Greek Science in the Long Run

Greek Science in the Long Run: Essays on the Greek Scientific Tradition (4th c. BCE-17th c. CE)

Edited by

Paula Olmos

CAMBRIDGE
SCHOLARS
PUBLISHING

Greek Science in the Long Run:
Essays on the Greek Scientific Tradition (4th c. BCE-17th c. CE),
Edited by Paula Olmos

This book first published 2012

Cambridge Scholars Publishing

12 Back Chapman Street, Newcastle upon Tyne, NE6 2XX, UK

British Library Cataloguing in Publication Data
A catalogue record for this book is available from the British Library

ISBN (10): 1-4438-3775-X, ISBN (13): 978-1-4438-3775-0

TABLE OF CONTENTS

ACKNOWLEDGEMENTS

The inspiration for this collection of essays came out of an international conference held Madrid in 2010, under the title "Twenty Centuries of Greek Science (4th c. BCE–17th c. CE)". This conference was made possible by generous financial support from the Spanish Ministry of Science and Innovation (research project: FFI2009-06957-E/FILO), University Carlos III of Madrid and the Department of Logic, History and Philosophy of Science at the Spanish National Distance Learning University (UNED). It gathered a highly distinguished interdisciplinary group of international scholars from the USA, United Kingdom, France, Italy, Germany and of course, Spain, in fields related to ancient history and culture and the history and philosophy of science. The event was a unique occasion to share interests and points of view in connection with the fascinating topic of the long-standing prominence of Greek models and traditions in the changing configuration of the sciences. It is therefore my pleasure to take this opportunity to express my deepest gratitude to *all* the participants, speakers as well as fellow organizers, in that conference– many of them, although not all, contributors to this volume. The gathering was undoubtedly the greatest achievement of my post-doctoral period at University Carlos III of Madrid, which is about to come to an end.

However, I would like to make special mention of three of these people, to whom, for different reasons, I should be particularly thankful. First of all, most certainly, Prof. Geoffrey Lloyd, who helped me more than anyone else with the organization of the conference. His academic prestige as well as his personal connections and generous mentorship of many of the participants made the event the best and most fruitful gathering possible, in both scholarly and personal terms. He was present at every session, took part in every discussion and was prepared and especially supportive when this publication started to take shape. In fact, he was the first to submit his contribution. Prof. Reviel Netz also played an exceptional role in the success of this volume, as he mentored and made possible my research term at Stanford University during the summer of 2011, when I was able to undertake editing of the text in a unique scholarly and collegial atmosphere that no doubt contributed to its improvement and completion. Finally, I must express my deepest gratitude and indebtedness to Prof. Antoni Malet, who made the decision to continue with his contribution to this volume despite very sad personal circumstances, doing us the honour of considering our project a means of solace, which I hope it was.

For reasons that exceed the limits of this particular achievement, my gratitude also goes out to all my colleagues at the Lucio Anneo Séneca Institute of Classical Studies, starting with its director, Prof. Francisco Lisi. For three years, they have offered me a brilliant example of scholarly excellence which, rather exceptionally in the academic world, has been made naturally compatible with an outstandingly collegial working environment.

Madrid, November 2011

INTRODUCTION: THE GREEKNESS OF SCIENCE

PAULA OLMOS

Throughout both classical and late antiquity, the European Middle Ages and most of our early modern period, systematic *technical* knowledge related to fields as varied as medicine, biology, mathematics, astronomy or even rhetoric and grammar repeatedly made reference to Greek sources, theories and discussions. In classical and post-classical Latin texts, medieval compilations or the Renaissance humanist return *ad fontes*–in part made possible through the teachings of Greek-speaking Byzantine émigré scholars–the content and legacy of Greek scientific production was esteemed and treasured. It was of course also reinterpreted, but not (not yet) discarded or much contradicted. If the dominance of the Aristotelian corpus as the basis of a broadly-construed scientific curriculum–albeit philosophically-driven–was a remarkable late medieval phenomenon with a rather prolific afterlife, more sustained indeed was the medical authority of a figure like Hippocrates, whose attributed corpus came to be mostly interpreted through the work of Galen, another Greek authority. Euclid and Archimedes were synonymous with advanced mathematics and although Cicero's textbooks were the main medieval source of information on technical rhetoric in the West, any attentive reader would note that what they really transmitted was the Latin translation of a systematic model based mainly on Hermagoras of Temnos and other Greek sources.

As is well known, the Greekness of technical literature and vocabulary was already an issue for early Latin writers such as Cicero, who tried to coin Latin terms equivalent to those used in Greek philosophy and rhetoric. However, for example, his *ratiocinatio* for *sullogismos* had very little impact, and he did not even try to come up with an alternative to *enthumema*. In the case of fields in which concept and object identification and naming were particularly pervasive, such as mathematics or rhetorical elocution–which included the study of an extensive and ever-growing list of "figures of speech" or *schemata lexeos*–Greek vocabulary remained the norm … until today.

The present volume is not intended to give a full account of the historical problem of the *longue durée* of the dominance of Greek science in itself. This is a complex topic that must be framed within a broader comprehension of the formation, development and survival of ancient literary culture as a whole, an issue to which some of the contributors to this volume are beginning to pay significant attention.[1] Our more modest aim is to offer a series of individual studies focused on different aspects, periods and fields related to the ongoing recycling of the Greek scientific legacy undertaken from the perspective of full awareness of this phenomenon.

When I began to make preparations for the international colloquium that was the starting point for this book and contacted different scholars about it, most of them seemed fascinated by the implications of the very title chosen, "Twenty centuries of Greek science". Nevertheless, I now clearly see that it was a somewhat restrictive one, if we take into account both the legacy of the Greek archaic period and the present-day usefulness of many Greek technical concepts (especially in fields like mathematics and rhetoric). In any case, the limits I set at the time and which are still present in the subtitle of this volume–4th century BCE to 17th century CE–remain useful in focusing our attention on the long period during which the Greekness of most of science was to a certain extent widely taken for granted.

The essays in this collection thus contain the response of a significant group of scholars to this call for attention to the way Greek scientific traditions enjoyed such an incredibly long reputation, while also displaying a kind of versatility that challenges any simplistic, dogmatic or *a priori* viewpoint about the meaning and social functioning of systematic knowledge. They revisit the different processes by which such doctrinal traditions originated, were transmitted and received within diverse socio-cultural contexts and frameworks. The concepts of continuity and discontinuity, *deuteronomic* or meta-textual recycling and contextualized originality therefore inform the various approaches presented here, while some of the contributions (see especially Lloyd, Netz and King) address them in a more direct way. Although the boundaries between scientific

[1] Reviel Netz is currently working on what he calls the *parameters* of ancient literary culture that, among other things, account for their preservation. I myself am working on the way the classical cultural legacy, and Greek science in particular, was filtered and *packed* for its future reception in the context of the crumbling world of late antiquity, a little studied period in the history of the sciences which is particularly crucial to the issues addressed in this volume.

fields and their denominations are rather problematic,[2] I have tried to structure the book in such a way that groups the different contributions into more or less coherent sections.

The first section includes six papers which address relatively general, method-focused and trans-disciplinary issues. They can be read in pairs as dealing first with method-centred caveats regarding our initial approach to Greek scientific traditions (Lloyd, Andò); then with the characteristically Greek–and specifically Aristotelian–concern with reasoning and argumentation in and of themselves (Vega, di Piazza) and finally, with the historically subsequent, *deuteronomic* compilation and reception (in different periods) of a comprehensive encyclopaedia of knowledge based on Greek sources (Olmos, Raschieri). The essays in the second section of the volume are devoted to arts and sciences based on numbers or quantification: mathematical or exact sciences in very general terms, with two original contributions to the study of Greek musical theory and its concepts, focused on very different authors of different periods and with very different aims (Tomasello, Tolsa), and two reflections on the historical processes of continuity, discontinuity, tradition and reception of Greek mathematics (Netz, Malet). The third section includes essays dealing with the arts and sciences relating to life and health. Here it is possible to distinguish two distinct blocks, the first exploring the broad social and intellectual spectrum of medical concerns (theoretical, philosophical) and practices (healing disciplines and professions) within ancient Greek culture (Cano, Macías), and the second looking at different aspects of the historical shaping (particularly undertaken by Galen) of a comprehensive, and to a certain extent forcefully coherent, idea of Greek medicine worthy of being passed along (Vegetti, King, Vélez).

Geoffrey Lloyd's paper, "Categorical Anachronisms and their Consequences for the History of Science", warns us about the perils of an anachronistic approach to the evolving fields and genres of Greek science that could create false expectations as to what we may find, for example, in Greek *mathēmatikē* or *harmonikē*. Following on from the fruitful comparative approach of his recent publications, he turns once more to his extensive knowledge of Chinese culture in order to explore significant cross-cultural differences in the classification of scientific fields. They thus appear as characteristic cultural products in themselves, rather than being dictated by the nature of their object.

[2] See Lloyd in this volume, although Netz has a somewhat different take on this issue, at least regarding the compactness of the field of the "exact sciences".

For its part, the exegetic review offered by Valeria Andò in her contribution, "Ancient Greece and Gender Studies", reminds us of the necessity to take gender issues into account–as the authors referenced by her have successfully done–in our exploration of the Greek literary and cultural realms. It will serve as a guide for anyone interested in this aspect of classical studies, which is necessarily intertwined with certain scientific practices and concerns, especially within the field of medicine, as the paper by Helen King in this same volume so keenly reveals.

The paper by Luis Vega, "The Field of Argumentation: From Aristotle to the Present Day", consciously exceeds the chronological limits initially set for this volume in presenting the particular case of Aristotle's broad and comprehensive approach to argumentation and its contemporary relevance. Aristotle's argumentative theories–contained in works like the *Topics*, the *Analytics* and the *Rhetoric*–although preserved and studied piecemeal for many centuries, seem to have required today's involvement with such issues to reveal all their potentialities. However, Vega shows how Aristotle's *approaches* essentially differ from the contemporary perspectives that have inherited the denominations of his triple legacy: logic, dialectic, rhetoric.

In "Stochastic Knowledge: *For the Most Part* and Conjecture in Aristotle", Salvatore di Piazza tries to reconstruct the meaning and philosophical import of the Aristotelian expression "ὡς ἐπὶ τὸ πολὺ" ("for the most part"), used with characteristic determinations (predications) applied to objects or states of the world, and opposed to both what is *necessary* and what is merely *by chance*. Di Piazza demonstrates the pervasive nature of this expression within the Aristotelian corpus–which is so abundant in self-referent methodological considerations–and seeks to establish its relationship with a kind of *conjectural* cognition that would be proper for wise (φρόνιμος) and reasonable people, capable of dealing with "truths *for the most part*" in a flexible, adaptable way.

My own contribution to this volume, entitled "*Euge, Graeculi nostri!* Greek Scholars among Latin Connoisseurs in Macrobius' *Saturnalia*", focuses on this significant text, a literary piece with encyclopaedic ambitions dating from the first third of the 5th century–and therefore belonging to the still poorly understood literary and scientific culture of late antiquity. The *Saturnalia* in fact reveals to us one of the main contentions of this volume, the Greek character attributed to systematic knowledge even within the context of a truly advanced and somewhat multicultural state within Latin culture. The text, in a Ciceronian mood, portrays a three-day gathering with highly social and partially *sympotic* or convivial erudite dialogues between members of the Roman elite and some

Greek scholars. It presents to us–mainly through its clever choice of characters–a series of confrontations and oppositions in which the tensions, as well as a certain sought-after harmony, between different approaches to advanced knowledge are brought to the fore. Among these approaches some are considered more Greek, which usually means more technical and systematic, than others.

To conclude the first, method-centred and metadisciplinary, section of the book, we have Amedeo Raschieri's contribution, "Giorgio Valla, Editor and Translator of Ancient Scientific Texts", focusing again on an author with encyclopaedic ambitions based on the Greek legacy, although this time situated at the latter extreme of our chronological range. Raschieri offers a very precise description of the significance of Giorgio Valla's publishing agenda, undertaken during the late 15th century in Northern Italy. This included a summary text aiming at compiling, in Raschieri's words, "the entire scientific and philosophical knowledge of the time" (*De expetendis et fugiendis rebus*), but also a vast programme of Latin translations of Greek scientific texts by, among others, Aristotle, Galen, Alexander of Aphrodisias, Aristarchus of Samos, John Philoponus, Nicephorus and Hypsicles. The works addressed such fields as rhetoric, mathematical and astronomical science, medicine and Aristotelian philosophy. Their presentations and extremely revealing preliminary material are the main focus of the paper.

The second section of the book, devoted to Greek traditions in the exact sciences, begins with Reviel Netz's useful and extensive panorama. Under the title "The More it Changes … Reflections on the World Historical Role of Greek Mathematics", this contribution covers the variations–continuities and discontinuities–endured by what Netz sees as the relatively compact and identifiable field of the Greek exact sciences, from pre-Socratic times up to the "late ancient synthesis" (4th to 6th centuries CE). This paper–along with Helen King's on Greek medicine–is a very serious response to the challenge posed by our initial call for reflections upon "twenty centuries of Greek science", providing a thought-provoking and insightful overview. After reviewing the different periods of creativity in Greek mathematics and exploring its genres and ways of expression, Netz concludes that, whereas Greek medicine and philosophy base their continuity on a canon of authors and schools, it is "formal style" that defines Greek mathematics, which can, once and again, avoid the mention of masters and individuals.

According to Netz's periodization, Marianna Tomasello's contribution can be seen as a case study on the *philosophical mathematics* of the 4th century BCE and the way authors like Plato or Aristotle engage in methodological reflections that make use of mathematical terminology and concepts–the terminology of music or harmonics in this case–albeit with significant cosmological repercussions. Thus, "Musical Terminology in Plato's Dialogues: The Image of Concord in the *Republic* and in the *Timaeus*" reviews Plato's understanding of the concept of σύμφωνος (what is concordant) and its crucial role in the theoretical comprehension of a kind of cosmology that, as is to be expected from this author, has ethical and political consequences. The mathematical basis of a cosmos which is initially perceived through its sensible phenomena–an allegedly Pythagorean idea–and its similarity to the also mathematical structure of the human soul, is conceived by Plato as suitable justification for his "teleological discourse" involving the harmony between the subject and the object of her knowledge.

Cristian Tolsa engages with a rather different aspect of Greek mathematical practice, so brilliantly explored in other publications by Reviel Netz: the use and meaning of diagrams. Medieval manuscripts have transmitted to us both the words and the graphical schemas of the original Greek mathematical treatises, but until recently, most attention has been focused on the words, while the diagrams have been deemed mere dispensable *representations* of the theorems and propositions allegedly developed and fully proved within the text. After the work of Netz and others, this is no longer acceptable and a whole world of new research on Greek mathematical diagrams has been opened up. In "On Ptolemy's *Harmonics* 2.4: Does the Text Refer to the Diagram?", Cristian Tolsa analyses a particular diagram belonging, according to part of the manuscript tradition–and thus assumed by modern editions–to the aforementioned treatise, but whose relationship to the text is rather problematic. His conclusion is that this particular diagram does not belong within the original treatise, but must rather be part of the commentary on it by Porphyry, a conclusion that in itself reveals the rather fascinating possibility of a visual commentary. The paper brings to the fore interesting aspects of the inner story of the texts within the long-standing tradition that is the main focus of this volume.

Antoni Malet's "Euclid's Swan Song: Euclid's *Elements* in Early Modern Europe" closes this section with a detailed review of the early modern reception of Euclid's *Elements*, ranging from the philologically-driven concerns of 16th-century editions of the text, to the conceptual and pedagogical criticism of its 17th-century editors and commentators, who tried to "amend" what they saw as the text's geometrical bias with clarifications and modifications that would make it more appropriate to the theoretical challenges of a general science of magnitude, especially in regards to the theory of ratios or proportions. His account ends with the changed context in which 18th-century authors such as Euler reinterpreted the content of the *Elements*, further confirmation of the lasting significance of Euclid's contribution.

After the section devoted to sciences based on numbers and quantification comes the vast field of sciences which examine and give an account of the diverse physiological aspects of living creatures. These are inevitably intertwined–something especially present in some of the papers (Macías, King)–with their most obvious practical side, concerning the health and physical condition of human beings.

As was the case with Tomasello's contribution on harmonics and cosmography, there is also room in this section for a paper on Plato's philosophical approach to and exploitation of scientific ideas and models in several of his dialogues, but most especially in the *Timaeus*, a work which because of its particular story, transmission and reception could in itself be considered one of the pillars of the long-standing tradition of Greek science. Jorge Cano's paper "Philosophy and Teleology: The Creation of the Marrow and the Head in Plato's *Timaeus*" addresses Plato's rationalistic treatment of the somewhat inferred anatomy and physiology of the organs allegedly associated with the body-soul relationship, i.e. the marrow and the brain. Cano explores the way Plato's contentions regarding the *encephalo-myeloid* system relate to the theories and speculations of other authors and schools, most especially Empedocles, Philolaus of Croton and the Hippocratic tradition, and concludes by restating the Platonic commitment to a general teleological scheme of divine creation and order: "As happens with arithmetic and geometry, other *mathemata*, medical physiology serves as proof and evidence of the mark of the demiurge underlying the entire universe."

With Sara Macías' "*Pharmaka*: Medicine, Magic and Folk Medicine in the Work of Euripides", we concentrate on matters relating to popular magic and folk medicine that have had a prominent place in recent literature. In this particular case, what is explored is the evidence of certain widespread ideas and popular prejudices involving these issues–and particularly the use of remedies with various properties (*pharmaka*)–in the literary work of Athens's most successful drama writer of his day, the enduring canonical author, Euripides. Reflecting the vastly *performative* culture of the Greek *polis*, with its high standard of citizen social participation, Euripides' tragedies bring to us expressions and images that were purportedly shared and understood by the laymen of the city, even as they also reflected certain controversies and conflicts within the technical and specialized realm of medical practice and literature.

This eminently technical and, by Hellenistic and Roman times, clearly professionalized medical field is the main focus of Mario Vegetti's contribution, "Galen on Body, Temperaments and Personalities". Galen's only somewhat successful efforts in building a coherent and unified theory of temperaments and humours that would also be the basis of a theory of personality, becomes Vegetti's object of study in what is also an account of the synthetic reworking of an earlier Greek legacy at the hands of the great Greek physician of a then Roman world.

The role of Galen in passing on his own synthesis of Greek medicine to the Western scientific tradition is also emphasized in Helen King's "Knowing the Body: Renaissance Medicine and the Classics". Before focusing on the very revealing and complicated reception of renowned classical case studies–involving moreover gender issues–in Renaissance medical and exemplary literature (that which compiles didactic stories, anecdotes and *exempla*), this paper offers us an insightful historical panorama of what would traditionally be considered Greek medicine. According to King's account, the concept of Greek medicine itself was an early construct whose canonical origins can be found in Hippocrates and which rather successfully resisted any temptation of fragmentation into different specialized practices–diet, surgery, drugs or gynaecology–by presenting itself as a unified field for an all-around practitioner. The Renaissance continued to revere the classical authors of Greek ancient medicine, while sometimes offering confused amalgamations of their case studies.

The final contribution to the volume is by Andrés Vélez: "The *Forum Vulcani* in the Work of Juan Huarte: Geographical Argument and Renaissance Medicine". The paper focuses on how renowned Spanish Renaissance physician Juan Huarte, author of the extremely well-received *Examen de Ingenios*, made use of the classical imagery associated with Pozzuoli. Located on the Gulf of Naples, this area of sulphuric emissions was traditionally linked, through mythological accounts, with the gates and halls of hell. Vélez's avowed aim is to explore the function of this and other geographical references as a basis for arguments based on location (*argumentum a loco*) in Renaissance medicine and science. Thus for Vélez, the preservation of the long tradition of classical, and particularly Greek, medicine involving itself with geographical knowledge in texts like *Airs, Waters and Places* (within the Hippocratic Corpus), as well as in Galen's writings, several Platonic dialogues, especially the *Timaeus*, and Aristotle's *Problemata*, is something that must be taken into account in the analysis of Juan Huarte's work and his way of employing erudite, typically humanist, scavenging of an ample encyclopaedic tradition for his medical goals.

This last paper, concerning the theories and writings of a renowned professional physician, while showing evident meta-disciplinary concerns, like those addressed in the first section of the volume, rounds off our journey through different aspects of the long tradition of the Greek sciences. It completes a book that has been put together with the aim of quite consciously attempting to convey what Greek science has meant in the long run.

PART I

METHOD, ARGUMENTATION AND TRANSDISCIPLINARY ISSUES

CATEGORICAL ANACHRONISMS AND THEIR CONSEQUENCES FOR THE HISTORY OF SCIENCE

GEOFFREY LLOYD

Most historians of science are well aware that the Greek term *mathēmatikē* is far from being the exact equivalent of our "mathematics", that the Greek term *phusikē* (the study of nature) is radically different from the modern understanding of "physics" and that "biology" as such is nowhere to be found in Greek antiquity.[1] If we want to grasp how Aristotle set about *peri zōōn historia* (the study of animals) we have to pay careful attention to his account of the different types of causes that are his principal target. Spreading our net wider, a slight acquaintance with classical Chinese texts reveals that their conceptual map of the major intellectual disciplines differs widely from that of the ancient Greeks and from our own. I shall be discussing one of their overarching categories, *shu shu*, 數 術 , "calculations and methods", in some detail in a minute. But for now it is enough to note that it spans both what we should think of as mathematical subjects (the study of numbers) and what we would class as physical ones, that is the explorations of the interactions between things that the Chinese discussed under the rubric of *wuxing*, 五行 , the five phases. Medicine provides another instance, where Greek *iatrikē* and Chinese *yi xue*, 醫 學 , represent very different ideas of the causes of diseases and what needs to be done to treat them, and widely different, in each case, from modern biomedicine. And many other examples could be given.

But this bewildering heterogeneity may lead us into two profound mistakes, each of which needs to be avoided like the plague. The first is to think that the differences between Greece and China (for instance) are such that no comparison is possible between them, that we are dealing with incommensurable conceptual systems. The second is to hold that in neither ancient civilisation is there anything at all close to (modern)

[1] See especially Cunningham (1988); Cunningham and Williams (1993).

science, which must therefore be considered a uniquely modern, specifically Western, phenomenon. So there is a second incommensurability in our way, this time between ancient and modern thought.

Let me first discuss, then, some of the complications of a cross-cultural study of "mathematics".[2] Our term "mathematics" is, of course, derived from the Greek *mathēmatikē*, but that word comes from the verb *manthanein*, which has the quite general meaning of "to learn". A *mathēma* can be any branch of learning, anything we have learnt, as when in Herodotus, 1 207, Croesus refers to what he has learnt, his *mathēmata*, from the bitter experiences in his life. So the *mathēmatikos* is, strictly speaking, the person who is fond of learning in general, and it is so used by Plato, for instance, at *Timaeus* 88c, where the point at issue is the need to strike a balance between the cultivation of the intellect (in general) and that of the body–the principle that later became encapsulated in the dictum "mens sana in corpore sano". But from the fifth century BCE certain branches of study came to occupy a privileged position as the *mathēmata* par excellence. The terms mostly look familiar enough, *arithmētikē*, *geōmetrikē*, *harmonikē*, *astronomia*, and so on, but that is deceptive. Let me spend a little time explaining first the differences between the ancient Greeks' ideas and our own, and secondly some of the disagreements among Greek authors themselves about the proper subject-matter and methods of certain disciplines.

Arithmētikē is the study of *arithmos*, but that is usually defined in terms of what we would call positive integers greater than one. Although Diophantus, who lived at some time in late antiquity, possibly in the third century CE, is a partial exception, the Greeks did not normally think of the number series as an infinitely divisible continuum, but rather as a set of discrete entities. They dealt with what we call fractions as ratios between integers. Negative numbers are not *arithmoi* and do not enter into consideration. Nor is the monad an *arithmos*: it is thought of as neither odd nor even. Plato draws a distinction, in the *Gorgias* 451bc, between *arithmētikē* and *logistikē*, calculation, derived from the verb *logizesthai*, which is often used of reasoning in general. Both studies focus on the odd and the even, but *logistikē* deals with the pluralities they form, while *arithmētikē* considers them–so Socrates is made to claim–in themselves. That at least is the view Socrates expresses in the course of probing what the sophist Gorgias was prepared to include in what he called the art of rhetoric, though in other contexts the two terms that Socrates thus

[2] This subject is the topic of a comprehensive collective volume, Robson and Stedall (2009).

distinguished were used more or less interchangeably. Meanwhile a different way of contrasting the more abstract and the more practical aspects of the study of *arithmoi* is to be found in Plato's *Philebus* 56d, where Socrates distinguishes the way the many, *hoi polloi*, use them from the way philosophers do. Ordinary people use units that are unequal, speaking of two armies, for instance, or two oxen, while the philosophers deal with units that do not differ from one another in any respect, abstract ones in other words. A gap thus opens up between the mathematical concepts that anyone might be expected to have, and a deeper, more sophisticated, approach, though that is here represented as depending not so much on advanced mathematical skills as on a philosophical grasp of the subject. We shall find other examples, in our ancient Greece, where it is the philosophers, as much as mathematicians themselves, who claim superior knowledge in the domain.

At the same time the study of *arithmoi* encompassed much more than we would include under the rubric of arithmetic. Like some other peoples the Greeks represented numbers by letters, where α represents the number 1, β the number 2, γ 3, ι 10 and so on. This means that any proper name could be associated with a number. While some held that such connections were purely fortuitous, others saw them as deeply significant. When in the third century CE the neo-Pythagorean Iamblichus claimed that "mathematics" is the key to understanding the whole of nature and all its parts (*On the Common Mathematical Science*), he illustrated that with the symbolic associations of numbers, the patterns they form in magic squares and the like, as well as with more widely accepted examples such as the identification of the main musical concords, the octave, fifth and fourth, with the ratios 2:1, 3:2 and 4:3. The beginnings of such associations, both symbolic and otherwise, go back to the pre-Platonic Pythagoreans who are said by Aristotle (*Metaphysics* 987b11, 1036b12) to have held that in some sense "all things" *are* or *imitate* numbers. Yet this is quite unclear, first because we cannot be sure what "all things" covers, and secondly because of the evident discrepancy between the claim that they *are* numbers and the much weaker one that they merely *imitate* them.

What about "geometry"? The literal meaning of the components of the Greek word *geōmetria* is the measurement of land. According to a well-known passage in Herodotus, 2 109, the study was supposed to have originated in Egypt in relation, precisely, to land measurement after the flooding of the Nile. Measurement, *metrētikē*, still figures in the account Plato gives in the *Laws* 817e when his spokesman, the Athenian Stranger, specifies the branches of the *mathēmata* that are appropriate for free citizens, though now this is measurement of "lengths, breadths and depths", not of land. Similarly in the *Philebus* 56e we again find a contrast between the exact *geōmetria* that is useful for philosophy and the branch of the art of measurement that is appropriate for carpentry or architecture.

Those remarks of Plato already drive a wedge between practical utility–mathematics as securing the needs of everyday life–and a very different mode of usefulness, namely in training the intellect. One classical text that articulates that contrast is a speech that Xenophon puts in the mouth of Socrates in the *Memorabilia*, 4 7 2-5. While Plato's Socrates is adamant that mathematics is useful primarily because it turns the mind away from perceptible things to the study of intelligible entities, in Xenophon Socrates is made to lay stress on the usefulness of geometry for land measurement and on that of the study of the heavens for the calendar and for navigation, and to dismiss as irrelevant the more theoretical aspects of those studies. Similarly Isocrates too (11 22-3, 12 26-8, 15 261-5) distinguishes the practical and the theoretical sides of mathematical studies and in certain circumstances has critical remarks to make about the latter.

The clearest early extant statements of the view privileging the theoretical come not from the mathematicians but from philosophers commenting on mathematics from their own distinctive perspective. What mathematics can achieve that sets it apart from most other modes of reasoning is that it is exact and that it can demonstrate its conclusions. Plato repeatedly contrasts that with the merely persuasive arguments used in the law-courts and assemblies where what the audience can be brought to believe may or may not be true, and may or may not be in their best interests. Philosophy, the claim is, is not interested in persuasion but in the truth.[3] Mathematics is repeatedly used as the prime example of a mode of reasoning that can produce certainty: and yet mathematics, in the view Plato develops in the *Republic*, is subordinate to dialectic, the pure study of the intelligible world that represents the highest form of philosophy. Mathematical studies are valued as a propaedeutic, or training, in abstract thought: but they rely on perceptible diagrams and they give no account of their hypotheses, but rather take them to be clear.[4] Philosophy by contrast moves from its hypotheses up to a supreme principle that is said to be "unhypothetical".

---

[3] This is a recurrent motif in Plato's *Gorgias* especially.

[4] The interpretation of the shortcomings of mathematics is controversial. See, for example, the articles in the collective volume edited by Anton (1980); Vlastos (1988); Lloyd (1991: ch. 14) and Burnyeat (2000). The *actual* role of the lettered diagram in Greek mathematical reasoning has been brilliantly explored by Netz (1999).

The exact status of that principle, which is identified with the Form of the Good, is highly obscure and much disputed.[5] Likening it to a mathematical axiom immediately runs into difficulties, for what sense does it make to call an axiom "unaxiomatic"? But Plato was clear that both dialectic and the mathematical sciences deal with independent intelligible entities.

Aristotle contradicted Plato on the philosophical point: mathematics does not study independently existing realities. Rather it studies the mathematical properties of physical objects (Lear, 1982). But he was more explicit than Plato in offering a clear definition of demonstration itself and in classifying the various indemonstrable primary premisses on which it depends. Demonstration, in the strict sense, proceeds by valid deductive argument (Aristotle thought of that in terms of his theory of the syllogism) from premisses that must be true, primary, necessary, prior to and explanatory of the conclusions. They must, too, be indemonstrable, to avoid the twin flaws of circular reasoning or an infinite regress. Any premiss that can be demonstrated should be. But there have to be *ultimate* primary premisses that are evident in themselves. One of Aristotle's examples is the equality axiom, namely if you take equals from equals, equals remain. That cannot be shown other than by circular arguments, which yield no proof at all, but it is clear in itself.

It is obvious what this model of axiomatic-deductive demonstration owes to mathematics. I have just mentioned Aristotle's citation of the equality axiom, which figures also among Euclid's "common opinions" (sometimes called "common notions") and most of the examples of demonstrations that Aristotle gives, in the first book of the *Posterior Analytics*, are mathematical. Yet in the absence of substantial extant texts before Euclid's *Elements* itself (conventionally dated to around 300 BCE) it is difficult, or rather impossible, to say how far mathematicians before Aristotle had progressed towards an explicit notion of an indemonstrable axiom (Knorr, 1975).

Of course the principles set out in Euclid's *Elements* themselves do not tally exactly with the concepts that Aristotle had proposed in his discussion of strict demonstration. Euclid's three types of starting-points include definitions (as in Aristotle) and common opinions (which, as noted, include what Aristotle called the equality axiom) but also postulates (very different from Aristotle's hypotheses). The last included especially the parallel postulate that sets out the fundamental assumption on which

[5] Modern discussion was opened up especially by Robinson (1953 [1941]: ch. 10). For a highly original study of Plato's ideal of the Good, see Burnyeat (2000).

Euclidean geometry is based, namely that non-parallel straight lines meet at a point. However, where the philosophers had demanded arguments that could claim to be incontrovertible, Euclid's *Elements* came to be recognised as providing the most impressive sustained exemplification of such a project (Mueller, 1981). It systematically demonstrates most of the known mathematics of the day using especially reductio arguments and the misnamed method of exhaustion. Used to determine a curvilinear area such as a circle by inscribing successively larger regular polygons, that method precisely did *not* assume that the circle was "exhausted", only that the difference between the inscribed rectilinear figure and the circumference of the circle could be made as small as you like. Thereafter the results that the *Elements* set out could be and were, treated as secure by later mathematicians in their endeavours to expand the subject. Mastery of the *Elements* came to be an essential prerequisite for those seeking to join the ranks of what became an increasingly self-conscious elite.

The impact of this development first on mathematics itself, then further afield, was immense. In statics and hydrostatics, in music theory, in astronomy, the hunt was on to produce axiomatic-deductive demonstrations that basically followed the Euclidean model. But we even find the second century CE medical writer Galen attempting to set up mathematics as a model for reasoning in medicine–to yield conclusions in certain areas of pathology and physiology that could claim to be incontrovertible (Lloyd, 2006: chh. 4 and 5). Similarly Proclus attempted an *Elements of Theology* in the fifth century CE, again with the idea of producing results that could be represented as certain.

This development is often held up as one of the glories of Greek rationality. Yet three points must be emphasised to put it into perspective. First, for ordinary purposes, axiomatics was quite unnecessary. Not just in practical contexts, but in many more theoretical ones, mathematicians and others got on with the business of calculation and of measurement without wondering whether their reasoning needed to be given ultimate axiomatic foundations (Cuomo, 2001).

Secondly, it was far from being the case that all Greek work in arithmetic and geometry, let alone in other fields such as harmonics or astronomy, adopted the Euclidean pattern. The three "traditional" problems, of squaring the circle, the duplication of the cube and the trisection of an angle were tackled already in the fifth century BCE without any explicit concern for axiomatics (Knorr, 1986). Much of the work of a mathematician such as Hero of Alexandria (first century CE) focuses directly on problems of mensuration using methods similar to those in the traditions of Egyptian and Babylonian mathematics by which, indeed, he may have been influenced (Cuomo, 2001; Tybjerg, 2004).

Thirdly, the recurrent problem for the model of axiomatic-deductive demonstration that the *Elements* supplied was always that of securing axioms that would be both self-evident and non-trivial. Moreover it was not enough that an axiom set should be internally consistent: it was generally assumed that they should be true in the sense of a correct representation of reality.

A lot more could be said about "mathematics" and the mathematical sciences in Greece, but it is time to introduce some of the complications that arise when we attempt to compare another great civilisation, China, and what they had to say on the subject. We can recognise their interest in numbers and shapes, and that looks familiar enough. But their maps of the relevant intellectual disciplines turn out to be very different from those of the Greeks and from our own. One of the two general terms for number or counting, *shu*, 數 , has meanings that include "scolding", "fate" or "destiny", "art" as in "the art of", and "deliberations" (Ho, 1991). The second general term, *suan*, 算 , is used of "planning", "scheming" and "inferring" as well as "reckoning" or "counting". The latter term figures in the titles of the two main canonical works on "mathematics" that date from a hundred years either side of the millennium, namely the *Zhoubi suanjing* (the mathematical canon of the Zhou Gnomon, Cullen, 1996) and the *Jiuzhang suanshu* (the Nine Chapters of Mathematical Procedures, Chemla and Guo, 2004).

When around the turn of the millennium the Han bibliographers, Liu Xiang, 劉向 , and Liu Xin, 劉歆 , catalogued all the books in the imperial library under six generic headings, one of these was *shu shu*, 數術 , that is, on one view, "calculations and methods". Its six sub-species comprise two that deal with the study of the heavens, namely *tianwen*, 天文 , (the patterns in the heavens) and *lipu*, 曆譜, (calendars and tables), as well as *wuxing*, 五行 (the five phases) and a variety of types of divinatory studies. The five phases provided the main framework within which change was discussed. They are named fire, earth, metal, water and wood, but these are not elements in the sense of the basic physical constituents of things, so much as processes. Fire, one text says,[6] is flaming upwards: water is soaking downwards. (Heraclitus might well have understood, but those like Aristotle who thought of fire as an elemental substance would have been lost).

[6] *Hong Fan* from the *Shang Shu*, see Karlgren (1950: 28 and 30); cf. Lloyd and Sivin (2002: 259-60).

So, on the one hand these Chinese inquiries deal with aspects of what we call astronomy and astrology and physics, the constitution of things, their correlations and associations, and their changes. But on the other the Chinese subsume all of these under a general category that straddles our "mathematics" and "physics". Moreover the aims of their more purely mathematical studies are strikingly different from those of the Euclidean tradition in Greece. We have direct evidence on the point in the work of a third century CE Chinese mathematician, Liu Hui, who wrote the first extant commentary on the *Nine Chapters*.

The key point, for our purposes, is that the notion of an axiom is absent from traditional Chinese mathematics, right down until the time of the arrival of the Jesuits in the sixteenth century. Rather the chief aim of Chinese mathematics, as Liu Hui describes it, was to identify the procedures that unify the subject and to extend its range. It is the *same* procedures, Liu Hui says, that provide the solutions to problems in different subject-areas. What he looks for, and finds, in such procedures as those he calls "homogenising", *qi*, 齊 , and "equalising", *tong*, 同 , is what he calls the guiding principles, *gangji*, 綱 紀 , of mathematics (*suan*, 算). In his account of how, from childhood, he studied the *Nine Chapters*, (91.6ff.) he speaks of the different branches of the study, but insists that they all have the same trunk, *ben*, 本 . They have a single source or principle, *duan*, 端 . The realisations and their categories, *lei*, 類 , are elaborated mutually. Over and over again the aim is to find and show the connections between the different parts of *suanshu*, extending procedures across different categories, making the whole "simple but precise, open to communication but not obscure". Describing how he identified the technique of double difference, he says (92.2) he looked for the essential points or characteristics, *zhi qu*, 指 趣 , to be able to extend it to other problems.

While Liu Hui is more explicit in all of this than the *Nine Chapters*, the other great Han classic, the *Zhoubi*, represents the goal in very similar terms. That text contains an interesting exchange between the Master and one of his Pupils on how "mathematics" holds the key to the understanding of all sorts of obscure phenomena (shades of Iamblichus, one might think), including especially astronomical ones, the height and distance and magnitude of the sun, for instance. But in the Chinese text the Master does not overwhelm the student with the claim to have demonstrated his results (quod erat demonstrandum, *hoper edei apodeixai*). Rather the Master insists that the student must internalise the procedures, to be so competent in them that he can get the results on his own (Lloyd, 2002: 51-2).

This parallelism between the *Zhou Bi* and Liu Hui shows that we are not dealing with some isolated, maybe idiosyncratic, point of view on the latter's part, but with one that represents an important, maybe even the dominant, tradition–at least among those in the forefront of Chinese mathematical speculation. "It is the ability to distinguish categories in order to unite categories" which is the key according to the *Zhoubi* (25.5). Again, among the methods that comprise the *Dao*, the Way, it is "those which are concisely worded but of broad application which are the most illuminating of the categories of understanding. If one asks about one category and applies [this knowledge] to a myriad affairs, one is said to know the Way" (24.12ff.).

I may now attempt to sum up what my rapid and selective survey suggests so far. The study of just two ancient mathematical traditions already brings to light certain generic similarities, but also some suggestive differences (Lloyd, 2009: ch. 2; Robson and Stedall, 2009). Axiomatization was a distinctive preoccupation of one Greek tradition. Faced with arguments as to the inadequacy of mere persuasion some mathematicians set about giving the subject a basis that would secure incontrovertibility. But if that was an ambition of some–certainly not all–Greek mathematicians, it is quite absent from classical Chinese mathematics, which is more concerned with exploring the connections and the unity between different studies, including those we consider to belong to physics and cosmology as well as those we classify as mathematics. Their aim was not to establish the subject on a self-evident axiomatic basis but to expand it by extrapolation and analogy.

Each of those two aims has its strengths and its weaknesses. The advantages of axiomatization are that it makes explicit what assumptions are needed to get to which results. But the chief problem was that of identifying self-evident axioms that are not trivial. The advantage of the Chinese focus on guiding principles and connections was to encourage extrapolation and analogy, but the corresponding weakness was that everything depended on perceiving the analogies, since no attempt is made to give them axiomatic foundations. It is apparent that there is no one route that the development of mathematics had to take, or should have taken. We find good evidence in these two ancient civilisations for a variety of views of its unity and its diversity, its usefulness for practical purposes and for understanding.

So what can we learn about our two basic questions from this foray into the "mathematics" of early civilisations? While there are considerable differences to which I have drawn attention, it is not that comparison is impossible. We can recognise interests in number theory in both civilisations. Despite a common view that the Chinese were no good at geometry, they investigated the circle-circumference ratio (π) just as the Greeks did. There are sophisticated investigations of the tricky problem of the volume of the pyramid that again involve the decomposition of the figures in question and recursive techniques that lead to better and better approximations to the desired result. In this instance both the conception of the problem (how to determine a particular volume) and the result are strictly comparable in Greece and China. There is this difference, however, that whereas the Greek (misnamed) "method of exhaustion" proceeds by an indirect proof (the volume cannot be greater, nor less, than the given formula, and so it must be equal to it) the Chinese use a direct method, getting closer and closer approximations and inferring that the series converges on the desired result.[7] But that difference in the methods of getting a result is far from legitimating any conclusion to the effect that the mathematics in the two civilisations are totally incommensurable.[8]

Once we take into account the whole gamut of Greek and Chinese investigations, the idea of dismissing Chinese work as somehow not proper mathematics at all can be seen to be mere Eurocentric prejudice. Of course many areas of the work of today's mathematicians are distinctly modern developments, and axiomatization itself underwent radical redefinition with Hilbert. But it is worth reminding ourselves that issues concerning the proper subject-matter and aims of mathematics continue to be hotly disputed in the twenty-first century. What mathematics studies and what knowledge it produces are questions that continue to be answered very differently by Platonists, constructivists, intuitionists, logicists and formalists (just to mention some of the principal divergent views). Those disagreements and the divergences in different historical mathematical traditions do not mean that there is no such subject as mathematics, after all: rather they suggest that what we can recognise as elements of the subject have been and continue to be pursued very

[7] See Lloyd (1996: 152-4, on the *Nine Chapters*, ch. 5: 167-8).

[8] It is important to distinguish between strong and weak versions of the incommensurability thesis. When Kuhn (1970 [1962]) introduced the notion, that was to draw attention to the undeniable point that radical shifts in the meaning of key terms took place as between Ptolemy, say, and Copernicus, or Aristotle and Galileo, or Newton and Einstein. But incommensurability has sometimes been used to suggest that there is a mutual unintelligibility across paradigms. Yet Galileo certainly thought he could understand Aristotle's views on weight and force, and modern interpreters proceed on the assumption that neither is totally beyond the reach of our understanding.

differently and with very different ambitions by different students of numbers and shapes.

But mathematics, some may think, is a comparatively easy subject in which to find answers to my two principal questions, of the possibilities of comparison between divergent ancient traditions and between them and modern work. So I turn now to the apparently much more difficult case of "science" itself.[9] Here there are no "actors" categories in ancient languages that provide a possible starting-point for comparisons across them and with our conception of science. Yet let me begin by reminding you of ongoing divergences in the latter. In the UK in one common usage "science" without qualification is taken to refer first and foremost to the "natural sciences". Of course psychology, anthropology, economics and the "social sciences" in general also stake a claim to be (proper) science, but the yardstick by which they are judged is provided by those natural sciences. Yet the other side of the English Channel, French "science", Spanish "ciencia", Italian "scienza" are not so restrictive, and in Germany "Wissenschaft" is certainly not limited to the "Naturwissenschaften".

How to define science, whether in terms of results, or methods, or aims, continues to be hotly contested. Any attempt to use results–the delivery of the truth–as the test is open to the objection that they, the results, are always subject to revision, never definitive. Some parts of what is accepted in 21st century chemistry, for instance, may look robust enough. But many of the fundamental problems in particle physics, in cosmology, even in parts of genetics, are, as is well known, still very far from resolved.

So rather than use results, most would focus on either methods or aims as the key criterion–and yet they yield very different views on our question. The invocation of the, or at least a, scientific method tends to favour the thesis of a Great Divide, which asserts that science has only been practised in the last 150 years or so (see, for example, Goody, 1977; Gellner, 1985). Science, on that view, may not require the entire gamut of the sophisticated gadgetry of modern laboratories, but it depends on such concepts as the mathematisation of physics, and on quantitative analyses generally, on the use of hypotheses and postulates, and especially on the experimental method. It is certainly not, then, a world-wide phenomenon to be found, to some extent at least, in any human society at any period.

But notions of correct methods have also been subject to fluctuation and the goal of being able to specify *the* scientific method, good for all

---

[9] See Cunningham (1988); Cunningham and Williams (1993), and cf. Lloyd (2009: ch. 8).

fields and purposes, recedes the more the variety of the actual practices of scientists comes to be appreciated. The alternative is to focus rather on the *aims* that science sets itself, principally the classification, prediction and explanation of the phenomena on the basis of a systematic collection of the data. That is not quite the way that Lévi-Strauss (1966) proceeded in advancing the notion of "concrete science"–or rather "la science du concret". But he, like Malinowski (1925), for one, before him, was enormously impressed by the detailed knowledge of the environment that is to be found in most societies across the world. The extremely detailed and complex plant and animal taxonomies reported from such societies as the Hanunóo, for example, will count as science, even though we are in no position to trace how that knowledge was first acquired and how it came to be consolidated.

Meanwhile so far as pre-modern societies go, to come closer to my own central interests, it seems absurd to deny the label "science" to the work of Mesopotamian, Chinese and Greek students of the heavens insofar as they aimed to build up the data that would enable them to regulate the calendar, that is to determine the relations between the solar year and the lunar month, and to predict and explain both lunar and solar eclipses. In those contexts their *aims* were surely strictly comparable with those of later investigators whom we have no difficulty in thinking of as astronomers. That is not to say that *all* the ancient interests in the heavens belong to a science of astronomy in our terms. Predicting the movements of the planets on the basis of observations of their periodicities was one thing, but saying what follows from their patterns for what is to happen to the fortunes of kings or states on earth is quite another. That too is prediction, of course, but not of celestial events, nor on the basis of observed periodicities.

Of course the broader view of what science includes has to make concessions. Once scientific research came to be carried on in purpose-built laboratories, that made a huge difference. Subjecting the data to statistical analysis is a phenomenon dating from the 17th century of which the same may be said.[10] But observation, measurement and classification as practised in modern science may all be represented as developments from activities that are as old as the human race. Even experimentation may be seen as continuous with, even though more elaborate and controlled than, very common trial and error methods deployed in all sorts of contexts by humans the world over. On that view the distinction

[10] The classic discussion of the emergence of the notion of probability is Hacking (1975). Hacking has further developed the notion of different styles of scientific reasoning or thinking, Hacking (1992; 2009).

between modern and pre-modern science is generally a matter of degrees of explicitness and of systematicity.

***

So my answers to the two questions I posed at the outset suggest a certain tension in my own position. Due attention has to be paid to the different ways different ancient investigators went about their work, the study of the stars and planets, that of animals and plants, diseases, their causes and cures, and more abstractly the geometrical properties of shapes. The answers are complex and vary between different areas of investigation. This is after all what we should expect, since the individuals and groups who did the work belonged to different institutions and to "professions" that were defined differently and occupied different niches in society. A fundamental methodological principle is that we should use *actors'* categories, not our own modern, observers' ones–as our starting-point. We have to ask what *their* aims were, their ideas about method, and about how their results should be checked and verified. But if that is a vital first step in the analysis of ancient work, that should not lead us to fall into the opposite mistake of asserting that–because *their* categories are unfamiliar and do not match ours–therefore we are dealing with totally *incommensurable* systems of thought. Getting the commonalities and then also the divergences (in aims, problematic, methods, and indeed results) is hard work. But that is what we should aim for, and if we do so with skill and perseverance, we should be able to throw some light on *why* science, in all its ramifications, developed in the different ways it did.

The first tendency (of eliding the differences) is repeatedly seen in those who write teleological history of science, assimilating ancient theories and interests to modern ones, as if the ancients conceived of a discipline that even approximates to our overarching "biology" or our "chemistry", let alone our "psychology". But the second tendency emerges in the blanket rejection of all ancient work as simply not "science" and the insistence that something called science only began in the 17th century. Evidently modern science makes use of many severely *modern* conceptual tools and techniques. Neither statistical analysis nor laboratory-based experimental science is represented in ancient China or in ancient Greece. Yet focussing on *aims* leads us towards a more pluralist understanding of investigations of the natural world and recognises that some ancient ambitions, in the matter of the systematic collection of data, with a view to classifying, predicting, explaining and understanding, are strictly analogous to modern ones. That is not to make a claim that science as such is a universal human phenomenon. But if we unpack its components, a case can be made that some of those correspond to basic cognitive endeavours. I would say observation certainly is, and I would add trial and error procedures and even measurement. The laudable aim of *avoiding* the use of anachronistic labels for ancient investigations (which is where I started) should not lead us into a different mode of Eurocentric chauvinism.

References

Anton, J.P., ed. (1980). *Science and the Sciences in Plato*, New York: Eidos.

Burnyeat, M.F. (2000). "Plato on Why Mathematics is Good for the Soul". In T. Smiley (ed.) *Mathematics and Necessity*, (pp. 1-81), *Proceedings of the British Academy* 103: Oxford University Press.

Chemla, K. and Guo Shuchun (2004). *Les Neuf chapitres: Le Classique mathématique de la Chine ancienne et ses commentaires*, Paris: Dunod.

Cullen, C. (1996). *Astronomy and mathematics in Ancient China: The Zhou Bi Suan Jing*, Cambridge: Cambridge University Press.

Cunningham, A. (1988). "Getting the Game Right: Some Plain Words on the Identity and Invention of Science", *Studies in History and Philosophy of Science* 19: 365-89.

—. (1999). "Aristotle's Animal Books: Ethology, Biology, Anatomy, or Philosophy?", *Philosophical Topics* 27 (1): 17-41.

Cunningham, A. and Williams, P. (1993). "De-Centring the 'Big Picture': The *Origins of Modern Science* and the Modern Origins of Science", *British Journal for the History of Science* 26: 407-32.

Cuomo, S. (2001). *Ancient Mathematics*, London: Routledge.

Gellner, E. (1985). *Relativism and the Social Sciences*, Cambridge: Cambridge University Press.

Goody, J. (1977). *The Domestication of the Savage Mind*, Cambridge: Cambridge University Press.

Hacking, I. (1975). *The Emergence of Probability*, Cambridge: Cambridge University Press.

—. (1992). "'Style' for Historians and Philosophers", *Studies in History and Philosophy of Science* 23 (1): 1-20.

—. (2009). *Scientific Reason*, Taipei: National Taiwan University Press.

Ho Peng-Yoke (1991). "Chinese Science: The Traditional Chinese View", *Bulletin of the School of Oriental and African Studies* 54: 506-19.

Karlgen, B. (1950). "The Book of Documents", *Bulletin of the Museum of Far Eastern Antiquities* 22: 1-81.
Knorr, W.R. (1975). *The Evolution of the Euclidean Elements*, Boston: Reidel.
—. (1986). *The Ancient Tradition of Geometric Problems*, Boston: Birkhäuser.
Kuhn, T.S. (1970). *The Structure of Scientific Revolutions* (1st. publ. 1962), Chicago: Chicago University Press.
Lear, J. (1982). "Aristotle's Philosophy of Mathematics", *Philosophical Review* 91: 161-92.
Lévi-Strauss, C. (1966). *The Savage Mind* (trans. of *La Pensée sauvage*, Paris: Plon, 1962), London: Weidenfeld and Nicolson.
Lloyd, G.E.R. (1991). *Methods and Problems in Greek Science*, Cambridge: Cambridge University Press.
—. (1996). *Adversaries and Authorities*, Cambridge: Cambridge University Press.
—. (2002). *The Ambitions of Curiosity*, Cambridge: Cambridge University Press.
—. (2006). *Principles and Practices in Ancient Greek and Chinese Science*, Aldershot: Ashgate.
—. (2009). *Disciplines in the Making*, Oxford: Oxford University Press.
Lloyd, G.E.R. and N. Sivin (2002). *The Way and the Word*, New Haven: Yale University Press.
Malinowski, B. (1925). "Magic Science and Religion". In J. Needham (ed.), *Science Religion and Reality* (pp. 19-84). London: The Sheldon Press.
Mueller, I. (1981). *Philosophy of Mathematics and Deductive Structure in Euclid's Elements*, Cambridge, Mass.: MIT Press.
Netz, R. (1999). *The Shaping of Deduction in Greek Mathematics*, Cambridge: Cambridge University Press.
Robinson, R. (1953). *Plato's Earlier Dialectic* (1st ed. 1941), Oxford: Clarendon Press.
Robson, E. and J. Stedall (eds.) (2009). *The Oxford Handbook of the History of Mathematics*, Oxford: Oxford University Press.
Tybjerg, K. (2004). "Hero of Alexandria's Mechanical Geometry". In P. Lang (ed.) *Re-inventions: Essays on Hellenistic and Early Roman Science* (pp. 29-56). Kelowna: Academic Printing and Publishing.
Vlastos, G. (1988). "Elenchus and Mathematics: A Turning-point in Plato's Philosophical Development", *American Journal of Philology* 109: 362-96.

ANCIENT GREECE AND GENDER STUDIES

VALERIA ANDÒ

Introduction

This paper deals with issues of methodology without focusing upon any particular area of Greek science. The intention is to analyse the specific contribution of *Gender Studies* and their underlying epistemology to the study of Greek texts. To begin with, the origin, use and definition of *gender studies* will be clarified.

Ever since the notion of *gender* was first elaborated in the 80s by the American historian Joan Scott, it has constituted an effective analytical tool. While sexual dimorphism appears to be determined by biological data and therefore an outcome so to say of "nature", gender on the other hand is the result of a social construction in that it is the cultural processing of the sexual differences inscribed in our bodies. In a certain sense, one could state that gender constitutes the development and theorisation of Simone de Beauvoir's famous statement "One is not born a woman, but becomes one,"[1] in which the insufficiency of biological data for the construction of a sexual identity is expressed. Therefore, the application of this notion in the field of historical/social and scientific study has revealed all its heuristic potential and methodological effectiveness. Joan Scott defines gender as follows: "*gender* is a constitutive element of social relationships based on perceived differences between the sexes, and gender is a primary way of signifying relationships of power."[2] As such gender has opened up modes of investigation which can penetrate deep into the social framework, the dynamics of social relationships and the tensions which come across the fabric of society. That is to say, polarity of gender has been considered as the key to interpret social history, institutions, ethical/political thought and scientific observation.

---

[1] Beauvoir (1999 [1949]: 325).

[2] Scott (1986: 1071). The notion had been earlier propounded within the field of sociological research by Oakley (1972).

Gender Studies have therefore ended up substituting or at least running in parallel with *women's studies*, studies of women considered as a single group to be analysed independently with the aim of restoring the presence, words and actions of women to history, from which they had been excluded for centuries. The standpoint of *women's studies* emerged from the period of post-modernism, and has been directly influenced by the feminist movement of the 70s and the emancipationist trends in political, social and economic life. Women, excluded from history, returned thanks to the libertarian and egalitarian thrust of feminism and became the object of individual studies which restored their roles, social functions, legal status, artistic subjectivity and creativity.

The risk of essentialism, inherent in the history of women, together with the blurring of the boundaries of the category of "sex" now free from the chains of biological determinism, and finally the weakening of militant feminism in politics, have permitted the category of "gender" to establish itself in academic studies. It has also become considered more inclusive and as such has given rise to a more reassuring neutrality: that is women are studied in dialectic tension with the opposing and historically hegemonic group of men, in the belief that "gender" is the determining factor in power dynamics, and therefore a valid investigative tool in the social sciences. Moreover, against the potential fixity of the category of sex there is the legal and changeable nature of gender, subject to historical change and changing social contexts. The passage from history of women to history of gender has therefore permitted the important passage from documentation to representation. That is, the analysis of the opposite poles male/female considered to be in constant tension and reciprocal exchange leads to the employment of data regarding social individuals ready for an overall interpretation of society as a whole and connected symbolic implications.

So far the category of gender continues to be used in research as an effective key to interpret all aspects of society[3] and to stimulate theoretical thought.

However, the American philosopher Judith Butler, who has concentrated all her theoretical research on *gender* has recently "undone" the idea. Within *queer theory*, which breaks down the rigid male/female dichotomy to open out to *transgender* sexuality and to different forms of sexual orientation, gender turns out to be the product of a continual and incessant work of *doing* and *undoing*, in the sense that it is necessary to undo it, deconstruct it, to avoid the risk of hypostatization and acceptance of

[3] Among the latest socio-historical contributions cf. Maruani (2005).

ethical/social norms. Moreover it is necessary to do it, construct it, to construct identity, even if it is fluid and continually evolving (Butler, 2004). Such deconstruction of the notion, which is certainly stimulating in philosophical and ethical terms, has not yet shown its effects in historical/social research in which, as has already been said, *gender* continues to be a useful research vehicle.

Gender studies have been and continue to be a fertile line of research in classical studies, especially in the field of the history, culture and texts of ancient Greece. Even here they have steadily replaced womens' studies though womens' studies have provided a far more well-founded theoretical epistemological framework. An interweaving between the two types of research sometimes makes distinguishing between them difficult: indeed in many cases the history of women in Greece is anyway *gendered* or *genrée*, perhaps because of the impossibility of ignoring it in the accounts of women which are almost all written by men. In the general silence of Greek women, save for a few sporadic voices which confirm the rule, it has been left to men to narrate, describe and codify the feminine according to rules and standards which reflect the social organisation of the relationships between the sexes.

As in all other areas of research, whether Anglo-American or European, and even those concerning the Greek world, gender studies have had various orientations and have focussed on different subjects and themes. While these studies are always in some way inspired by feminist politics, as well as Marxism, psychoanalysis, structuralism and post-structuralism, they are also influenced by the different ways the movements and currents of thought have been adopted in Europe and the United States.

Indeed, in the United States the university departments of women's studies have supplied a fully endorsed academic framework[4] within which to place the specialised study of women in antiquity. With this comes the radicalism that American feminism adopts at a political level: that is, a feminism far from the bourgeois, intellectual nature of white and largely heterosexual European feminism so criticised by American feminists, who on the contrary are receptive and welcoming not only of sexual difference but all the other differences such as skin colour or sexual orientation. Indeed, *Black Feminism* and *Lesbian Studies* are American disciplines, which surface even in studies about women in ancient times, with a strong political and cultural thrust, thus extending research to other marginal areas and other interests such as colonialism and inter-culturality.

---

[4] For the development of *Women's Studies* in the USA cf. Boxer (2001).

Given the academic success of Women's Studies it is therefore no coincidence that, shortly following on volume VI *Arethusa* (1973),[5] the first monograph on women in Greece and Rome by Sara Pomeroy appeared in the United States in 1975.[6] Then, again in the United States, a *reading* entitled *Feminist Theory and the Classics*[7] was published in 1993. This one, significantly dedicated "for all our sisters", falls within the theoretical framework of black and lesbian feminism in a spirit of sisterhood, that encompasses all women and their many differences.

Within the steady shift towards *Gender Studies*, the US approach to the Greek world has in some ways maintained a militant and political character towards discrimination and social exclusion.

As for Europe, these preliminary notes will simply mention that in France women's studies have always been contextually permeated by the notion of gender. Indeed in France these studies are on one hand linked to the writing of history by the *Annales* School and therefore influenced by social dynamics in all their complexity. On the other they have arisen from the Louis Gernet Centre of Comparative Research on Ancient Societies, founded by J.-P. Vernant and his disciples, in whose research one still feels the effects of Lévi-Strauss' structuralism as well as of social psychology and which is integrally inclined towards the reconstruction of social representation in all its forms, especially the gender relationship. For this reason French criticism has claimed a certain cultural priority with regard to the discovery and use of the notion of gender in spite of its declared genesis in an Anglo American setting.[8] Indeed, it was already in 1963 that the seminal article "Hestia-Hermès" by J.-P. Vernant[9] was published, which analysed these two Greek deities and identified their areas of expertise and consequent areas of activity, the interior and exterior worlds, which determine the social setting. Then in 1968 we have Vernant's well quoted statement "le mariage est à le fille ce que la guerre est au garcon,"[10] which marks the separation between different destinies and functions. Furthermore a large part of Nicole Loraux's work, which will be discussed more widely later, shows how much the feminine, because it is the product of a representation, is an effective research vehicle to reconstruct the masculine and approaches to social organisation

[5] In this volume cf. e.g. Arthur Katz (1984 [1973]).

[6] Pomeroy (1975). By the same author cf. Pomeroy (1984b).

[7] Rabinowitz and Richlin ([eds.], 1993).

[8] Cf. Offen (2006).

[9] Vernant (1965-1971 [1963]).

[10] Vernant (1968: 15).

in ancient Greece, particularly in the classical city.[11] And again, closely "genrée" is the *History of Women* edited by George Duby and Michelle Perrot, whose first edition was published in Italy. The introduction to the first volume, dealing with antiquity, which came out in 1990 and was edited by Pauline Schmitt Pantel, Duby and Perrot, explicitly states that their work follows twenty years of *Women's Studies*, in the historiographical tracks of the *Annales* and set amid the new interest in the centrality of *gender*.[12] In France, interest in this specific approach continues, fostered by the theories of the anthropologist Françoise Heritier and her research into male/female opposition.[13] More recently, in 2007 there is the French volume, *Problèmes du genre en Grèce ancienne*,[14] that collects different contributions from social history and literature. And, also in France, the online revue *Clio, Histoire, femmes et sociétés,* which deals with the social history "of gender", has been available since 1995.

It seems therefore that *gender studies* have been, and indeed continue to be, developed particularly in French and Anglo-American research centres, sometimes engaging in bitter debate but sometimes in a productive, scientific exchange of ideas.

Italy has made original contributions too, albeit not specifically declared as such. These are mainly in relation to scientific and philosophical ideas, rituals, sexuality and the theatre. Over the last few years a strong impetus has come from Germany where there have been individual works on specific subjects, which will be examined later. However, in 2002 the situation gained momentum when the proceedings of a meeting about gender studies in relation to classical antiquity were published, with the intent of interesting German academics.[15] There are also recent contributions from Spain.[16] Finally, in Greece, research has been conducted by Norwegian and Greek researchers with the Norwegian Institute for Classics, Archaeology and Cultural History in Athens. Through an examination of Greek literary texts and artistic works they have investigated the division of sexual roles as a social strategy, and also focused on the destabilisation forces that can provoke a crisis. This, they

---

[11] Cf. Loraux (2005).

[12] Duby and Perrot (1990: v-xvii).

[13] Héritier (1996; 2002).

[14] Sebillotte Cuchet and Ernoult ([eds.], 2007).

[15] Feichtinger and Wöhrle ([eds.], 2002).

[16] Cf. Zaragoza Gras (2007). The proceedings of a discussion about maternity have also been published in Riu *et al.* (2000-2001)

hold, is the case both in Ancient Greece and, most originally, in contemporary Greece.[17]

Having established this, an attempt will now be made to analyse works from different research settings and subjects. The sheer volume of this work is unmanageable. So, far from providing an exhaustive review,[18] only brief references will be made to highlight the heuristic nature, new perspectives and new questions which the notion of gender has permitted propounding, referring only to the studies which best exemplify the critical debate.

Social History

The main area in which research is located is history, in particular social history, because the notion of gender was originally elaborated in historical surveys. It is a field which in fact includes all the others which will be treated later, since the rituals, literature, biological ideas, philosophical thought and art are only cultural expressions of the society which has produced them and are therefore directly connected to ideology and dominant political culture. There is nevertheless a specific area of study which is directed at reconstructing sexual roles within the organisation of the city and therefore defining the history and anthropology of social relationships and power differentials within them. These studies are above all in the French and Anglo-American fields and manifest two different approaches, veering between studies which document the status of women and more markedly anthropological studies on social structure, on the imaginative world and different mentalities. Examples of the two types of research are, on one hand, the useful and rich collection of accounts about the lives of women in antiquity by Mary Lefkowitz and Maureen Fant,[19] and, on the other, the work by Sally Humphreys,[20] which has fuelled interest in identifying the spaces and relationship between public and private, between *oikos* and *polis,* and from there in the socio-political organisation as a whole. Her research, which concentrates mainly on classical Athens, also makes use of the cult of the tombs in an attempt to reconstruct the anthropology of the family through the way death is represented.

---

[17] Berggreen and Marinatos ([eds.], 1995).

[18] Over the years and even recently there have been several attempts to synthesise this research field. Among the latest, cf. Demand (2002); Bruit Zaidman and Schmitt Pantel (2007).

[19] Lefkowitz and Fant (1982).

[20] Humphreys (1983).

Both documentary and anthropological research have therefore multiplied, exploring the specific differences regarding women, the family and social fabric, both in different geographical areas, mainly Athens and Sparta, and different historical periods. Consider the monograph by Sarah Pomeroy on women in Hellenist Greece (1984b) and her more recent one on Spartan women (2002).

These studies are interesting because of the possibility through women to reconstruct their relationship with males and therefore the entire political fabric. This is the case in the study by Elke Hartmann (2002), in Germany. It deals with the three women who in the speech *Contra Neaeram*, the pseudo Demosthenes declares to be at the disposal of the Athenian man: the wife, the concubine and the courtesan (LIX 122). Their various functions are investigated by the author, who also states that in spite of their differences they all contribute to defining the male identity of the city.

In French research, the historian of the Greek world who in the course of her activities has constantly maintained a gender perspective is Pauline Schmitt Pantel. She is convinced that the relationship between the sexes is an indispensable lens to penetrate deep within the structures of social organisation and mentality. Schmitt Pantel has already been mentioned as editor of the first volume of *History of Women* dealing with antiquity. More recently, she collected her most significant papers in a single volume (2009), a work which covers over thirty years of research conducted with methodological rigour and focus. The author emphasises repeatedly the interest that the adoption of the notion of gender entails for the historian, particularly in antiquity, in that it can enrich and refine the feminist writing of history with other conceptual tools to extend thinking to forms of power other than the hierarchy between the sexes.[21]

After a survey of the history of gender studies regarding the Greek world, the book is organised around three main subjects: that is adolescents of both sexes whose initiating rituals reconstruct their function and the value that the city holds for them, the division of sexual space within the city, in particular of the *oikos*, and finally the *gender* of violence and heroism.

However, regarding the reconstruction of sexual roles in the city it is necessary to highlight the original contribution to the subject made in

---

[21] Cf. Schmitt Pantel (2009:162): "Poser la question du genre de façon théorique et globale permet à l'histoire féministe d'affiner et d'enrichir ces outils conceptuels, et en particulier de réfléchir au rapport entre la hiérarchie entre les sexes et d'autres formes de pouvoir". Also cf. pp. 23-48 for an analysis of women's studies and gender studies in the 80s and 90s.

earlier years by Nicole Loraux. She used analyses of historical, literary and mythological sources and actually developed a consistent and unitary piece of work on the organisation of the city and its world imagination, placing the division of the sexes right at the centre. Already her wide examination of funeral speeches had lead to the assertion that the properties of male speech construct and establish the city as a men's club (Loraux, 1981a). In *Les Enfants d'Athèna* (Loraux, 1981b) she reconstructs the myth of the autochthones Athenians through mythographic sources and drama, such as Euripides' *Hippolytos,* and *Lysistrata* by Aristophanes, once again construing them as expressions of a constant game of exclusion/inclusion between the two sexes.[22] In the article *Le Lit, La Guerre* (Loraux, 1991) as well as her thoughts on the notion of *ponos* and its uses, the author shows the superimpositions and reciprocal interferences between the representation of female pain at childbirth and male war.

In this area of research it seems that the latest trend is to challenge the rigid dichotomies seen in the work of Pauline Schmitt Pantel and Nicole Loraux, and to substitute reciprocal exchange and interrelation between males and females. Indeed recent work has rendered traditionally established distinctions even more problematic, above all that of *polis/oikos,* public/private, to be understood in sexually connotated opposition, since modern categories not applicable to the ancient world are recognised in it; areas of interchange and cross fertilisation have been identified between *oikos* and *polis*, between internal and external, between private and public, and therefore between feminine and masculine.[23] In my own work *L'ape che tesse* (Andò, 2005) the aim is to show how female knowledge, which grew in the *oikos,* enables the epic and stage heroine to have access to the public area of the *polis.* Moreover it is this same knowledge that became a political metaphor and paradigm in philosophical thought.

The trend to overcome opposition and develop marginality as expressions of autonomy had already been seen in the works edited by Giampera Arrigoni ([ed.], 1985) and Nicole Loraux ([ed.], 1993) respectively, and is the main thrust of a book by Claudia Montepaone (1999).

---

[22] Cf. also Loraux (1996).

[23] Wagner-Hasel (1988). And the collection of studies Polignac and Schmitt Pantel ([eds.], 1998). For the Hellenist-Roman age cf Frei-Stolba and Bielman ([eds.], 1998); Bielman (2002).

Rituals and myth

The works just quoted show that the privileged area occupied by women is in the sphere of the sacred and ritual. The organisation of religious rituals is a conspicuous aspect of social relations, and the problem regarding the presence of women has solicited questions which directly address not only the status of women, but more generally the entire social organisation. So what role did women have in the world of the sacred, the priesthood and sacrificial rites as well as in the numerous feasts reserved to them? This is an essential question, especially if one thinks that Greek religion is directly connected to the political sphere. And therefore, if women are excluded from political life, what function do they have in rituals in which their presence is planned, sometimes exclusively so? In substance, does female religious life consolidate male values in the city or–especially in the case of exclusively female rituals–can we speak of an alternative rituality and religiosity? As can be seen, we are posed with another dichotomy as regards religiosity.

The question is posed first of all regarding sacrifice, the act upon which the civic community consolidates its values and group cohesion. What therefore is the place reserved for women in ritual sacrifice? Is it not necessary to think that, in order to match with political life, they are excluded from sacrificial rites too? This is the response of Marcel Detienne in *La cuisine du sacrifice* (Detienne, 1979), in which, if indeed he really does identify rules of worship which enable women to have access to the sacred area of sacrifice and the meat banquet, they are seen as exceptions confirming the rule of the male monopoly in the sphere of bloodthirsty sacrifice and eating of flesh. He adds, moreover, that women are absolutely forbidden to hold the knife of the sacrificer and butcher. According to Detienne it is as if the city, which in its female feasts of Thesmophoria celebrates the reproduction of the entire civic body, must therefore avoid the danger that the women become the slaughterers, ready to lash out against men, as certain marginal myths suggest. The attractive idea in this study is that of having placed the presence of blood and its symbolism at the centre of the analogous relationship between politics and religion: namely, women would be excluded from pouring the blood of the sacrificial victim because they themselves bleed during menstruation. Political ideology and physiology are therefore interwoven here in the claim that women were excluded from the role of slaughterer. Detienne's position has been discussed and partially improved upon by Robin Osborne (2000), among others, who contests the homology between religion and politics which he sees as overestimated in the research. He claims that conversely, on the basis of the accounts, it is possible to state that women had access to the sacrificial area and, more importantly, ate the sacrificial meat. On the other hand, Giulia Sfamemi Gasparro (1991) states that intrinsic to the role of priestess would be the central function of sacrificer, something which would certainly distort Detienne's reconstruction of the situation. Matthew Dillon's position is analogous. He goes further saying that given the absence of conceptual distinctions between *polis* and temple, such a large participation of women in rituals is evidence against their exclusion from the city (Dillon, 2002).

A successful attempt at improving upon the idea of polarity which has been referred to is the work by Barbara Goff, *Citizen Bacchae* (Goff, 2004), one of the many studies dedicated to female ritual practices. Indeed, as the author demonstrates, if it is true that the civic context is dominated by patriarchs, it is also true that there are some areas of autonomy, albeit controlled ones: Goff observes that women seem to have a certain independence in ceremonies within the *oikos*, those for births, weddings and funerals, just as the rituals reserved for them at different ages, the Arreforia, the Brauronia ritual, the Adonis rituals and the Thesmophoria contribute to construct their gender identity and make them assume a kind of latent citizenship. This she claims is clearly shown by the voluntary sacrifice of virgins seen in the tragedies. Furthermore, there is poetry which, like Sappho's, rests upon worship practices of women only, while the rituals present in the tragedies imply the existence of a conflict between women and the city. Therefore we have a new perspective which strives to identify practices of gender construction in rituals albeit within the dominating male ideology.

A product of the same Californian university and equally original is the work by Susan Cole (2004) on ritual space, in which the quality *gendered* is highlighted: the landscape, cultivated fields, locations for politics and sacred space are analysed through detailed literary and epigraphic documentation. She starts from the first spatial organisation present in Hesiod's poems and continues with the reconstruction of the feasts and rituals of the city. The novelty of the book lies in the comparison of medical texts from which an image of the female body emerges, with its impurities and contaminating nature, a "ritual body", a body to fertilize, like the soil. This body is excluded from certain rituals or on the contrary called to enter the sacred space, in particular in Artemisian rituals, which are interpreted as being closely tied to the political community.

As regards the Artemis rituals at Brauron, the proceedings of a seminar in Urbino[24] are considered worth mentioning. They belong to that Italian output that doesn't claim to be "gender studies" but because of its tone and theme can be fully described as such.

There are many studies which focus on individual female rituals and clarify both their sense and function.[25] Inevitably, closely interwoven with these are the studies of the myths, where the perspective of gender has brought about new viewpoints and studies. One of the latest arguments is that provided by Lillian Doherty (2001), who after reviewing the different approaches to research into myths, including psychoanalysis, structuralism, post-structuralism and ritualism, shows how far myths subvert and reaffirm gender roles. The myths therefore represent a paradigm which, whatever method of analysis is adopted, leads towards an interpretation of the relationship between the sexes.

Some of Nicole Loraux's studies move in the same direction, even in this innovative and original field. In the previously quoted *History of Women,* her article "Che cos'è una dea?" (Duby and Perrot, [eds.], 1990: 13-55) goes to the roots of the conception of female divinities in the Greek pantheon and their function in the imaginative world, again facing, among other things, the historical-religious problem of the Great Mother Goddess, a phantom so common as to have become real. More generally, many of her studies, which sweep between different texts and epochs, tend to show how far the female is intrusively present in the construction of male heroic characters, such as warriors of the *Iliad* or Heracles.[26]

Literature: poetry texts and more

It is certainly difficult to isolate and distinguish an independent contribution from gender studies in relation to "literary" texts. And yet some attempts have been made. For example, to compensate for the substantial silence of women in antiquity, scholars have wanted to give voice to female expression, be they the few poetic voices of women or the voices of female characters which male writers speak about. This is the case in the collection of articles *Making Silence Speak. Women's Voices in*

[24] Gentili and Perusino ([ed,], 2002). Reference point on the religion of young Athenian girls is Brulé (1987).

[25] E.g. the *Halōa* festivities, cf. Patera and Zografou (2001). By re-examining the sources these scholars have scaled down the amount of obscenity and therefore the nature of the agricultural festival.

[26] Loraux (1991: chh. IV, V, VII).

Greek Literature and Society,[27] in which different contributions are dedicated to female lyric voices, the funeral epigrams for dead women and the women who emerge from Attic oratory.

Another interesting form of inquiry regards poetic *performances,* both choral and individual, in which gender has been effectively highlighted through the analysis of performance contexts, mainly at the thiasus and symposium.[28]

However, it is certainly Attic theatrical scripts, both tragedy and comedy, which have supplied most material, both in terms of quantity and prominence, to interpret gender. This is partly because of the paradoxical abundance of female characters on the Attic stage but mainly in consequence of the intrinsically politicised Greek theatre which imposed questions about ideology and social models conveyed in the drama. North American studies stand out in this sector, above all those by Froma Zeitlin and Helene Foley. In a 1996 collection of articles, Zeitlin closely examines the paradigmatic figures of Penelope and Pandora. She then dwells upon the dramas which best lend themselves to the reconstruction of social dynamics from gender relations such as the trilogies of *The Oresteia* and *The Danaides* (*Hecuba, Hippolytus, Ion*) and Aristophanes' *Thesmophoriazusae*.[29] Helen Foley has also collected and published her previous contributions on tragedy, which highlight in the first part the politicisation of the female tragic lament, in the second the dynamics in marriage relationships, and finally the numerous female characters who are "moral agents", of whom in spite of the apparent contradiction, she confirms there is real fear in Athenian life.[30] Laura McClure's *Spoken like a Woman, Speech and Gender in Athenian Drama* (1999) lies in the same area of research. She analyses female speech in five dramas: *Agamemnon, Hyppolytus, Andromache, Thesmophoriazusae* and *Ecclesiazusae*, through which she shows the hegemony of male speech on which the role of the citizen is constructed and civic ideology is founded. Research on sexual language specifically in the tragedy *Orestea* has been done by Simon Goldhill (1984).

Between the studies on rituality and those on literary texts lies the work of Laurie O'Higgins, *Women and Humor in Classical Greece* (2003), which after an analysis of cult obscenity focused on Iambe in the *Homeric hymn to Demeter,* concentrates on the women in Attic comedy.

[27] Lardinois and McClure ([eds.], 2001).
[28] Cf. Stehle (1997).
[29] Zeitlin (1996).
[30] Foley (2001).

However, among the analyses of drama texts from the perspective of gender one cannot but once again refer to Nicole Loraux, who has concentrated a large part of her research on the representation of the female on the Attic stage, both tragic and comic, to trace the social dynamics and mentality of the time. The works in question are: *Façons tragique de tuer une femme* (1985), *Les mères en deuil* (1990), and *La voix endeuillée* (1999). In the latter she shows that the female funeral lament is the counterpart of male funeral oration and that both have a public and political character.

Biology, Medicine: Body, Sexuality

Among the human sciences it has obviously been above all the studies on Greek medicine which have adopted the perspective of gender. In spite of the evidence of sexual dimorphism, which is mentioned insistently in the medical and biological texts, women's studies needed a push towards these non-literary texts and other sources thus far ignored by scholars. Thus emerged the first pioneering studies on ancient gynaecology intended to reveal its underlying ideology: e.g. the works of Aline Rousselle and Paola Manuli, which were both published in 1980. Since then there have been numerous studies giving credence to this perspective, mainly in Italy and among the Anglo-Americans, less so in France. What emerges is the strong character of ideological construction of the female body made by male medical discourse, which has represented females via the role of motherhood. For this reason female physiology and pathology are always directly linked to such a role, in the various events to which the female body is subjected at different ages and functions: virginity, pregnancy, childbirth and post partum.

After the papers by Geoffrey Lloyd in *Science, Folklore and Ideology* (1983), a series of monographs in England and the USA have further examined the nature of the female in Greek medicine. These include the work by Lesley Ann Dean-Jones (1994) on the female body in Greek science and that of Nancy Demand (1994) on maternity. In *Hippocrates' Woman* (1998), Helen King is credited with having picked out and followed the only thread which leads from the first "mythical" ideologised gynaecology to the Victorian age. The diachronic study on "hysteria" reveals its social construction, and the acute observations regarding menstrual blood, declared in the sources to be similar to that of the sacrificial victim, highlight the link that the Greek city endorses between the sacrificial function and the reproductive function of women. Work by

Lidie Bodiou (1999) is entirely concerned with female blood, both with regard to its symbolism and physio-pathological phenomena.[31]

Construction of the male is also necessary from a gender perspective. Jean Baptiste Bonnard (2004) takes steps to address this in *Le complexe de Zeus*, looking at the way in which the Greek world constructed and represented paternity. A peculiarity of the book is that of not only having used sources on biological and philosophical thought but also mythological and literary sources. From these Bonnard confirms the binary hierarchical logic between male and female, as suggested by the myths about paternal generation of children without a mother, for example the births of Athena, Dionysos and Aphrodite, and the embryological conceptions which, even if they relate the female contribution, never question the indispensability and superiority of male sperm.[32]

The study of the physiology and pathology of the body leads to the study of its management and social representation, particularly in the more intimate and personal sphere of sexuality. The brilliant research by Michel Foucault in the second and third parts of his *Histoire de la sexualité* (1984, *L'usage des plaisirs* and *Le souci de soi*), dealing with archaic and classical Greece and the first centuries of the Christian era respectively, have clarified how far sex is a product of social construction, the effect of dynamics of control and dominant relationships. Foucault's analysis is not from the perspective of the heterosexual/homosexual dichotomy, but that of active/passive, already present in *Greek Homosexuality* by Kenneth Dover (1978), considered apt to express sexual and political power of the adult male within the "realm of phallocracy" that is the classical city.[33]

Dover and Foucault have greatly influenced later studies on sexuality. There are those by Eva Cantarella (1988) in Italy, Halperin (1990) and Winkler (1990) in the USA,[34] as well as a collection of papers edited by Maria Wyke ([ed.], 1998), in which male and female bodies are explicitly and markedly interpreted as "parchments" on which each gender writes its own power discourse.

A different analysis is that of David Konstan (1994), who talks about "sexual symmetry" in the Greek novel, as well as that of James Davidson (1997), who challenges Foucault's dichotomy and phallocentricity by speaking of the important role of courtesans and prostitutes and of other pleasures of the body in classical Athens.

---

[31] Also cf. Jufresa ([ed.], 2000).

[32] On masculinity cf. also Foxhall and Salmon ([eds.], 1998a; [eds], 1998b).

[33] cf. also Keuls (1985).

[34] Cf. also the collection Halperin, Winkler and Zeitlin ([eds.], 1990).

These last two studies have focused attention on heterosexuality, neglected in the works previously cited, and have highlighted the risk that ideological factors and an excess of political militancy on the part of scholars could blur the object of inquiry to the point of identifying sexuality in Greece with homosexuality.[35]

Until recently, the main reference point on female homo-erotic experience was the excellent studies on the young female choirs by Claude Calame.[36] However, a recent book by Sandra Boehringer (2007) investigates the subject with depth of analysis and from an original perspective. She underlines the inexistence of "homosexuality". In the wake of American teaching and the Foucault school she asserts that to understand the construction of sexuality one has to refer to sexual practices, which develop and change over time in relation to the different social role assumed in different contexts.

Botany, practices of magic

Over the last few years the notion of gender has offered interesting results in the area of biology and also botany. Indeed an attempt to analyse and interpret the botanical texts of Theophrastus has been conducted to show how ancient society even projected cultural representation of male and female onto plants. Through stimulating comparisons with both Aristotelian biology and the ensemble of myths and rituals ascribable to the divinity of the pantheon it is thus possible to elaborate the symbolic value of vegetal species and the cultural function assigned to them in the social context and religious sphere.[37]

Animal Studies consists of the recent interest towards animals, analysed as the intersection of multiple cultural and symbolic combinations. Within this field we have the work by Cristiana Franco (2003) on the dog, in which the perspective of gender helps highlight the dog's substantial belonging to the female world, because it is an animal "without restraint", bearer of that *anaideia* which marks women in the Greek world.

Christopher Faraone (1999a; 1999b) has clarified the crucial importance of gender in the area of talismans and spells for the reconstruction of practices of love magic, namely distinguishing the sex of the performer of the practice from the person at whom it is aimed. His

---

[35] Cf. misgivings and advice on caution from Martos Montiel (2002). Very harsh criticism in Gourevitch (1999).

[36] Calame (1977; 1992, 2010; [ed.], 1983).

[37] Cf. Bretin-Chabrol and Leduc (2009). Cf also the previous article by L. Foxhall (1998).

analysis fully confirms the hierarchy between the genders emerging at a social level, even where roles are apparently inverted such as in the case of courtesans.

On the subject of magic the studies of Heinrich von Staden (1991) on female gynaecological medicine should be mentioned. Here the use of so-called unclean therapy, such as excrement are to be traced back to "homeopathic" magic, in that the contaminating nature of women's bodily functions leads to curing impurity with impurity.

Philosophy

The work *Madre materia. Sociologia e biologia della donna greca* (1983), from the Pavia school in Italy constitutes a kind of pivot between the studies on biology and those on philosophical thought. Indeed, Silvia Campese and Giulia Sissa show how according to Aristotle the male biological concept of the female is in perfect parallel with his political and ethical concept. This is in so far as the *adynamia* of the female body, incapable of reproduction without semen, corresponds to a deliberative faculty (*bouleutikon*) lacking in authority (*akyron*), which leads women to have need of the male *archon*, endowed with the right to control, for the sake of her *soterìa*. Later Giulia Sissa (1990) returned to the subject, extending this idea to Platonic thought. Here she emphasises the "appropriative" use of the female in particular in Plato's metaphors of pregnancy and childbirth to show the production of philosophical thought. Important in her study is the analysis of the notions of *genos* and *eidos* in both Aristotle and Plato, an analysis directed towards establishing the relationship between sexual difference and uniqueness of the *genos anthropinion.*

The recent work by Michael Kochin (2002) has a different structure. He analyses gender in the political works of Plato, where he sees a plan for real sharing between males and females in terms of justice. Spain has also supplied a recent overview on the conceptualisation of the female in Greek philosophy.[38]

Another way to understand the relationship between women and philosophy has been that of analysing and studying female philosophers,

---

[38] Pérez Sedeño E. ([ed.], 1994).

like Aspasia, Teanò and others about whom it is possible to have accounts.[39]

While on the subject of philosophy, the Italian philosopher Adriana Cavarero should be mentioned because of her completely different approach which follows in the wake of the theory of sexual difference. In *Nonostante Platone* (1990) she makes use of female figures, some of whom are in Plato, such as the Thracian maid or Diotima, to reconstruct the underlying symbolic maternal and female order that can be read between the lines. Work on the Greek world more generally seeks out archetypal power figures distinct from universalistic male speech and from a standpoint which permits women, traditionally uninvolved in philosophical *logos*, to recognise categories of thought which belong to their gender and themselves. Such is the case with Hannah Arendt's category of "birth". This is a philosophical perspective largely outside gender studies, still rooted in the feminism of equality and which in both France and Italy is unaffected by ideas of sexual difference, in spite of the philosophical practice of Luce Irigary and the Diotima community active in Italy.

Art and Archaeology

Gender studies regarding artistic monuments, ceramics, painting and sculpture "read" as texts, able to supply significant testimony about social structures and ways of thinking, reveal the marked effects of the different approaches already mentioned by the French and North American academics. The miscellaneous collection of articles *Pandora. Women in Classical Greece*,[40] which, in so far as it is a catalogue for an exhibition, is rich in illustrations and contains French and American contributions, clearly shows the dual methods of analysis of artistic monuments. The political militancy of radical North American feminism was explicitly revealed in the 1990s in the Archaeological Institute of America when studies directed towards the canons of Greek art revealed their dependence on hegemonic power structures and therefore the ideology underlying the depiction of bodies, in particular the female nude. Perhaps the most

[39] On Teanò and Aspasia see contributions by C. Montepaone and N. Loraux respectively in the collection Loraux ([ed.], 1993). An attempt at an overall summary in Pietra (1997).

[40] Reeder ([ed.], 1995).

militant and political example of this is the collection of articles *Naked Truths*.[41]

However, the French scholars François Lissarrague and Françoise Frontisi have different objectives. Starting with the images depicted in art, they aim at reconstructing the mental images that it is possible to hypothesise from specific subjects[42]. In this work the representation of the female and the times and places in life to which she is destined, for example the gynaeceum or marriage, is always analysed in relation to the male, although the authors concede that reality cannot truly be reflected or described in these pictorial images. In these studies there is evident challenge of notions which by then were held to be established, such as the participation of only heterosexuals and prostitutes in the male symposium. Such notions were unverifiable from the images, or even from the exclusive destination of space within the *oikos* reserved for women.

A full rediscussion of the life of Greek women through iconology is found in the book by Sian Lewis (2002), which is presented as an iconographic manual: all aspects of women's lives are analysed from the documentation offered by ceramics. The main objective in the book is to underline the distance between the images on one hand and real life and literature on the other. The vases were mainly produced for the foreign market so women are depicted in circumstances in which the city tended to value their role, such as religious rituals, funerals, the robing of warriors rather than other moments of their life. The same goes for domestic life where the depictions certainly show instances and objects of daily life, but lend more weight for example to working with wool rather than cooking, corresponding with the traditional and ideologically based sexual division of work. Moreover, it is difficult to identify the clear, distinct free/heterosexual to whom we are accustomed in the written sources, just as the traditional interpretation of female reclusion in the gynaecium contrasts with iconographical data.

In the same year, 2002, we have the work by Gloria Ferrari in which some fine research is conducted into "gender scenes" on vases, analysed as communicative events and using language theory. The research concentrates on the tension between male and female at times of life change for both sexes. It highlights that the passing of a young male into adulthood constitutes a sort of gender change in that he has to lay down female attributes to acquire *andreia*, while in nuptial rites the bride does not change gender but status, becoming a potential reproducer of citizens.

---

[41] Koloski-Ostrow and Lyons ([eds.], 1997); in particular see contribution by S. Brown (pp. 12-42).

[42] Cf. contributions by F. Lissarrague and F. Frontisi in Veyne ([ed.], 1998).

Attention to gender has produced particularly original and innovative results in archaeological research when excavations have been confronted with trying to identify an organisation which accounts for specific differences regarding gender in the necropolis, urban layout and domestic architecture.[43]

Conclusions

This rapid analysis has shown how much the adoption of the category of gender in studies of the Greek world over the last thirty years has enabled scholars to pose new questions and follow new lines of research. It is also true that in some cases, especially when ideology overruled and ended up determining the selection of sources, thereby obscuring them, the gender approach has shown limitations and potential risks for research.

I believe that gender can continue to be, to quote Joan Scott "a useful category for historical inquiry" if ideology is abandoned once and for all, without hypostatising and being inflexible regarding the female/male dichotomy. I would like to add Judith Butler's suggestion of *undoing* gender, when one is in danger of forcing the sources, locking them in simple and preconceived schemes. However, gender can still be a valid heuristic tool: that is, *doing* is also necessary, looking at the polarity between the sexual genders in their continual dialectic tension, continually shifting from one pole to the other, checking all the areas of intersection and exchange. The unambiguous traditional model of the hierarchy between the sexes is of course still an effective model to identify power structures even beyond the male/female dichotomy. In what it rejects and its anomalies, one will be able to find a much more complex and modulated reconstruction of Greek society in its different forms of expression, from politics to art, philosophical thought or science in its symbolic structures and its imaginative world. In the area of conflict and negotiation, loans and appropriation, resides the possibility of finding fresh answers to the ever new questions which we continue to pose the Greek texts.

---

[43] Cf. Schmitt Pantel (2009: 105-122); Morris (1999).

References

Andò, V. (2005). *L'ape che tesse. Saperi femminili nella Grecia antica*, Roma: Carocci.

Arrigoni, G., ed. (1985). *Le donne in Grecia*, Roma-Bari: Laterza.

Arthur Katz, M. (1984 [1973]). "Early Greece: The Origins of the Western Attitude toward Women". In J. Peradotto and J.P. Sullivan (eds.), *Women in the Ancient World. The "Arethusa" Papers*, (pp. 7-58), Albany: State Univ. New York Press.

Beauvoir, S. de (1999 [1949]). *Il secondo sesso*, (it. trans.), Milano: Il Saggiatore

Berggreen, B. and N. Marinatos, eds. (1995). *Greece and Gender*, Bergen: Norw. Inst. Athen.

Bielman, A. (2002). *Femmes en public dans le monde hellénistique, IV-I siècle av. J.-C.,* Paris: Sedes.

Bodiou, L. (1999). *Histoire du sang des femmes grecques: filles, femmes, mères. D'après les écrits médicaux et biologiques*, Dissertation Univ. Rennes II.

Boehringer, S. (2007). *L'homosexualité féminine dans l'Antiquité grecque et romaine*, Paris: Les Belles Lettres.

Bonnard, J.-B. (2004). *Le complexe de Zeus. Représentation de la paternité en Grèce ancienne*, Paris: Pubbl. de la Sorbonne.

Boxer, M. "*Women's Studies* aux Etats-Unis: trente ans de succès et de contestation", *Clio* 13: 211-238 http://clio.revues.org/index142.html

Bretin-Chabrol, M. and C. Leduc (2009). "La botanique antique et la problématique du genre", *Clio* 29. http://clio.revues.org/index9276.html.

Brown, S. (1997). "'Ways of Seeing' Women in Antiquity. An Introduction to feminism in classical archaeology and ancient art history". In A.O. Koloski-Ostrow and C.L. Lyons (eds.), *Naked Truths. Women, Sexuality, and Gender in Classical Art and Archaeology*, (pp. 12-42), London-New York: Routledge.

Bruit Zaidman, L. and P. Schmitt Pantel (2007). "L'historiographie du genre: état des lieux". In V. Sebillotte Cuchet and N. Ernoult, eds., *Problèmes du genre en Grèce ancienne*, (pp. 27-48), Paris: Pubblications de la Sorbonne.

Brulé, P. (1987). *La fille d'Athènes. La religion des filles à Athènes è l'éepoque classique. Mythes, cultes et société*, Paris: Les Belles Lettres.

Butler, J. (2004). *Undoing Gender*, New York-London: Routledge.

Calame, C. (1977). *Les choeurs de jeunes filles en Grèce archaïque,* Roma: Edizioni dell'Ateneo.

—. (1992, 2010). *I Greci e l'eros, Simboli, pratiche e luoghi*, Roma: Laterza.

—. ed. (1983). *L'amore in Grecia*, Roma-Bari: Laterza.

Campese, S. (1983). "Madre materia. Donna, casa, città nell'antropologia di Aristotele". In S. Campese, P. Manuli and G. Sissa, *Madre materia. Sociologia e biologia della donna greca*, (pp. 13-79), Torino: Boringhieri

Campese, S., P. Manuli and G. Sissa (1983). *Madre materia. Sociologia e biologia della donna greca*, Torino: Boringhieri.

Cantarella, E. (1988). *Secondo natura. La bisessualità nel mondo antico*, Roma: Editori Riuniti.

Cavarero, A. (1990). *Nonostante Platone. Figure femminili nella filosofia antica*, Roma: Editori Riuniti.

Clio, Histoire, femmes et sociétés (1995-). http://clio.revues.org

Cole, S.G. (2004). *Landscapes, Gender and Ritual Space. The Ancient Greek Experience*, Berkeley-Los Angeles: Univ. of California Press.

Davidson, J. (1997). *Courtesan and Fishcakes. The Consuming Passions of Classical Athens*, Hammersmith: Fontana Press.

Dean-Jones, L.A. (1994). *Women's Bodies in Classical Greek Science*, Oxford: Clarendon Press.

Demand, N. (1994). *Birth, Death and Motherhood in Classical Greece*, Baltimore-London: Johns Hopkins Univ. Press.

—. (2002). "Gender Studies and History: Participation and Power". In S. M. Burstein, N. Demand, I. Morris and L. Tritle, *Current Issues and the Study of Ancient History*, (pp. 32-43), Claremont: Regina Books.

Detienne, M. (1979) "Violentes Eugénies. En pleines Thesmophories, des femmes couvertes de sang". In M. Detienne and J. P. Vernant (eds.), *La cuisine du sacrifice en pays grec*, (pp. 183-214), Paris: Gallimard.

Dillon, M. (2002). *Girls and women in classical Greek religion*, London-New York: Routledge.

Doherty, L.E. (2001). *Gender and the interpretation of classical myth*, London: Duckworth.

Dover, K.J. (1978). *Greek Homosexuality*, London: Duckworth.

Duby, G. and M. Perrot, eds. (1990). *Storia delle donne. L'antichità* (a cura di P. Schmitt Pantel), Roma-Bari: Laterza.

Faraone, Ch. A. (1999a). "The construction of gender in ancient Greek love magic". In D.R. Jordan, H. Montgomery and E. Thomassen (eds.), *The world of ancient magic*, (pp. 279-282), Bergen: Norw. Inst. at Athens.

—. (1999b). *Ancient Greek Love Magic*, Cambridge Mass.: Harvard Univ. Press.

Feichtinger, B. and G. Wöhrle, eds. (2002). *Gender Studies in den Altertumswissenschaften: Möglichkeiten und Grenzen*, Trier: Wissenschaftlicher Verl.

Ferrari, G. (2002). *Figures of Speech. Men and Maidens in Ancient Greece,* Chicago-London: The University of Chicago Press.

Foley, H.P. (2001). *Female Acts in Greek Drama*, Princeton-Oxford: Princeton Univ. Press.

Foucault, M. (1984a). *L'usage des plaisirs,* Paris: Gallimard.

—. (1984b). *Le souci de soi*, Paris: Gallimard.

Foxhall, L. (1998). "Natural sex: the attribution of sex and gender to plants in ancient Greece". In L. Foxhall and J. Salmon (eds.), *Thinking Men. Masculinity and its Self-Representation in the Classical Tradition*, (pp. 57-70), London-New York: Routledge.

Foxhall, L. and J. Salmon, eds. (1998a). *When Men were Men. Masculinity, Power and Identity in Classical Antiquity*, London-New York: Routledge.

Foxhall, L. and J. Salmon, eds. (1998b). *Thinking Men. Masculinity and its Self-Representation in the Classical Tradition*, London-New York: Routledge.

Franco, C. (2003). *Senza ritegno: Il cane e la donna nell'immagine della grecia antica*, Bologna: Il Mulino.

Frei-Stolba, R. and A. Bielman, eds. (1998). Special Issue "Femmes et vie publique dans l'Antiquité gréco-romaine", *Études de Lettres* 1.

Gentili, B. and F. Perusino, eds. (2002). *Le orse di Brauron. Un rituale di iniziazione femminile nel santuario di Artemide*, Pisa: ETS.

Goff, B. (2004). *Citizen Bacchae. Women's Ritual Practice in Ancient Greece*, Berkeley-Los Angeles: Univ. of California Press.

Goldhill, S. (1984). *Language, sexuality, narrative, the Oresteia*, Cambridge: Cambridge Univ. Pr.

Gourevitch, D. (1999). "La sexualité de l'Antiquité. Essai à propos de pubblications récents", *L'Antiquité classique* LXVIII: 331-334.

Halperin, D.M. (1990). *One Hundred Years of Homosexuality and Other Essays on Greek Love*, New York: Routledge.

Halperin, D.M., J.J. Winkler and F. I. Zeitlin, eds. (1990). *Before Sexuality. The construction of the Erotic Experience in the Ancient Greek World.* Princeton: Princeton Univ. Press.

Hartmann, E. (2002). *Heirat, Hetärentum und Konkubinat im klassischen Athen*, Frankfurt am Main: Campus Verlag.

Héritier, F. (1996). *Masculin/Féminin. La pensée de la différence*, Paris: Odile Jacob.

—. (2002). *Masculin/Féminin II. Dissoudre la hiérarchie*, Paris: Odile Jacob.

Humphreys, S.C. (1983). *The Family, Women and Death. Comparative Studies*, London: Routledge.

Jufresa, M., ed. (2000). Special Issue "Mujer y medicina en el mundo antiguo", *Arenal*, 7/7.

Keuls, E.C. (1985). *The reign of the phallus. Sexual politics in ancient Athens*, New York: Harper & Row.

King, H. (1998). *Hippocrates' Woman. Reading the Female Body in Ancient Greece*, London-New York: Routledge.

Kochin, M.S. (2002). *Gender and rhetoric in Plato's political thought*, Cambridge-New York: Cambridge Univ. Press.

Koloski-Ostrow, A.O. and C.L. Lyons, eds. (1997). *Naked Truths. Women, Sexuality, and Gender in Classical Art and Archaeology*, London-New York: Routledge.

Konstan, D. (1994). *Sexual Simmetry. Love in the Ancient Novel and Related Genres*, Princeton: Princeton Univ. Press.

Lardinois, A. and L. McClure, eds. (2001). *Making Silence Speak. Women's Voices in Greek Literature and Society*, Princeton: Princeton Univ. Press.

Lefkowitz, M.R. and M.B. Fant (1982). *Women's Life in Greece and Rome,* London: Duckworth.

"Les voies traversières de Nicole Loraux: une helléniste à la croisée des sciences sociales", Special Issue *Espaces Temps-Clio* 87-88 (2005).

Lewis, S. (2002). *The Athenian Women. An Iconographic Handbook*, London-New York: Routledge.

Lloyd, G.E.R. (1983). *Science, Folklore and Ideology. Studies in the Life Sciences in Ancient Greece*, Cambridge: Cambridge Univ. Press.

Loraux, N. (1981a). *L'invention d'Athènes. Histoire de l'oraison funèbre dans la «cité classique»*, Paris: Mouton.

—. (1981b). *Les enfants d'Athéna. Idées athéniennes sur la citoyenneté et la division des sexes*, Paris: Maspero.

—. (1985). *Façons tragique de tuer une femme*, Paris: Hachette

—. (1990). *Les mères en deuil*, Paris: Seuil.

—. (1991). *Les expériences de Tirésias. Le féminin et l'homme grec*, Paris: Gallimard.

—. (1996). *Né de la Terre*, Paris: Seuil.

—. (1999). *La voix endeuillée Essai sur la tragédie grecque*, Paris: Gallimard.

—. ed. (1993). *Grecia al femminile*, Roma-Bari: Laterza.

Manuli, P. (1980). “Fisiologia e patologia del femminile negli scritti ippocratici dell’antica ginecologia greca”. In M. D. Grmek (ed.), *Hippocratica. Actes du Colloque hippocratique de Paris (4-9 sept. 1978)*, (pp. 393-408), Paris.

Martos Montiel, J.F. (2002). “Luces y sombras de los *women and gender studies* en el campo de los estudios clásicos”. In I. Calero Secall and M. de los Ángeles Durán López (eds.), *Debilidad aparente, fortaleza en realidad: la mujer como modelo en la literatura griega antigua y su proyección en el mundo actual*, (pp. 237-247), Málaga: Universidad de Málaga.

Maruani, M., ed. (2005). *Femmes, genre et sociétés: l'état des savoirs*, Paris: La Découverte.

Mattioli, U. *et al* , eds. (1991). *Donna e culture*, Genova: Marietti.

McClure, L. (1999). *Spoken like a Woman, Speech and Gender in Athenian Drama,* Princeton: Princeton Univ. Press.

Montepaone, C. (1999). *Lo spazio del margine. Prospettive sul femminile nella comunità antica*, Roma: Donzelli.

Morris, I. (1999). “Archaeology and gender ideologies in early archaic Greece”, *TAPhA* 129: 305-317.

O’Higgins, L. (2003). *Women and Humor in Classical Greece*, Cambridge: Cambridge Univ. Pr.

Oakley, A. (1972). *Sex, Gender and Society*, London: Temple Smith

Offen, K. (2005). “Le *gender* est-il une invention américaine?”, *Clio* 24: 291-304. http://clio.revues.org/index4702.html

Osborne, R. (2000). “Women and Sacrifice in Classical Greece”. In R. Buxton (ed.), *Oxford Readings in Greek Religion*, (pp. 294-313), Oxford: Oxford Univ. Press.

Patera, I. and A. Zografou (2001). “Femmes à la fête des Halôa: le secret de l’imaginaire”, Clio 14. http://clio.revues.org/index102.html

Peradotto, J. and J.P. Sullivan, eds. (1984). *Women in the Ancient World. The “Arethusa” Papers*, Albany: State Univ. New York Press.

Pérez Sedeño, E., ed. (1994). *Conceptualización de lo femenino en la filosofia antigua*, Madrid: Siglo XXI.

Pietra, R. (1997). *Les Femmes philosophes de l'Antiquité gréco-romaine,* Paris-Montréal: l'Harmattan.

Polignac, F. de and P. Schmitt Pantel, eds. (1998). Special Issue “Public et privé dans les cités grecques”, *Ktema* 23.

Pomeroy, S.B. (1975). *Goddesses, whores, wives, and slaves,* New York: Schoken Books.

—. (1984a). *Selected Bibliography on Women in Classical Antiquity*. In J. Peradotto and J.P. Sullivan (eds.), *Women in the Ancient World. The*

"Arethusa" Papers, (pp. 315-72), Albany: State Univ. New York Press.

—. (1984b). *Women in Hellenistic Egypt from Alexander to Cleopatra*, New York: Schocken.

—. (2002). *Spartan Women*, New York: Oxford University Press.

Rabinowitz, N.S. and A. Richlin, eds. (1993). *Feminist Theory and the Classics*, New York-London: Routledge.

Reeder, E., ed. (1995). *Pandora. Women in Classical Greece*, Princeton: Princeton University Press.

Riu, X. *et al* (2000-2001) "Les mares dels grecs", *Quaderns Catalans de Cultura Clàssica*, 16-17: 25-145

Rousselle, A. (1980). "Observation féminine et idéologie masculine; le corps de la femme d'après les médecins grecs", *Annales ESC* 35/5: 1089-1115.

Schmitt Pantel, P. (2009). *Aithra et Pandora. Femmes Genre et Cité dans la Gréce antique*, Paris: L'Harmattan.

Scott, J.W. (1986). "Gender: A Useful Category of Historical Analysis", *American Historical Review* 5/91: 1053-75

Sebillotte Cuchet, V. and N. Ernoult, eds. (2007). *Problèmes du genre en Grèce ancienne*, Paris: Pubblications de la Sorbonne.

Sergent, B. (1984). *L'homosexualité dans la mythologie grecque*, Paris: Payot.

Sfameni Gasparro, G. (1991). "Ruolo cultuale della donna in Grecia e a Roma: per una tipologia religiosa". In U. Mattioli *et al* (eds.), *Donna e culture*, (pp. 58-121), Genova: Marietti.

Sissa, G. (1983). "Il corpo della donna. Lineamenti di una ginecologia filosofica". In S. Campese, P. Manuli and G. Sissa, *Madre materia. Sociologia e biologia della donna greca*, (pp. 81-145), Torino: Boringhieri.

—. (1990). "Filosofie del genere: Platone, Aristotele e la differenza dei sessi". In G. Duby and M. Perrot (eds.), *Storia delle donne. L'antichità*, (pp. 58-100), Roma-Bari: Laterza.

Staden, H. von (1991). "Matière et signification. Rituel, sexe et pharmacologie dans le Corpus hippocratique", *L'Antiquité classique* 60: 42-61.

Stehle, E.M. (1997). *Performance and Gender in Ancient Greece. Nondramatic Poetry in its Setting*, Princeton: Princeton University Pr.

Vernant, J.-P. (1965-1971). "Hestia-Hermès". In *Mythe et pensée chez les Grecs. Etudes de psychologie historique*, Paris: Maspero. (First publ. *L'Homme. Revue française d'Anthropologie* 3 (1963): 12-50).

—. (1968). *Introduction* to *Problèmes de la guerre en Grèce ancienne*, Paris: Mouton.

Veyne, P., ed. (1998). *Les mystères du gynécée*, Paris: Gallimard.

Wagner-Hasel, B. (1988). "Das Private wird politisch. Die Perspektive 'Geschlecht' in den Altertumswissenschaften". In U.A.J. Becker and J. Rüsen (eds.), *Weiblichkeit in geschichtlicher Perspektive*, (pp. 11-50), Frankfuurt am Main: Suhrkamp

Wike, M., ed. (1998). *Parchments of Gender. Deciphering the Body in Antiquity*, Oxford: Clarendon Press.

Winkler, J.J. (1990). *The Constraints of Desire. The Antropology of Sex and Gender in Ancient Greece*, New York: Routledge.

Zaragoza Gras, J., ed. (2007). *La construcció social del gènere a Grècia*, in *Invisibilitat i poder*, Tarragona: Arola

Zeitlin, F. I. (1996). *Playing the Other. Gender and Society in Classical Greek Literature*, Chicago-London: Univ. Chicago Press.

THE FIELD OF ARGUMENTATION: FROM ARISTOTLE TO THE PRESENT DAY*

LUIS VEGA

Aristotle's paternity of argumentation studies

Let us recall the famous statement from the final summary in *On Sophistical Refutations* in which Aristotle himself claims paternity of the study of reasoning:

> Moreover, on the subject of rhetoric there exists much that has been said since long ago, whereas on the subject of reasoning [*peri tou sullogidsesthai*] we had nothing else of an earlier date to speak of at all, but were kept at work for a long time in experimental researches (*SR*, 184a9-184b3).

As is widely known, this statement has been misinterpreted in two ways. On one hand, some have understood that Aristotle was claiming that before him no one had engaged in rational discourse, or at least, that no one had taken an interest in this type of discourse. This incorrect interpretation contributed to Locke's irony in saying that the task of making human beings rational had been entrusted to Aristotle, as well as the disproportionate reaction of Enriques, who considered Aristotle to be a sort of compiler and editor of the knowledge of logic developed by the earlier tradition of mathematics and dialectic.[1] On the other hand and at virtually the opposite extreme, historians of logic have taken this statement *pro domo sua* as the founding act of the discipline of formal logic.

For my part, as I have already suggested, I read this as Aristotle attributing to himself the founding of or initial steps in the area of

---

* English translation by Nedra Rivera Huntington. Paper supported by a grant by the Spanish Ministry of Economy and Competitiveness (FFI2011-23125).

[1] Vide J. Locke (1689: B. IV, Ch. xvii, § 4): "But God has not been so sparing to men to make them barely two-legged creatures, and left it to Aristotle to make them rational", F. Enriques (1949 [1922]: 8-9).

discourse studies which is today encompassed by the field of argumentation theory. It is related to formal and informal logic, dialectic, rhetoric, fallacies and even hints of other fields or disciplines such as those we now call 'public deliberation' or 'critical thinking'.

For example, Aristotle mentions a variety of uses of critical thinking, a sort of generic discipline covering training in and acquisition of discursive abilities and skills. To give an indication of this, we might mention the distinction between two different outcomes of education: (a) the first-order subject or substantive knowledge provided by the science in question; and (b) certain second-order critical training: an educated person should be capable of forming a fair judgement about the quality of an exposition in any branch of general knowledge, not just a particular field, and should be capable of judging whether the discourse is correct irrespective of the truth or falsity of the propositions in question. There are also specific references to a characteristic of people who have been trained and educated, namely: their capacity for discernment with regard to any sort of proof and in any field. Thus, Aristotle states that the false idea that everything must be demonstrated is due to "want of education, for not knowing which things require demonstration and which do not is simply a want of education [*apaideusia*]" (*Metaphys*. 1006a5-8). And in another context he insists:

> It is a mark of an educated man to look for precision in each field of inquiry just so far as the nature of the subject admits; it is evidently equally foolish to accept a merely persuasive account from a mathematician and to demand scientific proofs from a rhetorician (*NE* 1094b24-27).

Now then, here I propose to consider Aristotle's paternity of the field of argumentation studies by looking at two fundamental issues implied in the statement quoted above. First, in what sense is Aristotle's contribution to argumentation studies 'foundational' and therefore in what regard does it constitute a basic precedent for the later development of the field? And second, how far does this presumed founding extend?

With regard to the first question, at this point I would like to state that, unlike what we have already seen in the particular case of logic, Aristotle's paternity of the specific study of argumentative discourse has not, to my knowledge, been refuted. In fact, his *Analytics*, *Topics* and *Rhetoric* are still acknowledged as classic sources for three traditional perspectives within the field of argumentation: logic, dialectic and rhetoric.

With regard to the second question, at this point, I would also like to put forward two general theses:

(a) Aristotle does not offer an argumentation theory. (In fact, if it is any consolation, we should acknowledge that the *theory* of argumentation, in the sense of a body of established, comprehensive and systematically organized knowledge, remains something still sought today).

(b) What is more, if we accept certain of Aristotle's aims, such as the programme of logical validation envisaged in the *Analytics*, not only is there no such theory, but neither can there be one.

A wide range of superficial or circumstantial reasons may be put forward to support this second thesis. For example, we should remember the cases of arguments which cannot be reduced to the syllogistic format for the purpose of validation: from those which contain n-adic or embedded predications which are completely unrelated to the logic that underlies predication in *On Interpretation* or the *Analytics*, to practical reasoning or "syllogism". However, there is a deeper reason with a broader scope: the inability of any system of formal validation, in this case the syllogistic system of *Prior Analytics*, to serve as a basis for a theory of argumentation.

Naturally, we must consider these issues and the very attribution of paternity to Aristotle not in the eyes of God or some substitute (let us say the presumed essence of argumentative discourse or the genuine and pristine nature of Aristotle's *Organon*), but with our own eyes. We must therefore take the current situation in the field of argumentation as a starting point for tackling the issue of the Aristotelian origins of our study of argumentation. Incidentally, as in Aristotle's time, this is not today a stagnant discipline, but rather an interdisciplinary field of second-order studies on discourse which are open to various perspectives and also expanding (with moveable and sometimes uncertain inner limits and outer boundaries).

One aim of this comparison will be to identify some similarities and commonalities, as well as some disparities and differences, between Aristotle's contributions and the current state of argumentation "theory".[2] It would perhaps be wise to indicate here those which may be most relevant. And so, among the similarities or shared aspects are the three perspectives or dimensions of the analytic and normative study of argumentation already indicated: logic, dialectic and rhetoric. And of the disparities and differences, the following two are worth mentioning at this point: (i) the problem of how these dimensions or perspectives are interrelated, regarding which Aristotle seems to adopt a descriptive and

[2] Allow me the stylistic licence of using "theory" to abbreviate and expedite references to the studies and disciplines (logic, dialectic, rhetoric, etc.) which make up the field of argumentation today.

"naturalistic" position, giving very little consideration to the integration of theory and explanation which we demand today; (ii) the fundamental role in evaluating argumentation–or at least discursive proofs–which Aristotle assigns to formal and systematic logic in the *Analytics*, even putting it forward as the criterion for validation. It currently has only a limited role, in part negative, as a standard for invalidation, and in part *ad hoc*, as a tool for exploring new fields, for example, rebuttable inference.

There are of course other differences resulting from the historical and cultural development of the uses of discourse. Examples include how to handle new types of fallacies, such as "political" or parliamentary, or the case of the modern addition of a fourth perspective to the three classical ones, that of the so-called public sphere, which encompasses the uses of discourse in discussing and resolving matters of public interest and domain within the framework of processes such as deliberation. In Aristotle, this paradigm in particular, i.e. public deliberation, is not yet seen as its own distinct entity (as it moves between the monologic and dialogic levels and among the planes of rhetoric, ethics and politics). However, I cannot at this point linger on these specializations and developments, and so I will focus my attention on the general case of the three traditional perspectives.

The classical perspectives of argumentation: logic, dialectic, rhetoric

I have already indicated that I consider Aristotle's paternity of argumentation studies to be a given: in particular, the perspectives of logic, dialectic and rhetoric–for example, through the *Analytics*, *Topics* and *Rhetoric,* respectively. In the face of this traditional and nearly unanimous attribution–in the face of this, shall we say, historiographic *endoxon*–I believe that in principle the burden of proof lies with those who seek to refute it.

Let us move directly to recalling, even if only very briefly, the terms of reference for a comparison between Aristotle's ideas and our own. As I have already stated, I will not start at the beginning of the story, with Aristotle, but rather at the end, with the present day. I will begin with our current view of the field of argumentation, followed by some comparisons and contrasts with early examinations and explorations by Aristotle.

Today *to argue*, within our particular field, can generally be understood to mean accounting and giving reason for something to someone or before someone in order to obtain their understanding and agreement. As this is a discursive and deliberate activity, it is carried out by an agent with a specific intention–in particular, to account or give reason for something to someone–and with a certain purpose–in particular, to lead the recipients of the discourse to agree to or accept what is being proposed. According to this, argumentation consists of an action or interaction, in person in a conversation or frozen in a text, which involves not only an agent, but also real, potential or imaginary recipients of the discourse. In addition, it assumes a certain understanding, and even certain express or tacit agreement, between them. As an extreme case, the agent may "dialogue" and argue with himself or, at the opposite end of the spectrum, believe that he is addressing humanity as a whole. In any event, argumentation can be considered either the activity of arguing or its product. An argument would be a basic unit of discourse, either within the framework of the activity of arguing, or in the context of the respective texts produced. How units of this type and their components (e.g. premises, conclusion, inference markers) are handled may reflect various pragmatic suppositions, such that, for example, they may be considered complex acts of speech or expressions of commitment.

Based on this pragmatic infrastructure of argumentation discourse, the study of argumentation can introduce various dimensions or aspects regarding both the formation and consideration of lines of argument and arguments. Since the 1980s, it has been customary to differentiate between the following aspects:

(**a**) the argument as **product**, consisting of the complete or enthymematic expression–e.g. in textual exposition–of an argument or series of arguments;

(**b**) argumentation as argumentative **interaction**, which could in turn be understood

- (**b.1**) as **procedure,** e.g. as regulated confrontation between arguments and counter-arguments, or between the roles of questioning and responding;
- (**b.2**) as **process**, e.g. as an interaction between people or the effect of one person acting on others, in person or through a "pre-recorded" format;

(**c**) argumentation as a **social-institutional phenomenon** which takes place within or between social groups in public spaces of discourse, in a variety of forms such as, for example, public deliberation, negotiation, jury deliberations and parliamentary debate.

The first three–(**a**) and (**b.1, b.2**)–expressly recall the three classic or traditional perspectives of logic, dialectic and rhetoric. They can be outlined as follows:

Perspective	Most relevant aspect of argumentation
LOGIC	*products* ≈ textual arguments * basic form: < premises - nexus of inference - conclusion > * traditionally orientated towards monologue * a paradigm: a conclusive proof (demonstration)
DIALECTIC	*procedures* ≈ interactive and dynamic argumentation * rules of the debate (roles of proponent-opponent) * orientated towards regulated dialogue and interaction * a paradigm: reasonable debate
RHETORIC	*processes* ≈ persuasive / dissuasive processes * resources and strategies to induce adoption of a belief, attitude or action * orientated towards dialogue and interpersonal interaction * a paradigm: persuasive discourse

A variety of different types of factors have converged in the modern formation and development of argumentation: historical, interdisciplinary and cultural. Historical factors refer to the influence of certain contributions dating from the middle of the last century, which were later acknowledged and accepted as "foundational" (e.g. Perelman & Olbrechts-Tyteca, 1958; Toulmin, 1958). Disciplinary factors are in turn related to the institutional and academic framework of logic, the new rhetoric, communication sciences, and even other spheres which might be thought of as more removed, such as philosophy, psychology and law, as well as certain emerging disciplines like informal logic and critical thinking. Lastly, cultural factors take the form of the metaphors used as sources of inspiration or guides, and are of greater interest in the context of comparison between the ancient and modern worlds. Thus, viewing argumentation as construction and argument as the resulting edifice is consistent with choosing the logical point of view regarding the basis or soundness of an argument. On the other hand, viewing argumentation as a standardized contest leads one to consider the rules of the discussion, or if you will, the "laws and arts of war", in a manner suited to the point of view of dialectic. And viewing argumentation as the (re-)presentation of a case on a stage or before an auditorium is clearly similar to the rhetorical perspective on the effectiveness and efficiency of discourse. As these are ultimately somewhat trans-cultural metaphors regarding argumentation, their consideration invites one to seek them out in the Aristotelian corpus. To give an example: in *On Sophistical Refutations*, eristic argument is called "dirty fight" [*adikomakhia*], in the context of a debate/contest where victory is sought at any cost (*SF* 171b23-25).[3]

Otherwise, they are not mutually exclusive, but rather complementary, approaches to argumentation in many cases support each other and even overlap, although one may take on greater importance than the others depending on the type of argumentation employed, the framework and the context in which they are used.

We shall now look at some of Aristotle's significant contributions to these three dimensions or perspectives, albeit in a succinct and simplified fashion.

Some of Aristotle's contributions
The perspective of logic

The idea of syllogism and its function in the logical validation of arguments is a fundamental part of this perspective. This is not a simple and transparent notion. First of all, the relation between this term and *sullogidsomai* (reason, summarize, infer, argue) may lend itself to misunderstanding by involving the psychological and cognitive aspects of a process of reasoning, the linguistic aspects of statements structured in an inferential format, and the discursive aspects of a line of argument. In addition, as is widely known, *sullogismos* has several meanings in Aristotle's texts on logic. It can have the general meaning of the discursive product of reasoning, inference or a line of argument. But it also has two specific uses: one, as $syllogism_1$, that is, informal deduction in general; the other, as $syllogism_2$, the technical sense of canonical deduction or deductive schema belonging to the syllogistic system. Here I will consider these two specific senses.

Let us recall that *syllogism* is understood to mean a discourse in which certain things being assumed, through what is established, something other than what is established must also result (*Top.* 100a25-27); a discourse which begins with questions stated in such a way that, through what is established, something different from what is established must follow (*SR*

---

[3] Other trans-cultural, and shall we say "trans-historical", metaphors are those related to spatial images which refer to either the traditional conceptual polarization of opposites, or to the diagrammatic representation of arguments. Let us recall, for example, Aristotle's notion of contraries in *Categories*: "for they say that those things are contraries which, within the same class, are separated by the greatest possible distance" (*Cat.* 6a17-19). Apart from that, some "technical" terms in the presentation of the syllogistic system in *Prior Analytics* come from diagrams or graphic representations.

165a1-2). That is, we are operating within the context of *syllogism*$_1$, understood as a deduction/product or a deductive argument which meets these characteristic conditions:

(*i*) the conclusion must follow from the premises adduced;

(*ii*) the conclusion is a proposition which is different from every one of these premises;

(*iii*) the conclusion inherently follows from them.

According to (*i*), in principle this is a valid deduction based on a relation of logical consequence between the premises and the conclusion. According to (*ii*), this relation is not reflexive–unlike our standard idea of logical consequence. And according to (*iii*), it involves the premises being highly relevant to the conclusion–once again in contrast with our standard conception of formal consequence (e.g. the *ex contradictione quodlibet* rule: from a contradiction, any other proposition follows, does not apply to syllogistic deduction).

The paradigm of syllogistic deduction is an illative text made up of explicit statements which are in turn broken down into general terms that act as subjects and monadic predicates. However, it still allows for certain variants, both basic and derived or applied. The basic variants are those already mentioned:

(a) Syllogism$_1$ or syllogism in the specific sense of an informal yet logically valid deduction: the conclusion must follow from the premises, such that it could not be false as long as the premises are true.

(b) Syllogism$_2$ or syllogism in the technical sense of constituting a mode or schema of one of the three figures which make up the syllogistic system in *Prior Analytics*, especially the system of categorical syllogism in *PrA*. I 4-7. This is a canonical deductive schema whose logical validity is evident in itself or is demonstrated by reduction to obviously valid schemata.

Relations between these basic forms appear to be governed by these aims:

1/ Soundness: any canonical syllogism is a valid deduction, or in other terms, any syllogism$_2$ meets the specific conditions of a syllogism$_1$.

2/ Completeness or completion: any valid deduction can be validated in terms of a canonical syllogism or a chain of canonical syllogisms. In other words, the validity of any syllogism$_1$ can be confirmed by expressing it in, or reducing it to, the terms of a syllogism$_2$.

The claim that the syllogistic system is sound does not present further problems. However, the same cannot be said of the claim that it is complete. Aristotle makes an attempt at a proof based on an *a priori* assumption that it is necessary for any conclusive deduction to establish that a certain predicate does or does not belong to certain subject, either in a universal sense or in a particular sense (*PrA*. I 23, 40b17-19). Nonetheless, he himself is aware that this is a programmatic claim with which certain true deductions do not agree, such as, for example, some mathematical proofs. This point makes it necessary to take into account other types of argumentation or deductive proofs, which would make it possible to consider applied or derivative variants. These applications can be divided into three main types: monologic, dialogic and practical.

Use in the monologue form is fundamentally to demonstrate that something is or is not really the case. This is a "syllogism that makes something known" by way of either direct demonstration or indirect demonstration or refutation by means of reduction to the impossible. This is the apodictic paradigm of *Posterior Analytics*. However, Aristotle himself still acknowledges more or less strict and rigorous variants of *apodeixis,* depending on the sphere of application in question. Therefore, not only are there categorical or modal demonstrations, but also others which are functional and retroductive. Apart from that, it is also necessary to recognize the hypothetical syllogism which truly deduces a conclusion not based on previously assumed or established theses, but rather from an agreement or a hypothesis, e.g. in refutations or proofs by reducing an assumption to an impossible consequence.

Applications in the world of dialogue are much more dynamic. To start with, it encompasses several functional variants: didactic, dialectic, peirastic, and eristic (cf. *SR* 165b39). The paradigm in this case is the dialectical syllogism which begins with questions posed by a questioner/opponent to a respondent/proponent; it progresses through plausible propositions or dialectical premises and is based on a certainty or a commonplace, which applies in all or most cases. Another important variant is the rhetorical syllogism, the enthymeme, which begins with characteristic premises: probable propositions (referring to what happens most of the time or to what is likely to happen), or those related to signs–if they are necessary signs, the argument is conclusive; if the signs are not necessary, such as panting being a possible sign of fever, the argument is refutable such that even with true premises, e.g. observing that a person is panting, the conclusion may turn out to be proven false; it is possible to demonstrate that the person in question does not have a fever (cf. *Rhet*. 1357b19-21).

A third variant worth mentioning here is the practical or deliberative syllogism, which concerns putting the measures which will lead to the desired end into practice. Generally speaking, a practical syllogism is made up of a primary or major premise which establishes (i) a rule, (ii) a value, or (iii) a design; a minor premise in which the agent himself states that he has an involvement in the case; and a conclusion which presents the appropriate action. The following may serve as examples of the variants:

(i) One must do what one needs. Now then, I need an article of clothing, a cloak. Therefore, I make myself that article of clothing, a cloak.
(ii) All sweets are worth tasting. Now then, what I have before me is sweet. Therefore, I taste it.
(iii) I want to achieve X. For this purpose, the method I deem appropriate and find feasible is Y. Therefore, I do Y.

It is important to bear in mind that this reasoning is carried out by an agent with the intention of taking some action. This is the root of its distinctive traits: using the first person and sometimes confusing the conclusion with the actual implementation.

Based on these suppositions, we can consider the key point in Aristotle's treatment of argumentation from the perspective of logic: the evaluation of an argument, or more precisely, how it is validated or invalidated insofar as it moves within the framework of a deductive paradigm. However, it is first advisable to clear up a misunderstanding in traditional critiques which accuse Aristotle's claim that all valid deductions can be validated through syllogism of naïve reductionism. In contrast to what these critiques generally suppose, the function of the syllogism is not to provide a particular method of reasoning: the idea is not to make the canonical syllogism a fixed type of deductive argument, intended to replace or shape deduction in general. The idea is to present the ways in which any correct deduction may ultimately be reduced for the purposes of validation, thus serving the purpose of testing the validity of any argument called into question. And in these terms, it is intended to act not only as a negative standard, for invalidation, but also a positive one, for validation or confirmation. Therefore, ultimately, Aristotle's aim is not radically different from the aims of our standard logic–at least during the second third of the 20th century–i.e. to establish the norms of rational discourse and the standard for validating any true deductive proof. In these terms, Aristotle represents a clear precedent, although his logic/formal resources are rather limited.

This point leads us to an issue which is still being discussed today, the relation between formal logic and argumentation "theory", by means of the crucial case of validation, or more generally speaking, of evaluating arguments by virtue of their logical form.

First, it is advisable to note that Aristotle is considering a formal process of validating and challenging arguments, but not a formalized process like that which our languages and systems of formal logic can apply today. The syllogistic system does not have an effective method of determining logical forms; rather it has functional resources such as schematic letters and pronominal "variables".[4] It therefore ultimately reveals a technical limitation in terms of validation or invalidation.

In addition, it is advisable to recognize that the properties of being a deduction (syllogistic) and being valid/invalid are formal. In contrast, the properties of being a proof, a counter-proof or refutation, or a sophistic refutation are not formal: they are not maintained/transmitted through the logical form.

This difference has serious repercussions when it comes to the evaluation of arguments by virtue of their logical form. Let us look at this:

(a) Validation relies on identifying or confirming the formal properties of the argument: whether it is deductive, whether it is valid or invalid–or in the case of propositions, whether they are a logical truth or a contradiction. These properties are transmitted and maintained through the logical form, so that in effect they do not validate or invalidate a given argument, but rather a type of argument, i.e. all those arguments that have the same form. Thus, transmitting/maintaining the form means, for example, that if A is a deductive argument, any argument with the same

---

[4] In Aristotle it is possible to find several uses of variables in this sense. For example, they are used: 1) in the role of deictic/diagrammatic letters, as was the tradition in geometry: "Suppose the Lines A and B have been drawn to the centre. If then one should assume that the angle AΓ is equal to the angle BΔ ..." (*PrA*. 41b15-17: reference to the proof that the angles at the base of an isosceles triangle are equal.); 2) in the role of nominal abbreviations. E.g.: "If magnitude ΛBΓ is composed of the indivisibles A, B, Γ, each corresponding part of the motion ΔEZ of Ω over ABΓ is indivisible" (*Phys*. 231b23-25). Apart from that, there are also uses which combine 1/ and 2/, e.g. in *Meteorology*, 375b10 ff., describing the semicircular shape of a rainbow; 3) in the role of schematic letters and a kind of pronominal variables–marked by the neuter gender (*to A*...)–within the framework of Aristotelian logical analysis. After having turned to trivial or vacuous examples of propositions or terms (e.g. "every pleasure is good", "no pleasure is good", "if some man is animal"), they are introduced without prior warning at *PrA*. 25a15 ff.: "First then take ... A and B. If no B is A, neither can A be B." Use combined with resorting to trivial examples or terms, e.g. "If A were motion, B animal and that by which [*eph'ho*, epi + dative] Γ this man ..." (*PrA*. 30a30).

logical form as A is deductive; and that if A is valid (or invalid), any argument with the same logical form as A is valid (or respectively, invalid).

(b) Now then, the most important properties of argumentative discourse are not transmitted/maintained in a similar way through their logical form: being a proof (or a sound argument), being a counter-proof (refutation), being an apparent or failed proof or counter-proof (a fallacy). If a deduction is a proof (or a demonstration), not every deduction with the same logical form constitutes a proof (a demonstration). Therefore, if a deduction is a counter-proof or refutation, not every deduction with the same form constitutes a refutation. And consequently, if a counter-proof is apparent and failed, that is: it consists of a sophistical refutation, not every counter-proof with the same form is equally specious or fallacious.[5] In short, pure formal logical analysis is incapable of explaining and giving an account of the most important and characteristic discursive properties of argumentation, such that it cannot constitute or even lead to a suitable and inclusive "theory" of argumentation.

The perspective of dialectic

Let us recall the opening statement of *Topics*:

> Our treatise proposes to find a line of inquiry whereby we shall be able to reason from opinions that are generally accepted about every problem propounded to us, and also shall ourselves, when standing up to an argument, avoid saying anything that will obstruct us (100a18-21).

According to this statement, the purpose of Aristotelian dialectic is in essence to construct and fine-tune a method of preserving "pragmatic soundness" in any proposed debate. There is no lack of precedents in this respect, including the Socratic process in Plato's early dialogues. However, in Aristotle's hands the applications of the dialectic method are more varied than this statement might lead us to imagine: It can have a, shall we say, gymnastic use, for practice and training, or a discursive use in casual encounters. At the same time it can be used to put known theses to the test, as well as to make critical philosophical or scientific contributions, for example in the sense of resolving doctrinal aporias

[5] On this point, it is equitable and essential to observe that the idea of "formal fallacy" is not Aristotle's. It has a relatively modern origin and perhaps first came from Whately (1826). In any event, as we have seen, it seems to be a classification which is difficult to justify.

according to the maxim of minimal mutilation with regard to the plausible propositions in the sphere under consideration (*NE* 1145b2-7).

In the dimension of dialectic we find three characteristics which distinguish it from the canonical logic of the perspective discussed above:

(a) the focus of attention is on common informal argumentation, as a rule deductive, but how it is identified and evaluated does not always meet the schematic and validation criteria of the canonical syllogism;

(b) it takes the form of a dialogic discourse such that it involves an argumentative interaction;

(c) paradigmatically, the interaction is carried out by contrasting discursive roles or figures, whose behaviour is regulated by certain rules of the game or certain conventions regarding reputation or standing. The figures in question are the opponent, who asks or questions a position regarding the subject of the debate, and the proponent, who responds or supports the position in question. The aim of the opponent is to refute this position by leading the proponent to an inconsistency in this respect. It is odd that in this standardized context, Aristotle does not seem to be aware of the importance of allocating the burden of proof.

The propositions in play are generally plausible premises or dialectical propositions. The Aristotelian notion of plausibility is especially interesting in view of its pragmatic and social description, which might not be expected in the light of the logical and analytical context mentioned above: "those opinions are plausible which are held by all men or by most men or by the wise, and among these, either by all, or by most, or by the most notable and illustrious of these" (*Top*. 100b21-23, 104a8-15). I will return to this point later. Additionally, "a dialectical proposition is one which thus holds in several instances and to which no objection is forthcoming" (157b32-33; cf. 158a4-5: "either no objection whatever appears or at least not any on the surface"). This idea also seems to have a modern guise: could a traditional dialectical proposition resemble what today we understand to be a sustainable yet defeasible proposition within the framework of reasoning by default?

The dialectical elements *par excellence* are the *topics* which lend their name to Aristotle's work on dialectical method. They seem to function as templates–e.g. in *Topics* itself–or as a stock–e.g. in the *Rhetoric*–of arguments which can be constructed based on more or less plausible or acceptable certainties. In the first case, they could be taken as a precedent for our "argumentative schemes". The following examples provide a clue as to their variety and meaning:

(α) "Courage is a virtue. Therefore, cowardice is a vice" (*Top.* 113b32-34). [That is: If S is P, then the opposite of S is the opposite of P.]

(β) "A man who strikes his father also strikes his neighbours" (*Rhet.* 1397b17-18); "If the gods are not omniscient, certainly human beings are not" (1397b13-15). [That is: If the least likely thing holds, then the most likely holds *a fortiori*.]

(γ) "If one of two things is an end, and the other is not, the former is the greater good" (1364b3-5); "A greater number of goods is a greater good than a smaller number, if that small number is included in the first" (1363b19-20); "If A plus B is greater than C and A minus C is greater than B minus C, then A is greater than B" (1363b34).

(δ) "And [one must] look and see if, supposing the one to be the same as something, the other also is the same as it; for if they are not both the same as the same thing, clearly neither are they the same as one another" (*Top.* 152a31-32). [This could be considered the negative correlative of "common notion 1" in Euclid's *Elements*.]

Thus, classifications become important for the purpose of descriptive systematization, where the topics constitute above all a repertoire of common resources and commonplaces. In particular, Aristotle recognizes the following:

1. Those which encompass relations of opposites (opposites; negation; privation-possession; relatives).
2. Those which encompass relations of coordinates (e.g. proceed fairly-be fair).
3. Those which encompass relations of more / less / equal.

With regard to evaluating the perspective of dialectic, it is advisable to note that the schemata of the *Topics* consist of deductive forms which are valid, but which cannot be validated with reference to the syllogistic system of the *Analytics* (*PrA.* I 4-7). This also refutes the claim of canonical completeness or completion which we have already considered in the discussion of the perspective of logic above. In its place, other considerations related to formal logic seem to be at work, in which taxonomic interest converges with relations of "material' implication". As happens in Aristotle's handling of the so-called predicables, which analyzes relations of predication from a dual approach: intensional and extensional. This results in the following variants:

(α) essential and coextensive predications with regard to the subject, which constitute the conceptual *definition* of the subject;

(β) essential but non-coextensive predications, which determine its *genus*;

(γ) non-essential but coextensive predications, which determine its peculiar *property*;

(δ) non-essential and non-coextensive, or indeterminate, predications, which distinguish an *accident*.

Beyond what we have discussed, is it possible to locate in Aristotle any sort of nascent dialectic theory or specific systematization? Of course, there is nothing resembling the syllogistic system of the *Analytics* which we find in the perspective of logic. But there are a number of cues which could perhaps be developed and systematized. I will give two possible examples.

a) One would be the case of material consequence or inference in contexts of dialectical and rhetorical argumentation. Along these lines, Dick (2000) has suggested reconstituting the *topos* as a binary relation which, alone or in combination with other relations and predicates, serves to build deductive arguments. We might then think of a certain parallelism between the implicative forms of a canonical syllogism S, and of a topical syllogism S':

S: *If P, and P implies Q, then Q*, where the implication consists of a logical or analytical formal consequence.

S': *If P, and T(P,Q), then Q*, where the implication is a material consequence. Thus, for example, if courage is a virtue and the correlation of its opposites is valid (such that if S is P, the opposite of S is the opposite of P), then cowardice is a vice; or if political supremacy is an aim and revenge is not, and the commonplace with greater value for the end is valid, then political supremacy is a greater good and preferable to revenge.

b) Another example would be the handling of plausibility or the plausible [*to endoxon*]. This does not consist of a semantic property of a proposition or by extension, of an argument, but rather in the light of his description, a pragmatic, approximate and social relation, as we have already seen: "Those opinions are *endoxa* which are held by all men or by most men or by the wise, and among these, either by all, or by most, or by the most notable and illustrious of these (*endoxois*)" (*Top*. 100b1-3). For example, the common belief that one ought to do good for friends (104a24); the more unusual opinion (of some sophists) that not everything that exists is either created or eternal (104b25-28) or even Empedocles' assertion that the elements of bodies are four (105b16-18).

Regarding this, in his commentary on the *Topics,* Alexander of Aphrodisias specified:

> The difference between plausible and true is not that what is plausible is false (for some plausible opinions are also true), but is in the judging. With what is true, judging is on the basis of the fact which this truth has reference to, for when this accords with it, then it is true. But with what is plausible, the judging is not based on facts but on the audience and on their suppositions about the facts (*On Top*. 19.22-27).

One no less important feature of plausibility is its gradual nature: the assessment proceeds by degrees or levels of correlative plausibility/implausibility, where the greatest degree or the highest level corresponds to the opinion of all, as we can deduce from the order given in Aristotle's description. This point ultimately remains somewhat problematic. However, it is at least possible to identify the following principles:

Let α be a dialectical proposition and α' its opposite:

- Tacit assumption of correlation: If α is more/less plausible (or implausible), then α' is so much more/less implausible (or plausible).
- Assumption of correction: one must start with what is more plausible or what is as plausible as the case under consideration will allow (159b8-15, 161b19 ff., 183b7).

So that:

- Principle of correction: he who argues correctly proves what has been put up for discussion based on premises which are no less plausible than the conclusion; otherwise, his line of argument will not be sound (160a15-16).

All the same, Aristotle's handling of this matter beyond this point is not very promising with regard to a complete conceptualization or "theorization" of plausibility in terms of *endoxa*.[6]

As a coda to our discussion of the perspective of dialectic in Aristotle, it is not out of place to turn our attention to the study of fallacies or more precisely, *sophistical refutations*. Firstly, their study serves as an appendix to the *Topics* and secondly, reconsidering fallacies has also been a determining motive in the development of our "theory" of argumentation since the 1970s.

Let us recall the characteristics of the syllogism: deducing a conclusion based on premises, such that (*i*) the conclusion necessarily follows from them; (*ii*) the conclusion is a separate proposition; (*iii*) the conclusion inherently follows, or in other words, each and every one of the premises, but only those, are relevant.

In turn, a *refutation* [*elenchus*] is in this context a syllogism which leads to the contradiction of the conclusion in question (*SR* 165a3-4), that is, a deduction of the contradictory proposition to the thesis maintained by the addressee or respondent in the dialectical debate. In addition, in light of the purpose of the *Topics* (100a17-21) cited above, this function of refutation or counter-argument may be understood as the deduction of either (*a*) a proposition contradictory to the thesis held by the respondent, or (*b*) a proposition which is inconsistent with the premises assumed by him, in accordance with an Aristotelian variant of *ad hominem* argumentation.

A *sophistical* refutation is, to summarize, an apparent refutation: it seems to meet the conditions of a refutation without actually doing so. For example, a *petitio principii* would not meet (*ii*), just as a fallacy of false cause–or a false attribution of cause–would not meet (*iii*). They would therefore be failed deductions or counter-proofs; they would also be sophistical if they appeared to give or gave the opposite impression. However, this type of argumentation can also be found in other cases, judging by the statement: "I use sophistical refutation and deduction to refer not only to what seems to be a deduction or refutation, but is not, but also to what while being so, is only apparently appropriate to the case" (*SR*, 169b20-23). This may occur, for example, if non-geometric arguments are used in geometry. We should also add the case of eristic reasoning: according to the *Topics*, "eristic reasoning is that which is based on things which appear to be plausible, but are not, and also on what

---

6 Vega (1998) includes a more detailed discussion of the possibilities and limitations of Aristotle's handling of plausibility (*endoxa*) and its importance in the modern world.

while seeming to be a deduction, without being one, is based on things that are plausible or appear to be so" (100b16-18). Thus it is also sophistical.

A sophistical refutation is, in short, that which seems to be based on (1) plausible propositions to reach a conclusion, (2) deductively and (3) in a relevant manner, (4) regarding the proposition contradictory to the thesis brought into question, but in reality fails in one or more of these respects (1)-(4), and thus is no more than a failed or apparent counter-proof.

In this handling of fallacious argumentation there seems to be a tacit assumption of correlation or correspondence as a counterbalance: every failed and apparent proof or counter-proof is the reverse of a genuine one, at least in the sense that any failure, defect or breach of a genuine refutation establishes a corresponding apparent refutation. This supposition has gone on to have a noteworthy history. In our time, it remains active in dealing with fallacies not only for scholarly purposes involving cataloguing in manuals, but also for the theoretical purposes of detection and explanation: thus it is expressly assumed, for example, by a school of thought as influential as pragma-dialectics. However, this supposition gives rise to debatable conclusions. From this perspective, being fallacious means a lack of virtue; it is violating one of the conditions which define a good argument. As such, only a bad argument could be fallacious, or worse yet, any bad or faulty argument would be a fallacy. Nonetheless, there can also be fallacious or sophistical uses of good arguments, as Aristotle himself acknowledged, and in any event, it is not advisable to remove the distinction between *good*, *bad* and *fallacious* arguments.

The perspective of rhetoric

As is widely known, Aristotle considered rhetoric to be the "antistrophe" or correlate of dialectic (*Rhet.* 1354a1) in the sphere of persuasion or generating convictions. The enthymeme constitutes the rhetorical syllogism (1356b5), as well as the "flesh and blood" of the persuasion (1354a15; cf. also 1355a3-15). However, in contrast, he also conceived it as an offshoot or branch of dialectic and of that practical knowledge of character and ethics which it would be right to call politics (1356a25-27). Nevertheless, rhetoric must not be confused with or subsumed under politics (1356a28). Additional statements in one sense or the other may be made without shedding further light on the matter of their mutual relation. "Rhetoric does not belong to a single definite class of subjects, but is as universal as dialectic; it is useful and its function is not simply to succeed in persuading, but rather to discover the most relevant means of convincing in each case" (1355b10-13). "It is without a doubt a branch of dialectic and similar to it, as we said at the outset, as neither rhetoric nor dialectic is the scientific study of any one separate subject, but rather simple faculties for providing arguments" (1356a30-35, cf. also 1359b13-14).

Let us conclude, provisionally at least, that rhetoric and dialectic are partner arts involving constructing and providing arguments, although with different areas of specialization and purposes:

- dialectic specializes in the methods of dialogical confrontation, as well as its employment in investigating principles and testing opinions;
- rhetoric, in turn, specializes in the means of generating convictions among the audience which the speaker is addressing, through different kinds and varieties of proofs.

Although dialectic seems to arrogate certain precedence to itself through the image of rhetoric as a schema or offshoot, rhetoric retains its distinguishing features and discursive autonomy. For example, the types of technical proofs, resources of the rhetorical presentation itself and rhetorical competence of the speaker determine: the *ethos*, which depends on the moral character of the speaker; the *pathos* which gives the audience a certain mental disposition; and the *logos*, or argumentative discourse, to the extent which it proves or tries to prove something to someone (1356a1-4).

Another characteristic discursive resource is the *enthymeme*, the most powerful of the proofs by persuasion (1355a8), which consists not so much of a truncated syllogism, as later tradition has insisted, as a syllogism which is based on probabilities [*eikota*, what happens most of the time or as a general rule] and signs [*semeia*], whether necessary and conclusive [*tekmeria*] or not necessary [*anonyma* or lacking a specific name] (1357a37-b7). In *Prior Analytics*, *eikos* is also defined as "plausible premise [*protasis endoxos*], what men know to happen or not to happen, to be or not to be, for the most part thus and thus, is a probability, for example, 'envious men hate' or 'those who are loved show affection'" (70a4-7).

According to a proposal by Brunschwig (1994), dialectic is to individual or conversational use of language what rhetoric is to the public use of discourse. We can make use of this proposal to note an interesting inclination towards the deliberative genre. Aristotle says, for example:

> rhetoric [draws upon] the regular subjects of debate. The function of rhetoric is to deal with such matters as we deliberate upon without arts or

> systems to guide us, in the hearing of persons who cannot take in at a glance a complicated argument, or follow a long chain of reasoning (*Rhet.* 1357a1-5).

The relevance of this reference to deliberation and discourse in public lies in certain aspects which are significant with regard to the current situation in this field. Two in particular may perhaps deserve special attention: one is related to how deliberation itself is understood; the other refers rather to the relation between classical rhetoric and the modern socio-institutional perspective of public discourse.

In principle, deliberation may be considered (1) a rhetorical genre, (2) the practice of prudence in an ethical context, and (3) a form of public discourse in the socio-political sphere. It was the first two considerations which most attracted the attention of the ancient Greeks, Aristotle in particular. In terms of its status as a rhetorical genre, the first known text is the *Rhetoric to Alexander*, which speaks of deliberation as a genre of public discourse [*genos demegorikon*] that takes place within the framework of a discussion in assembly concerning a decision and may utilize both persuasive and dissuasive effects in this respect (*RhA.* 1421b1). Aristotle's *Rhetoric* echoes this idea in the terms of the "deliberative genre" [*genos sumbouleutikon*]: "Deliberative speaking [*sumbouleusis*] urges us either to do or not to do something, as one of these two courses is always taken by private counsellors, as well as by men who address public assemblies about a common matter" (*Rhet.* 1358b7-9). It is a discourse which points to the future: it deals with something to come, whether to advise it or to prevent it. In sum, its characteristic purpose ranges from the advisable to the detrimental: he that advises recommends what he deems best; he that dissuades tries to move away from what he deems worst.

With regard to the second consideration as the thoughtful practice of prudence, this is precisely the meaning which Aristotle gives to deliberation [*bouleusis*] in the context of the *Nicomachean Ethics*. For example: "It seems to be the mark of a sensible man to be able to deliberate well about what is good and advantageous for himself, about what sorts of things conduce to the good life in general" (*NE.* 1140a25-27). In addition, "No one deliberates about things that cannot be otherwise, or about things he has no possibility of doing" (ibid. 30-31). In short, deliberation ultimately amounts to practising prudence "as a true, reasoned and practical state of capacity to act with regard to the things that are good or bad for man" (1140b4-5). However, he does not concern himself with aims or purposes, but rather with the appropriate means of achieving or fulfilling these. Accordingly, it is a matter of thoughtful and meditative consideration of the conditions surrounding the decision to undertake an action with the intention of achieving the goal or aim one has set. Therefore, in this regard, what becomes important is its personal and practical dimension, e.g. its nature as practical reasoning, compared to the public and rhetorical dimension of the deliberative genre. However, there may also be a certain connection between the two, for example, insofar as one can engage in deliberation in private as a way of advising oneself with regard to some future option available to one.

Today we also understand deliberation as a form of public discourse, an interactive and collective form of argumentation which weighs up and manages information, options and preferences in order to jointly make a decision or practical determination in a responsible and reflective way. It occurs within social frameworks of discussion regarding issues of common interest, which may be resolved by means of reasons which are public knowledge. The ancient Greeks did not of course lack experiences in this regard–some as famous as the debate concerning the fate of Mytilene recounted by Thucydides in his *History of the Peloponnesian War*, III, §§ 36-49. However, like the later Roman orators, they were understood as the simple projection of rhetoric onto the public arena of politics and public or deliberative assemblies.

It is true that deliberation, in its modern sense, has a certain connection to the two previous meanings: it encompasses an interactive and relatively complex form of practical argumentation, along the lines of (2), while at the same time dealing with questions of common interest for the agents involved in the discussion, along the lines of (1). However, the fact that the object is a collective decision or determination, to be taken within and by a group, together with the associated conditions and complications–e.g. coordinating how the participants take part in the discussion or the various forms of collective decision-making–introduces substantial differences between this new meaning of deliberation and the other two traditional meanings, to such an extent that its new form constitutes not only a difference of degree, but also of scale. And so, despite efforts to reduce this new dimension to rhetoric, for example, in line with Kock (2009), the socio-institutional perspective of public discourse is today acknowledged and credited with having its own identity. It is clear that this may include and accommodate relevant considerations from the classical perspectives, particularly dialectic and rhetoric–and through these, the logic involved. However, it cannot be reduced to any one of these or to them as a whole. Among other reasons, because the rationality of collective determinations in frameworks of institutional decision-making does not conservatively, linearly and directly follow on from the rationality of individuals and the process of aggregating individual decisions.

Lastly, it is worth mentioning another significant difference between our public uses of discourse and those envisaged by traditional rhetoric: that introduced by online discourse and the new virtual or electronic public existence which results from the mixed language of the computer, which lies somewhere between spoken and written. Let us note that a *face-to-face* public existence not only implies physical and total communication, direct and in person, but also a certain degree of exposure, in the sense of being on view to another person, as well as a certain amount of precaution in response to feeling oneself in their gaze. In this regard, Aristotle recalls an ancient proverb: "shame is in the eyes [of others]" (*Rhet.* 1384a34). This complex idea lends itself to other considerations, such as: (i) Shame implies a certain awareness of exposure and personal risk, an awareness which generally encourages self-restraint. (ii) Finding oneself in public creates a framework which operates as a stimulus and motivation to contribute to the course of the conversation while observing the appropriate conventions, with the aim of obtaining esteem and recognition; this is what we might call "the civilizing role of hypocrisy." (iii) In addition, it favours adopting attitudes and assuming responsibility within the argument, such as the obligation to take responsibility for one's own opinions and proposals in the face of objections or against other opinions and alternative proposals. In contrast, an *electronic* public existence is based on a virtual presence and fosters other codes of conduct. For one thing, it refers to a "presence" which is transcribed and read on a computer screen, not felt or seen. This presence also refers to a persona and to a performance/representation, rather than a personal agent; it is enough to note that the only identifying feature of the interlocutors is the IP address of their device's online interface. Therefore, in short, what we have is a public existence which would be considered "shameless" according to Aristotle's proverb.

Conclusions

Having reached this point, it would be advisable to summarize some of the key dissimilarities and differences between current argumentation "theory" and its founding father in so many aspects, Aristotle.

For a start, at the very level of conceptualization and theorization, we notice that Aristotle shows little interest in issues related to the theoretical explanation of discursive analysis–except when referring to a case of syllogistic grammar or logic. In particular, the absence of a relatively integrated or comprehensive "argumentation theory" becomes apparent: it does not even exist as a programme or search horizon, in contrast with the concerns and proposals of a significant number of modern theories and specialists in this field of study. Aristotle is rather more interested in classifications and descriptive systematization. The classifications reveal a somewhat naturalistic handling of discourse. In the *Topics*, these systematizing ambitions are summarized as the aim of finding a method capable of dealing with dialectic situations or confrontations without this involving being inconsistent within oneself or deceiving others; and in the *Analytics*, as the aim of presenting the system of logic underlying scientific demonstration. On the other hand, despite the paradigmatic role assigned to the syllogism–that is, to conclusive deduction–Aristotle shows signs of being especially sensitive to the relation between the forms of argumentation and proof and their framework and context of use, for example, with regard to both the substantive relevance of the proofs and the suitable degree of rigour.

Other notable dissimilarities and differences are related to the problems posed by the interrelation among the traditional perspectives and, in particular, to the role of logic. Firstly, in the Aristotelian texts, the relations between dialectic and rhetoric are not always stable and clear. In addition, formal or informal logic, that is, analytic or dialectic, seems destined to play a canonical role, except when the public use of discourse makes it advisable for the speaker not to linger on rigour and rules. In this regard, it has been suggested that there was a transition from the priority initially given to dialectic–as the *Topics* shows–to the priority later bestowed on logic–as the *Analytics* indicate (cf. Allen 2007). At this point, it is not possible to go into this complex question, one which is both hermeneutic and chronological. However, it would not be out of place to recall some of Aristotle's statements in this regard: "The consideration of syllogisms of all kinds, without distinction, is the business of dialectic, either of dialectic as a whole or of one of its branches" (*Rhet.* 1355a8-10). "We must now state that it is not only dialectical and demonstrative syllogisms that are formed by means of the aforesaid figures, but also rhetorical syllogisms and in general any form of persuasion [*pistis*], however it may be presented" (*PrA.* 68b9-14). Still, it would be necessary to point out the standard reservations, known or overlooked by Aristotle himself, as in the difficult case of a syllogism based on a *reductio ad absurdum* hypothesis or the irreducible case of practical reasoning which occurs in the first person and concludes with an action. However, what it is not possible to accuse Aristotle of is an excessive or irrelevant reductionism using the syllogistic pattern. The canonical syllogism does not constitute a specific method of reasoning or arguing, but rather the form to which any valid deductive reasoning may be transferred for the purpose of validating it. Let us consider the case of grammar as an analogy: grammar is not a way of talking, but rather the authority to which any expression or utterance may be referred in order to determine whether it is correct.

I will conclude with a brief reference to the current approach to the relations among the three traditional dimensions or perspectives. First, it is of interest to recall that in this regard we are the heirs not only of Aristotle, but also to a greater extent of the particular history of these approaches, which soon became independent disciplines–logic, rhetoric, and their unstable relations of proximity. Today, these perspectives and the relations between them can also be viewed in different ways. For example, according to Cohen (2001), they are three dimensions which are logically independent in the light of their respective criteria: logical cogency, adequate fulfilment of the rules of dialectic, rhetorical effectiveness. According to Johnson (2000), dialectic is presupposed by logic in the sense that it is a necessary condition of the logical cogency of an argument that it is dialectically adequate, at least in part, and responds properly to the counter-arguments it gives rise to. Whereas in the eyes of pragma-dialectic (cf. e.g. van Eemeren and Grootendorst, 1992), the opposite relation exists, as logical correctness is assumed in the dialectical rules of reasonable debate. In any event, Blair (2003) leans towards pairing logic and dialectic based on how good the argument is, at least in the negative sense: α is not a good argument if it violates a rule of logic or ignores the characteristic cogency of a logical relation or rule of inference; α is not a good argument if it does not properly meet the demands of dialectical confrontation.

Nonetheless, apart from these variations, the prevailing opinion can be summarized in the following three points:

a/ The three perspectives are required, in view of the constituent and regulative characterization of good argumentation, such that none is self-sufficient or may seek to exclude the others. On the other hand, as a whole, neither do they cover the broad field of argumentation today, as indicated by the later addition of a fourth socio-institutional perspective which is irreducible to the three original perspectives.

b/ The priority or greater relevance given to one over the others depends on the framework and context of the debate.

c/ The criteria related to logical and dialectical quality, on one hand, and estimations of rhetorical efficiency or effectiveness, on the other, are normally considered not only different, due for example to the normative aspects of the former and the empirical qualities of the latter, but also rather independent of each other–as the success of certain fallacies indicates. In any event, the quality of a line of argument does not necessarily mean that the effect will be persuasive, nor does a persuasive effect necessary indicate the quality of a line of argument.

References

Allen, J. (2007). "Aristotle on the disciplines of argument: Rhetoric, Dialectic, Analytic", *Rhetorica*, XXV/1: 87-108.

Blair, J.A. (2003). "Relation among Logic, Dialectic and Rhetoric". In F.H. van Eemeren *et al.* (eds.), *Anyone Who Has a View: Theoretical contributions to the study of argumentation*, (pp. 91-107), Dordrecht: Kluwer.

Brunschwig, J. (1994). "Rhétorique et Dialectique, *Rhétorique* et *Topiques*". In D.J. Furley and A. Nahams (eds.), *Essays on Aristotle's Rhetoric* (pp. 57-96), Princeton: Princeton University Press.

Cohen, D.H. (2001). "Evaluating arguments and making mate-arguments", *Informal Logic* 21: 73-84.

Dyck, E. (2002). "Topos and enthymeme", *Rhetorica* 20/2: 105-117.

Eemeren, F. H. and Grootendorst, R. (1992). *Argumentation, Communication and Fallacies*, Hillsdale, NJ: Lawrence Erlbaum Associates.

Enriques, F. (1949 [1922]). *Para la historia de la Lógica*, Madrid: Espasa Calpe.

Johnson, R. H. (2000). "More on arguers and their dialectical obligations". In H.V. Hansen and C.W. Tindale (eds.), *Argument at the Century's Turn. Procds. 1997 OSSA Conference*, Brock University, St. Catharines, ON, Canada. CD-Rom.

Kock, C. (2009). "Choice is not true or false: The domain of rhetorical argumentation", *Argumentation* 23: 61-80.

Locke, J. (1985 [1689]). *An Essay concerning Human Understanding*, Oxford: Clarendon Press.

Perelman, Ch. and Olbrechts-Tyteca, L. (1958). *Traité de l'argumentation. La nouvelle rhétorique*, Paris: Presses Universitaires de France.

Toulmin, S.E. (1958). *The uses of argument*, Cambridge: Cambridge University Press.

Vega Reñón, L. (1998). "Aristotle's *endoxa* and plausible argumentation", *Argumentation*,12: 95-113.

Whately, R. (1826). *Elements of Logic*, London: B. Fellowes.

STOCHASTIC KNOWLEDGE: *FOR THE MOST PART* AND CONJECTURE IN ARISTOTLE

SALVATORE DI PIAZZA

Introduction

The entire *Corpus Aristotelicum* is permeated with occurrences of the fundamental notion of ὡς ἐπὶ τὸ πολὺ, usually translated as "for the most part". Although this notion is derived from the common language of ancient Greece, Aristotle often uses it in a technical sense thus it seems to be a key-concept of his theoretical framework. In the following pages, we aim to analyze the strong relationship between what Aristotle calls "*for the most part* objects" and the so-called conjectural or *stochastic*[1] knowledge; in addition, we hope to shed light on how and why the notion of *for the most part* plays such a central role in Aristotelian thinking, in particular in relation to the two classical philosophical notions of truth and rationality.

The basic idea on which I have founded my enquiry is that, within the dense philosophical debate on conjectural knowledge, Aristotle's contribution was, in my opinion, often underestimated,[2] probably because the notion of *for the most part* was undervalued. I believe, though, that it is possible to track an approach that "rehabilitates the conjectural knowledge and the intelligence that proceeds by way of deviations" (Detienne and Vernant, 1974: 247) through all of Aristotle's works and, above all, I think that the Aristotelian point of view can provide interesting theoretical indications for the contemporary debate on these matters.

The method and the object

The whole Aristotelian theoretical framework revolves around a fundamental methodological principle, that is to say the strong relationship

[1] We will see the real meaning of this adjective further on.
[2] A partial exception concerns Detienne and Vernant (1974).

between the object that has to be investigated and the method with which this object is investigated. See, for example, the first chapter of the first book in the *Nicomachean Ethics* where Aristotle states:

> Our discussion will be sufficiently adequate if it has as much clearness as the subject-matter admits of, for precision is not to be sought for alike in all discussions [...]. We must be content, then, in speaking of such subjects and with such premisses to indicate the truth roughly and in outline, and in speaking about things which are only *for the most part* true and with premisses of the same kind to reach conclusions that are no better. [...] For it is the mark of an educated man to look for precision in each class of things just so far as the nature of the subject admits; it is evidently equally foolish to accept probable reasoning from a mathematician and to demand from a rhetorician scientific proofs.[3]

As mentioned above, this is a well-known passage where both the role "the nature of the subject" has in the cognitive process and the need for a *flexible* approach are made clear: we cannot expect a single type of knowledge for every kind of question or object. This epistemological flexibility is also confirmed by a series of terms used in the quotation: ἱκανῶς (sufficiently), ἀγαπητὸν (we must be content), ἐπιδέχεται (it admits), ἀποδέχεσθαι (to accept). All of these terms point out to the same approach, according to which the idea of *adapting* to the circumstances is central and we should never expect more than what the φύσις τοῦ πράγματος, the nature of the subject, allows.

We find a similar idea in another passage, this time from the *Metaphysics:*

> Thus, some people do not listen to a speaker unless he speaks mathematically, others unless he gives instances [...]. And some want to have everything done accurately, while others are annoyed by accuracy [...]. Hence one must be already trained to know how to take each sort of argument [...]. The minute accuracy of mathematics is not to be demanded in all cases, but only in the case of things which have no matter.[4]

And again, in the *incipit* of the *On the soul:*

> It might be supposed that there was some single method of inquiry applicable to all objects. [...] But if there is no such single and general method for solving the question of essence, our task becomes still more

[3] *Nicomachean Ethics*, 1094b 11-28. Translation slighlty modified by me.
[4] *Metaphysics*, 995a 5-17.

> difficult; in the case of each different subject we shall have to determine the appropriate process of investigation.[5]

When Aristotle speaks of φύσις τοῦ πράγματος, the nature of the subject, he probably refers to two elements in particular: the nature of the subject, of the matter, in a literal sense, and the goal that is typical to that subject. In this way, these three elements (method, subject, goal) interact with each other, in order to stabilize the most suitable method. According to Aristotle, in fact, there is an inextricable interaction between these three elements, so that the method is bound to the subject, just as the nature of the subject conditions the goal which it aims for. We consider this aspect as really innovative, especially in comparison with a still fashionable trend that aims to impose a single and universal method without considering the specificities and the differences of the types of objects being addressed.

The for the most part

In the passage from the *Nicomachean Ethics* we quoted above, the notion of ὡς ἐπὶ τὸ πολὺ, *for the most part*, plays a really central role. The over two hundred occurrences of ὡς ἐπὶ τὸ πολὺ, homogeneously distributed throughout the whole *Corpus Aristotelicum*, clearly show what an important role this notion plays within Aristotelian philosophy. Despite the fact that it is "foremost an expression of the ordinary Greek language" (Mignucci, 1981: 186), in Aristotle's corpus it often has a technical meaning.

In the attempt to summarize the technical definition of *for the most part* described in Aristotelian thinking, we could say that it represents one of the types of the contingent (τὸ ἐνδεχόμενον), precisely that type that distinguishes itself from the chance (αὐτόματον or τύχη) or from the indefinite (ἀόριστον), or from the accidental (κατά συμβεβηκός) (which represent the other type of contingent) and, obviously, from what is always (ἀεὶ) or what is necessary (ἀναγκαῖον) or what is common to all of the cases (παντὶ ὑπάρχειν) (which represent, instead, the pole opposite to what is contingent). This definition needs a preliminary clarification, since both poles between which the *for the most part* is encompassed, can be represented by elements that, in this context, we can consider equivalent even though they not always are. In fact, depending on which elements we take into account, *for the most part* may have a modal, temporal or

[5] *On the soul*, 401a 13-19.

quantificational value.[6] In the first case, the *for the most part* would be the central element of the triad "necessary (ἐξ ἀνάγκης)–for the most part–accidental (or fortuitous) (κατά συμβεβηκός or αὐτόματον or τύχη)"; in the second case, instead, the triad would be "always (ἀεὶ)–for the most part–rarely (ἐπ' ἔλαττον)", and, finally, in the third one, "universally (καθόλου)–for the most part–in a few cases (ὀλίγοι)."

I admit with Mignucci that:

> we cannot choose between the temporal interpretation and the quantificational interpretation, since both are possible depending on the context. [...] Actually in everyday language, the expression ὡς ἐπὶ τὸ πολὺ generally denotes a plurality of *cases*, where the cases are, according to the context, the individuals in a field of thought or instants in time or some other things (Mignucci, 1981:186).

In relation to this last element, a passage in the second book of the *Rhetoric* is particularly explicit: "[The *for the most part*] may take place in two ways, from consideration either of the time or of the facts."[7] Furthermore it is necessary to specify that in the *Corpus Aristotelicum* the ὡς ἐπὶ τὸ πολὺ sometimes plays a temporal role (or a quantificational one) and sometimes a modal role. The matter, although interesting for other approaches, is of little interest in our case. Aristotle, in fact, as it has been correctly noted, associates judgments of frequency (not to be taken as a theory of probability), with the modal ones. Thus, "'always is associated with '(by nature and) of necessity', 'for the most part' with 'by nature (but falling short of necessity)', and 'rarely' with 'incidentally'" (Judson, 1991: 82). And again:

> these two ideas, "the universal" and "the necessary", are closely interconnected, as are those of "the particular", "the accidental", or "the contingent", for universal attributes are treated as those which belong necessarily (*ex anangkēs*) to their subjects, and the universal is also treated as "the always and anywhere", *to aei kai pantakhou*, (de Ste. Croix, 1992: 25).

Since Aristotle does not always distinguish between these aspects, when we refer to the *for the most part* we generally intend what is included between what is always or universal or necessary, on one side, and what is rare (from a both quantificational or temporal point of view) or accidental on the other side. In addition, the same explicit definitions that

---

[6] On this topic see the detailed analysis provided by Mignucci (1981).

[7] *Rhetoric*, 1402b 35-37.

we find in the *Corpus Aristotelicum* show that this missed distinction lingers. Let's read, for example, one of the classical definitions of the *for the most part*, in the first book of *Prior Analytics*:

> Having made these distinctions we next point out that the expression "to be possible" [the contingent, A/N] is used in two ways. In one it means to happen *for the most part* (ὡς ἐπὶ τὸ πολὺ) and fall short of necessity [...] or generally what naturally belongs to a thing [...]. In another sense the expression means the indefinite, which can be both thus and not thus [...] or generally what happens by chance.[8]

As we will also see in the other occurrences, this is not a proper *positive* definition, but rather a *negative* definition that tells us, above all, what the *for the most part* is not. The *for the most part*, refers, in fact, to what is neither necessary nor accidental and consequently refers to all that is included between these two poles. But, because they are at the antipodes, they contain a very ample and variegated conceptual slice that ranges from the borders of the *necessary* to the borders of the *accidental.*

Another passage, this time from *Topics*, backs up our idea that the difference between the various interpretations of *for the most part* is not really important in Aristotle In this case in particular, the modal interpretation and the quantificational one are treated as equals:

> Some things occur of necessity, others *for the most part* (ὡς ἐπὶ τὸ πολὺ), others however it may chance; [...] For if a necessary event has been asserted to occur *for the most part* (ὡς ἐπὶ τὸ πολὺ), clearly the speaker has denied an attribute to be universal which is universal, and so has made a mistake: and so he has if he has declared the *for the most part* (ὡς ἐπὶ τὸ πολὺ) attribute to be necessary: for then he declares it to belong universally when it does not so belong. [...] For the contrary of a *for the most part* (ὡς ἐπὶ τὸ πολὺ) attribute is always a comparatively rare attribute.[9]

Here, as in the *Prior Analytics*, Aristotle defines the category of *for the most part*. He defines this category as being on one hand in opposition to what is necessary or to what belongs to everybody, and, on the other hand to what is accidental or rare, without specific logical details: what is necessary, therefore, happens in all cases and, symmetrically, what happens in all cases is necessary. *Fortuitous* and *rare* are also both treated as equivalents, hence supporting what has previously been illustrated,

---

[8] *Prior Analytics*, 32b 4-13. Translation slighlty modified by me.

[9] *Topics*, 112b 1-12. Translation slighlty modified by me.

which is that the *for the most part* objects can be treated equally from the modal, temporal and quantificational point of view.

In the sixth book of the *Metaphysics* we find another passage that stresses that in Aristotle there is a technical use of the expression *for the most part* and there is no distinction between the three approaches (modal, temporal, quantificational): "Therefore, since not all things either are or come to be of necessity (ἐξ ἀνάγκης) and always (ἀεὶ), but, the majority of things are *for the most part* (τά πλεῖστα ὡς ἐπὶ τὸ πολὺ).[10]

This passage is also interesting for another reason; by speaking of the *for the most part* "things," or "objects", Aristotle highlights that being *for the most part* is an *ontological* and ineradicable condition. Elsewhere,[11] Aristotle also underlines this particular interpretation and this undoubtedly represents a notable difference compared to the previous philosophical tradition. In fact, while admitting that it is not even possible to improve and modify the knowledge of *for the most part* objects from a theoretical point of view, it is tantamount to consider that this is not a *limit* of human knowledge, rather there is an *objective* and unchangeable situation which makes it necessary to identify a more suitable cognitive method.

Considering what we have discussed up to now, considering the position of the ὡς ἐπὶ τὸ πολὺ between two opposite poles, this notion refers to two aspects that are only apparently in opposition, but that in reality are only complementary: the regularity and the variability. To say that something happens *for the most part*, in fact, implies that there is a sort of regularity; this regularity, nevertheless, since it is neither necessary nor accidental, *tends to* happens again, but without the guarantee that it might not happen and, above all, without the possibility to quantify and *manage* this regularity: we know that there is a regularity that admits exceptions, but we don't know how often these exceptions happen. We certainly know, on the one hand, that this regularity cannot be absolute, because otherwise it would be within the realm of necessity, and, on the other hand, that the number of exceptions to the regularity, although variable, cannot be such to replace such regularity with accidentality. This regularity, therefore, may show a certain lack of homogeneity. If it is true, as Mignucci says: "the deviations in comparison to the rule [...] always have the character of rarity" (Mignucci, 1981:203), it is also true that the concept of *rarity* itself is so *fluid* that it is impossible to categorize in a rigorous way.

---

[10] *Metaphysics*, 1027a 8-10. See also, for example, *Metaphysics*, 1064b 32-36.

[11] In *Nicomachean Ethics*, 1094b 21, for example, Aristotle speaks explicitly about "*for the most part* objects", Cf.: *Metaphysics*, 1027a 20-26.

Certainly, as we will see below, the regularity is not such as to impose a forced choice (as happens in the case of what is necessary), rather this is the real specificity of the *for the most part*: the fact that it may be known and forecasted (otherwise it would be pure chance) but not in an algorithmic way. It is for this reason that the Aristotelian concept of *for the most part* has the characteristic of being continuous. We use this adjective to highlight the fact that there aren't different types of *for the most part* that are easily separable and distinguishable from one another. For example, a *for the most part* doesn't exist in the ethical-rhetorical field that is radically different from the *for the most part* in the field of physics. Rather it is more opportune to speak about a certain gradualness within the interpretation of the concept; there are not discrete *for the most parts*, but, rather, a continuous *for the most part* that ranges from one pole to the other without real interruptions.

The coexistence of these two aspects, regularity and variability, in the concept of *for the most part* is demonstrated by the fact that when Aristotle defines *for the most part*, he sometimes associates it with what is variable (when he identifies it as a type of contingent), and sometimes, on the contrary, with what is regular, when he associates it to *always* and to states of which there is ἐπιστήμη (science), i.e. of the *always* and of the *for the most part.* Actually, by saying, as Aristotle does, that the world is divided into what is necessary and in what is contingent and then by defining *for the most part* as a type of the contingent, it is tantamount to underline the difference between the *for the most part* and the necessary, in other words, what happens with a regularity that does not admit exceptions. Consequently, just because the *for the most part* is something that can be otherwise, in a certain sense it is more similar to τύχη, chance, than to what is necessary and this inexactitude becomes a constitutive and unavoidable element of the *for the most part*.

It is also true, on the other hand, that Aristotle often tends to underline the similarities between *always* and *for the most part*, by highlighting the differences between these two concepts and the τύχη. This especially happens when he uses the well-known formula according to which science, ἐπιστήμη, is of what is *always or for the most part*, underlining, therefore, elements of affinity, most importantly regularity as opposed to accidentality.

The *Metaphysics* and the *Prior Analytics* are the two works where this expression occurs more often. In the sixth book of *Metaphysics* we read:

> For all science is either of that which is always or of that which is *for the most part*. (For how else is one to learn or to teach another? The thing must be determined as occurring either always or *for the most part*, e.g. that

> honey-water is useful for a patient in a fever is true *for the most part*.) But that which is contrary to the usual law science will be unable to state.[12]

And again, in the eleventh book: "For all science is of that which is always or *for the most part*, but the accidental is in neither of these classes."[13] This formula is mentioned twice, both in the *Prior Analytics* and in the *Posterior Analytics*–even if not as explicitly as in the *Metaphysics*–and, in the latter, *for the most part* is replaced with πεφυκὸς, *according to nature*, *natural*, that, as we will see, can almost be considered as a synonym in many contexts:

> There is no knowledge (ἐπιστήμη) by demonstration of chance conjunctions; for chance conjunctions exist neither by necessity nor *for the most part* [...]. Now demonstration is concerned only with one or other of these two;[14]

> Science (ἐπιστήμη) and demonstrative syllogism are not concerned with things which are indefinite [...]; but they are concerned with things that are natural (τῶν πεφυκότων).[15]

First of all, it should be noted that, in this context, the use of the term ἐπιστήμη does not have to be understood in a *strong* sense, that is to say, as an absolutely certain knowledge of what is unchangeable, because there may be ἐπιστήμη of what is *for the most part* as well. However, when Aristotle says that there is a science of what is *always* and of what is *for the most part*, does he really mean that *always* and *for the most part* can be treated as equivalent, and therefore that a unique type of science exists that deals indifferently with what is *always* and what is *for the most part*? Or does he mean that the two forms of science exist, and that one deals with *always* and one deals with *for the most part*?[16] If we accept the first

[12] *Metaphysics*, 1027a 20-26.
[13] *Metaphysics*, 1065a 4-6.
[14] *Posterior Analytics*, 87b 19-22. Translation slighlty modified by me.
[15] *Prior Analytics*, 32b 18-20.
[16] Among the proposals formulated in order to provide a solution, we briefly recall Reeve's and Anagnostopoulos' analysis. Reeve (1992) chooses the solution according to which there are two different types of science, "a pure science" concerning what is always and a "plain science" dealing with *for the most part*. What isn't convincing in this analysis is the attempt to maintain, nevertheless, *necessity* as a fundamental element in order to define science. For this reason, Reeve is obliged to hypothesise two different types of necessities according to the object in cause. Instead, Anagnostopoulos (1994) considers that necessity is not a

hypothesis we minimize excessively the difference between *always* and *for the most part* and in this situation we are forced to consider the fallibility intrinsic to the regularity of the *for the most part* as if it were irrelevant. This does not seem to be acceptable. Surely the other proposal sounds more reasonable, according to which in the expression *always or for the most part* we have to different elements, *always* and *for the most part*, each of which are the object of a type of *scientific* investigation. This formula, therefore, implies that there is also a science of *for the most part*, but it is not the same science as that of *always*, it is another type of science. In this situation, *for the most part* remains independent from *always*. In this way we no longer require necessity as a fundamental condition of science and we can use a weaker concept, *regularity*. In fact, the statement that what unites the two types of science is necessity, in one case absolute and in the other case partial, would not only be difficult to accept (it is difficult to imagine a partial necessity), but above all, would not really explain the nature of the *for the most part*. However, it is altogether another matter to consider *regularity* as the common element between *always* and *for the most part* and as the condition of their being subjected to a scientific investigation. In fact, it is absolutely legitimate to speak of a regularity that does not admit exceptions (in this case, it would coincide with necessity) and of a regularity that admits exceptions (this would be, obviously, the regularity typical of the *for the most part*). From this perspective, the concept of *for the most part* wouldn't be misrepresented and would maintain its second characteristic, variability.

A wider notion of conjecturality: the stochasticity

Considering the relationship between method and object developed above, the question now is: which, therefore, is the most suitable method for the type of object we have just delineated, that is, the *for the most part*? Concerning the characteristics of *for the most part* objects, the most adequate type of knowledge is the conjectural one or, even better, the *stochastic* one. The verb στοχάζεσθαι, from which the adjective

---

necessary condition for Aristotelian science and in *for the most part* there isn't, indeed, a necessity but instead an *inexactitude,* or *flexibility*. However, if such way of reasoning is rather convincing up to this point, the scholar's conclusion is less so. He states that it is possible to eliminate the inexactitude of *for the most part* sentences. Anagnostopoulos' arguments have already been criticised by others (Winter 1997).

στοχαστικός is derived, is usually translated as "to conjecture".[17] We can say, in general terms, that the verb στοχάζεσθαι refers to a non-algorithmic cognitive activity, which is not limited to performing pre-arranged procedures for reaching a goal, but which necessarily elaborates a hypothesis, conjectures and devises in order to reach the target. When we speak of a *stochastic* attitude or intelligence, therefore, we are referring to this type of knowledge and approach to objects that constantly presents itself throughout Aristotle's works.

While the studies on conjectural knowledge have historically been associated with the notion of sign, in Aristotle the analysis of *stochastic* knowledge rarely concerns, in an explicit manner, the notion of sign and instead, more often refers to the wider concept of *for the most part*.[18] In all modern languages, in fact, we define the disciplines which belong to the so-called evidential paradigm[19] as conjectural. The typical examples of conjectural disciplines are those of investigation, of medicine, of historiography etc., that is to say, those disciplines which show an evident link between conjecturing and the existence of signs. Instead, the activity of στοχάζεσθαι, in addition to such evidential disciplines, incorporates others activities that we instinctively would not hold conjectural: such as music, archery, gardening etc. These activities, as said above, are not explicitly related to the σημεῖα, signs, but, nevertheless, we can consider them conjectural since they don't present an element of certainty that must be followed. In order to reach a result, instead, some traces must be used: as Plato said in *Filebo*, the musician doesn't harmonize the cords through measure but through the conjecture that is made based on practice (μελέτης στοχασμῷ) (*Filebo*, 56a 3-4), and he must use such signs because the matter he deals with is always variable.

In a certain sense, it could be said that *stochastic* knowledge is to the *for the most part*, as conjectural knowledge is to the notion of sign, with the consequence that there isn't a total coincidence between conjectural and *stochastic* knowledge. The second one is conceptually wider than the first one. Therefore, *stochastic* intelligence is the ability to manage these

---

[17] For the etymology of the verb see Allen (1994). From case to case, we consider that it is possible to identify at least other four semantic nuances beside the conjectural one: to aim for the target (Piazza, 2004); to aim for something; "to try (to do) something"; "to approximate to something". From the verb στοχάζεσθαι, mainly in the post-Aristotelian period comes the notion of τέχνη στοχαστική. See Isnardi Parente (1961).

[18] The sign itself–since only *for the most part* indicates what it is sign of–can be integrated into the broader category of *for the most part* objects.

[19] Ginzburg (1979).

types of objects, in which the *for the most part* shows regularity as well as variability. The fact that Aristotle sees conjectural knowledge as playing the central role does not prevent him from recognizing the intrinsic *fallibility* of this form of knowledge. It would be better to say, nevertheless, that Aristotle not only does not obscure this aspect, but above all, he does not believe that this possibility of failure could devalue this cognitive approach, given the characteristics of the objects to which this type knowledge is applied. Rather, Aristotle uses the notion of fallibility to point out another aspect that is complementary to it, that is, that not everyone is able to manage *for the most part* objects and, therefore, to use *stochastic* intelligence in a suitable and profitable way. Aristotle is so convinced of the importance of this ability, that he holds those who can use it, even though they may always be subject to the possibility of error, as excellent, σπουδαῖος (*Nicomachean Ethics*, 1109a 20-25).

What makes *for the most part* objects knowable in a *stochastic* way is really the conjunction of their characteristics, their being at the same time both regular and varying, at times tending more toward the necessary, at times nearer to the accidental. If there is the *for the most part*, in fact, "it is necessary that there is the accidental" (*Metaphysics*, 1027a 10-11) and consequently anyone who wants to get a suitable knowledge of the *for the most part* has to know how to *manage* the possibility of deviation, the risk of an intervention from chance, and the impossibility of applying *tout court* the rule he knows.

The φρόνιμος, the wise man, precisely because he knows how to adapt the rule to singular cases and because he knows just as much about the universal as about the particular, is the only one, in a certain sense, that is able to make the *for the most part* as regular as possible, without being able, nevertheless, to totally foresee the intervention of the chance element, although he might still always take it into account. The passage from the third book of the *Nicomachean Ethics*, 1112a 34-b 10, clarifies this idea according to which the φρόνιμος is the most suitable individual for managing the *for the most part*: the specific ability of the φρόνιμος is making decisions and the decisions can be made only on what is *for the most part* and not necessary, consequentially then, the more uncertain the decision is (or, the less regular the *for the most part* is), the more able the φρόνιμος should be in order to carry out his task. The *for the most part* is the most suitable object for *stochastic* intelligence because it is regular (therefore knowable by science, not by chance) but with exceptions (therefore fallible and not algorithmic).

That's why the two aspects of the *for the most part* are both necessary. If there were no variability, there wouldn't be any need for the ability to adapt the rules to the single cases, or of eventually elaborating a new rule, since just the algorithmic application of the rule would effectively be enough. If there were not a sort of regularity and everything were absolutely accidental it would not be possible to manage these accidental elements with rational procedures and *stochastic* intelligence would lose the rationality that Aristotle constantly recognizes in it. When we cannot automatically apply the rules and we need to decipher signs, we do it because, in a sense, these signs have a past that we know, because we have observed that they referred *for the most part* to something else or, if these are new signs, because we treat them as similar to others signs we consider regular. We trust, therefore, that these signs, on which we ground our conjectures, can lead us to making the correct decision, because we believe that a background regularity exists, because, despite the unavoidable chance component of the case, we trust and we believe, as Aristotle says, that "the future *for the most part* resembles the past" (*Rhetoric*, 1394a 7-8).[20]

An alternative to the opposition between rationalism and irrationalism: the *for the most part* truths

The Aristotelian analysis of the notions of στοχάζεσθαι and *for the most part*–the way, above all, in which Aristotle connects these two notions, and particularly the methodological principle according to which there is no *single* method that can be suitable for any type of object–is a fundamental aspect of Aristotelian thought and, at the same time, is one of the more interesting theoretical resources that can be useful in contemporary debates. Facing the issue of knowledge "through conjectures and approximations" in this way allows Aristotle to avoid *hierarchical* evaluations of the various forms of knowledge.

The general tendency in (but not exclusive to) a large part of contemporary philosophy, is different. With such a variety of objects to know, in fact, just two alternatives seem to exist: either the method is *forced* and is applied to objects for which it is not suitable, or the object is *forced*, modified, in order to make it suitable to the method. Both these positions reveal an intellectual violence that not only seems to be ethically improper, but that also does not seem to be useful, since in this way the original inadequacy of method and object clearly comes out.

[20] Translation slighlty modified by me.

The Aristotelian proposal allows a much wider epistemological setting, since exactness is not always the correct criterion to evaluate a type of knowledge and there is not only one sound type of knowledge. In this way, *stochastic* knowledge is not second-hand knowledge and the *for the most part* objects are not second-hand objects. The clearest example that being *for the most part*, in the Aristotelian perspective, does not indicate a condition of inferiority or negativity is given by the fact that all natural objects are *for the most part, and* that nature itself, φύσις, has a *for the most part* trend.

In some cases the adjective πεφυκὸς and the expression ὡς ἐπὶ τὸ πολὺ seem to be used as synonyms, beginning with the passage from the *Prior Analytics* we have quoted before: "In one it means to happen for the most part (Τὸ ὡς ἐπὶ τὸ πολὺ) and fall short of necessity [...] or generally what naturally (τὸ πεφυκὸς) belongs to a thing."[21] We can constantly find this idea in the *Corpus*, in several works that differ from one another quite a bit, and not necessarily always with the use of the adjective πεφυκὸς but also with equivalent expressions such as φύσει or κατὰ φύσιν (according to nature.)[22] But it wouldn't be correct, either, to say that in the φύσις there is only space for the *for the most part* and not for the *always*. In the first book of the *Rhetoric*, for example, Aristotle clearly says that in φύσις there is space for both and, above all, confirming what we said before, that their unifying *trait* is regularity: "Things which are the result of nature (φύσει) are all those of which the cause is in themselves and regular; for they turn out always, or *for the most part*, in the same way."[23]

We also find something similar in the *Physics* (198b 34-36) and in the *Eudemian Ethics* (1247a 31-32) so that, even though we cannot speak of a total identification between ὡς ἐπὶ τὸ πολὺ and πεφυκὸς (or φύσει or κατὰ φύσιν), we have to recognize nevertheless that, according to Aristotle, it is, above all, the element of the *for the most part* that characterizes φύσις. The clearest evidence comes from the works that address biology, where Aristotle speaks of nature *in action.* It is no by coincidence, in fact, that in these works we find the highest frequency of the expression ὡς ἐπὶ τὸ πολὺ, both when it is used to describe natural phenomena, and when Aristotle makes some theoretical clarifications on

---

[21] *Prior Analytics*, 32b 5-8. Translation slighlty modified by me. In *Prior Analytics*, see the almost analogous paragraph 25b 14-15.

[22] See also Ferejohn (1991: 120-121).

[23] *Rhetoric*, 1369a 35-b 2. Translation slighlty modified by me.

the statute of φύσις. In the *On the parts of animals* we find the following methodological warning:

> But in all our speculations concerning nature, what we have to consider is the general rule; for that is natural which applies either universally or *for the most part* (ἢ γὰρ ἐν τῷ παντὶ ἢ τῷ ὡς ἐπὶ τὸ πολὺ τὸ κατὰ φύσιν ἐστίν).[24]

In *On the generation of animals* this question is particularly interesting because, when Aristotle approaches the *for the most part* aspect of the φύσις from a theoretical point of view, he explicitly refers to what escapes regularity and he does it talking in terms of *monstrosity* or *morbidness:*

> Nor is it at all absurd that the catamenia should still continue after conception [...] This, however, is a *morbid* phenomenon, wherefore it is found only in a few cases and then seldom, whereas it is that which happens *for the most part* that is according to nature; [25]

> For the *monstrosity* (τὸ τέρας) belongs to the class of things contrary to nature, not *any and every kind* of nature, but nature in her *for the most part* (ὡς ἐπὶ τὸ πολὺ) operations; nothing can happen contrary to nature considered as eternal and necessary, but we speak of things being contrary to her in those cases where things *for the most part* (ὡς ἐπὶ τὸ πολὺ) happen in a certain way but may also happen in another way;[26]

> Hence such cases seem to be more *monstrous* (τερατώδη), because they are contrary to the *for the most part* (ὡς ἐπὶ τὸ πολὺ) and customary rule;[27]

> While women are suckling children the catamenia does not occur *according to nature* (κατὰ φύσιν), nor do they conceive; if they do conceive, the milk dries up. This is because the nature of the milk and of the catamenia is the same, and nature cannot be so productive as to supply both at once; if the secretion is diverted in the one direction it must needs cease in the other, unless some violence is done contrary to the *for the most part* (ὡς ἐπὶ τὸ πολὺ). But this is as much as to say that it is contrary to nature, for in all cases where it is not impossible for things to be otherwise

[24] *On the parts of animals*, 663b 27-29. Translation slighlty modified by me.
[25] *On the generation of animals*, 727b 26-30. Translation slighlty modified by me.
[26] *On the generation of animals*, 770b 9-13. Translation slighlty modified by me.
[27] *On the generation of animals*, 772a 36-37. Translation slighlty modified by me.

> than they generally are but where they may so happen, still what is *for the most part* (ὡς ἐπὶ τὸ πολὺ) is what is *according to nature* (κατὰ φύσιν).[28]

Because nature is *for the most part*, some exceptions occur and there are cases where regularity and the *for the most part* show their fallibility. The fact that Aristotle, as we have seen in these passages, takes into account these exceptions by using terms such as *monstrous*, *morbid* etc. it is an evident sign that according to him the *for the most part* aspect of nature is strongly characterized by regularity and that exceptions, although they are always possible, are considered as a rarity. However, at the same time, such rarities (see in particular the first passage) aren't *absurd* because it is not absurd that the regularity of the *for the most part* shows its fallible side, since "the possibility that things are otherwise from what they are, which is part of the φύσις [...] admits the possibility of deficiency and error" (Isnardi Parente, 1966: 177).

The regularity, or the frequency with which phenomena occur that is not generally introduced as "deriving, *a posteriori*, from the verification of phenomena, but as a guaranteed *a priori* resulting from the regularity of φύσις, and from the constancy of the structures of the *ousiai* nature itself consists of" (Vegetti, 1971: 100), is not such as to obscure the presence of the fortuitous, of the accidental, or of the anomalous. Nature's anomalies correspond, therefore, to the *triads* of *for the most part* we described at the beginning, the *rare*, the *accidental*, or the *few*. They are the other face of the regularity of nature just as *rare*, *accidental* and *few* are the other face of *for the most part*. Once more, therefore, even though one of the two aspects of the *for the most part is* highlighted, in this case regularity, the other aspect is not completely darkened, so that the idea of distinguishing between the two types of *for the most part* seems, yet again, to be inadequate according to the Aristotelian perspective.

In light of these observations, therefore, it is possible to state that:

> within the Aristotelian paradigm, naturalness and variability are not incompatible characteristics. [...] The naturalness, in action and in production, defines an area of invariability within which a multiplicity of different and unique but isomorphous events can happen (Lo Piparo, 2003: 124).

Saying that a biological *for the most part* also exists, in the sense that nature also shows itself being *for the most part*, is yet another evidence of what we have often repeated, that this characteristic (being *for the most*

[28] *On the generation of animals*, 777a 12-21. Translation slighlty modified by me.

part) is not eliminable and does not depend on the cognitive limits of man, but is a necessary condition of anything dealing with matter and movement.

Therefore, it is not a characteristic which only belongs to the fields of ethics and rhetoric, two areas usually relegated to an inferior level in the hierarchy of knowledge; it is, rather, a way of being that exists where there is movement and matter, where nature itself is included, regardless of whether physics is a theoretical science (*Metaphysics*, 1025b 25-28) and whether ethics and rhetoric are, instead, a practical and a poetic science respectively. In this way, it is also clear that not only one type of rationality exists that is useful in cognitive processes, the algorithmic or mathematic ones, and, consequently, the classical opposition between "rationalism" and "irrationalism" becomes inadequate: there are various types of rationality, all legitimate and all sound, depending on the objects and the goals involved, and *stochastic* knowledge, far from being mere divination, is based on a strong rationality. A type of knowledge that is founded on signs, which by definition are *for the most part*, neither can nor must aim for the rigor of the exact sciences. Nevertheless, it can be rigorous within the limits that its object allows. If it did not do so, and only one method would be used for all the objects and if a type of *inadequate* accuracy was expected, according to Aristotle, it would not appear to possess enough παιδεία, education. The *stochastic* knowledge also has an accuracy that, however, is different from algorithmic knowledge; various forms of accuracy exist, so that the necessity of choosing among a strong or weak scientific status vanishes. The inevitable consequence of this approach is that there are not only one, but various *types* of truth: there are certain truths which concern what is *always* and *necessary*, but there are also *for the most part* truths,[29] that is to say, truths that can be reached by *stochastic* means. These truths cohabitate with fallibility but they are the only ones this approach accepts. They are, above all, those really suitable for objects of this kind.

---

[29] See also Piazza (2008).

References

Primary Sources

Aristotle (1923). *Metaphysics*, translated by W. D. Ross, Oxford: Clarendon Press.

—. (1908). *Nicomachean Ethics*, translated by W. D. Ross, Oxford: Clarendon Press.

—. (1912). *On the Generation of Animals*, translated by A. Platt, Oxford: Clarendon Press.

—. (1882). *On the Parts of Animals*, translated by W. Ogle, London: K. Paul, French & Co.

Aristotle, (1984). *On the Soul*, translated by J.A. Smith. In J. Barnes (ed.). *The Complete Works of Aristotle*, Vol. 1, Bollingen Series LXXI-2, Princeton: Princeton University Press.

Aristotle (1928). *Posterior Analytics*, translated by G. R. G. Mure. In W. D. Ross (ed.), *The works of Aristotle*, London: Oxford University Press.

Aristotle (1944). *Prior Analytics*, translated by A. J. Jenkinson, Cambridge: Harvard University Press.

Aristotle (1941.) *Topica*, translated by W. A. Pickard-Cambridge. In R. McKeon (ed.), *Basic Works of Aristotle*, New York: Random House.

Aristotle, (1925). *The 'Art' of Rhetoric*, translated by J. H. Freese, London: The Loeb Classical Library.

Secondary Sources

Allen, J. (1994). “Failure and Expertise in the Ancient Conception of an Art”. In T. Horowitz & A.I. Janis (eds.), *Scientific Failure* (pp. 81-108), London: Rowman & Littlefield.

Anagnostopoulos, G. (1994). *Aristotle on the Goals and Exactness of Ethics*, Berkeley: University of California Press.

De Ste. Croix, G.E.M. (1992). “Aristotle on History and poetry (*Poetics*, 9, 1451a36-b11)”. In A. Oksenberg Rorty (ed.), *Essays on Aristotle's Poetics* (pp. 23-32), Princeton: Princeton University Press.

Detienne, M. e Vernant, J.-P. (1974). *Les ruses de l'intelligence. La mètis des Grecs*, Paris: Flammarion. (Italian ed, (2005). *Le astuzie dell'intelligenza nella antica Grecia*, Roma-Bari: Laterza).

Ferejohn, M. (1991). *The origins of Aristotelian science*, New Heaven and London: Yale University Press.

Ginzburg C. (1979). "Spie. Radici di un paradigma indiziario". In AA. VV. (eds.), *La crisi della ragione* (pp. 80-112), Torino: Einaudi.

Judson, L. (1991). "Chance and 'Always For the Most Part' in Aristotle". In L. Judson (ed.) *Aristotle's Physics: A collection of Essays* (pp.73-100), Oxford: Clarendon press.

Isnardi Parente, M. (1961). "TECHNE", *La parola del passato. Rivista di studi antichi*, vol. XVI: 257-296.

—. (1966). Techne. *Momenti del pensiero Greco da Platone a Epicuro*, Firenze: La Nuova Italia.

Lo Piparo, F. (2003). *Aristotele e il linguaggio. Cosa fa di una lingua una lingua*, Roma-Bari: Laterza.

Mignucci, M. (1981). "'ὡς ἐπὶ τὸ πολὺ' et nécessaire dans la conception aristotélicienne de la science". In E. Berti (ed.) *Aristotle on Science. The "Posterior Analytics", Proceedings of the Eight Symposium Aristotelico Held in Padua from September 7 to 15, 1978* (pp. 173-203), Padova: Antenore.

Piazza, F. (2004). "Prendere la mira e cogliere nel segno. L'*eustoxia* nella filosofia aristotelica", *Forme di vita*, 2-3: 95-103.

—. (2008). *La retorica di Aristotele. Introduzione alla lettura*, Roma: Carocci

Reeve, C.D.C. (1992). *Practices of Reason: Aristotle's Nicomachean Ethics*, Oxford: Clarendon Press.

Vegetti, M. (1971). "Introduzione" ad Aristotele, *Ricerche sugli animali*, Torino: Utet.

Winter, M. (1997). "Aristotle, *hōs epi to polu* relations, and a demonstrative science of ethics", *Phronesis*, 42: 163-189.

EUGE, GRAECULI NOSTRI! GREEK SCHOLARS AMONG LATIN CONNOISSEURS IN MACROBIUS' *SATURNALIA*

PAULA OLMOS

Introduction

As is well known, Macrobius' *Saturnalia* (ca. 430 CE)[1] features a monumental three-day-long convivial[2] and highly erudite dialogue, which takes place in Rome during the festivities mentioned in its title–December 17th-19th–and brings together a mixed assembly of "leading members of the Roman nobility and other learned men" ("Romanae nobilitates proceres doctique alii congregatur", *Sat.* I.1.1.).[3] As in any other ancient literary dialogue with doctrinal (philosophical or otherwise) ambitions, the choice of characters is a key issue here and most editors of the text acknowledge this fact by offering, within their introductions, detailed lists of *dramatis personae* and commenting extensively on the fictional and historical particulars of the people portrayed in it. For, following the genre's conventions,[4] all the participants in Macrobius' dialogue represent real, historical, and in most cases, rather well-known, people,[5] all of them (with the possible controversial exception of the two younger ones) dead

---

[1] This date, today almost universally accepted, used to be considered polemical. Known as "Macrobius' late dating hypothesis", it was established by Cameron (1966), as opposed to the traditional assumption that placed his writings around the end of the 4th century.

[2] More will be said later about the just partially convivial nature of the *Saturnalia*.

[3] Unless otherwise stated, all *Saturnalia* quotations in both Latin and English are taken from R. Kaster's recent edition and translation (Kaster [ed.], 2011).

[4] Macrobius mentions Cicero's dialogues (*Sat.* I.1.4.) and Plato's *Symposium* (*Sat.* I.1.3.) in his Prologue. We also have to take in account that the tradition of the symposium genre had grown and expanded considerably since Plato, with both Greek and Latin authors. See Navarro Antolín ([ed.], 2010: 14-22).

[5] See Kaster ([ed.], 2011: xxvi-xxxvi).

by the time of its writing.[6] In various significant instances, moreover, as has been effectively argued, they are related by blood to Macrobius' own circle and contemporary friends.[7]

The historical identification of Macrobius' protagonists and the particular details of their age and circumstances place the dialogue's dramatic date in the 380s, perhaps during the Saturnalia festivities of the year 382.[8] This means that Macrobius would be writing his work something like fifty years after its dramatic date–another classical convention very close to Cicero's models (*De oratore*, *De re publica*)–and most surely in a virtually all-Christian, post-Theodosian Rome. The traditional assumption that made Macrobius a pagan militant adversary of Christianity–a hypothesis that made somewhat more sense together with the also traditional earlier dating of his works to the 390s–has been rather contested in recent years and ultimately rejected by Cameron (2011) and Kaster (2011). In any case, whether publicly pagan or Christian, Macrobius–again according to Cameron (2011) and earlier, to Marinone ([ed.], 1967)[9]–would not have counted among the *committed* ones and his love of classical lore and antiquarian interests should in fact be considered compatible with either faith.

However, returning to the dialogue's participants and their literary characterization–which is the main focus of this paper–most scholars, and the author himself,[10] emphasize the polarity between the two main and

---

[6] See Kaster ([ed.], 2011: xxxvi). For a detailed discussion of the possibility that Avienus or Servius, the youngest characters in the *Saturnalia*, could still have been alive around the time of its writing, see Cameron (2011: chap. 7, 231-272).

[7] "The interlocutors of the *Saturnalia* reflect less the original circle of Praetextatus in the pagan 380s than their descendants, the circle of Macrobius himself, in the Christian 430s" (Cameron, 2011: 261).

[8] Kaster ([ed.], 2011: xxiv), Cameron (2011: 232). The argument for its exact placement in 382, advanced by Cameron, has to do with this being the eve of the withdrawal of state subsidies for civic/pagan rites by Emperor Gratian. It would mean that these were the last Saturnalia festivities before the progressive hardening of anti-pagan policies, when the characters gathered by Macrobius would still have felt rather comfortable in their pagan faith and civic offices.

[9] Cameron (2011: 270): "My own feeling is that Macrobius could not have written about the minutiae of pagan sacrifice in the distanced, casual way he does at the time he does if he had been a committed pagan himself." Marinone ([ed.], 1967: 24): "Nel nostro scrittore la fede religiosa è cosi poco combattiva, direi quasi così tepida, che trova spiegazione soltanto in un deliberato silenzio nei confronti degli avversari."

[10] In addition to the already mentioned passage (*Sat.* I.1.1.), see e.g. *Sat.* I.2.5.: "quod doctissimis procerum ceterisque nuper apud Vettium Praetextatum".

most evident categories they belong to: Roman *proceres* who were members of the senatorial elite vs. foreign (mostly Greek) professional experts specializing in different subject areas (namely philosophy, rhetoric or medicine). The Latin noblemen in the dialogue number seven, of which three are clearly above the others in rank and position (Vettius Agorius Praetextatus, Quintus Aurelius Symmachus and Nicomachus Flavianus, the three hosts to the three-day gathering). Another two, blood relatives and similarly noble, stand out for their antiquarian knowledge and taste (Rufius Albinus and Ceacina Albinus). There are also an elite youngster eager for learning (Avienus) and a mature nobleman with rather unpleasant and sometimes offensive manners and a contrasting suspicious attitude towards learning, who comes uninvited but ultimately plays a major role in the development of the dialogue (Evangelus).[11] The plebeian foreigners number four: three Greeks (Eustathius, Eusebius and Dysarius) and one Egyptian (Horus). The clear exception to this main division would be the twelfth guest,[12] Servius the Latin grammarian, portrayed here in his youth, just after opening his own school of grammar. He represents what is necessarily the most *Latinized* of the arts and disciplines: the art of language (in the end, a particular language) itself. Kaster (1980, 1988) has commented extensively on the ambiguous social position of grammarians among their elite contemporaries and the modesty (*verecundia*) that was

[11] The intriguing character of Evangelus, of which more will be said later, has been interpreted by some as a Christian stereotype, mainly because of his name. Ramelli, for example, favours this interpretation (Neri and Ramelli, [eds.], 2007: 7 ff), which was also present in Davies's introduction to his English translation of the *Saturnalia* (1969: 11-13). However, this possibility was somewhat contested by other scholars (Flamant, 1977; Scarpa, [ed.], 1981) and has ultimately been rejected by both Kaster (2011: xxxii-xxxiv) and Cameron (2011: 245), who says this is "a major misunderstanding of the *Saturnalia*".

[12] Twelve guests take part in the three-day gathering and banquets–a number, according to *Sat.* I.7.3., which is the sum of the number of Muses (the original nine invited participants) plus the Graces (the uninvited three who appear later, Evangelus, Dysarius and Horus). The gatherings are arranged by a subset of the group (Preatextatus, Rufius Albinus, Avienus, Symmachus, Caecina Albinus and Servius) on the previous eve (*Sat.* I.2-5.). To these main characters we must add another two, who appear only in the Prologue: Postumianus, a Latin noble advocate originally among the nine invited but in the end too busy to attend (sending his Greek friend, the rhetorician Eusebius, in his place), and Decius Albinus (the son of Caecina), to whom Postumianus will recount the story of the banquet as told to him by Eusebius–in a complex "reported speech plot" similar to Plato's *Symposium*, where Apollodorus tells a friend how he had previously told Glaucon about the banquet as recounted to him by Aristodemus.

expected of them as a matter of course. That most grammarians (as a profession in great demand for basic literacy) would belong to a lower rank does not mean, however, that a further distinction between ordinary grammarians and grammatical/literary scholars (and Servius would be one of the latter) would not be socially acknowledged. Nonetheless, it seems that a subtle sense of competition could always be felt between these scholars and the elite amateur connoisseurs particularly interested in literary studies–an issue discussed in depth by Keulen (2009) in reference to Aulus Gellius' anecdotes and settings in *Noctes Atticae*, one of the *Saturnalia*'s most obvious sources.

Here, however, following the main theme of the present volume, we are particularly interested in the performance of the Greeks within the dialogue, taking special account of their role as professional experts in various arts and sciences. As we have already mentioned, there are three definitely Greek characters in the *Saturnalia*: Eustathius, the philosopher (a formally invited and most respected guest), Eusebius, the rhetorician (accepted in lieu of the invited Roman advocate Postumianus) and Dysarius, the physician (uninvited). The Egyptian Horus, formerly a pugilist and presently living as a Cynic philosopher (also uninvited), is usually considered part of this group of foreigners. Yet as we will see, his role in the dialogue (at least in the extant parts of the dialogue) is rather different from that of the others.

Over the course of this paper, we will seek to describe the way these three foreigners interact with their Latin hosts and the other guests, the role they play in the assessment of particular areas of knowledge, the degree of respect (from most of those present) or scorn and abuse (from the caustic/satiric character Evangelus) they receive, and how these different attitudes and cues shape the dialogue's development. We will also see how, even though most of the characters agree to present a sample of their expertise in public that of the Greeks seems to be of a definitely different nature, namely more systematic and professional. Finally, we will see as well how the differences among them (and among their different areas of knowledge) will be made part of the dialogue's topics (towards its final climax), in a renewed late-antique presentation of the Greek agonistic culture in which methodological issues and inter-scientific controversies are naturally foregrounded.

However, our interest in the Greeks will necessitate our first taking a closer look at the actions of Evangelus–the speaker, by the way, of the "Euge, Graeculus noster!" (VII.9.26.) which I have borrowed and slightly modified for my title. The reason for this is that, as we will see, most contributions by the Greek experts are mediated by remarks, comments or complaints from Evangelus, while at the same time, he is the one that mentions, time and time again, the specific Greekness of either the knowledge involved in the dialogue or the modes of exposition employed. The adjective "Greek" and its cognates will constantly be used by this character in an apparently derogatory but very distinctive way that calls for comment.

The Plotted Erudition and Technical Content of the *Saturnalia*

Although Cameron (2011: 353-398) has effectively challenged certain traditional images of the late Roman elite as particularly dedicated to erudite issues and classical lore in their *otium*, emphasizing instead their political and mostly economic occupations,[13] if we are to tackle Macrobius' text, we must say that, idealized or not, such is the image he conveys about the particular gathering he devises.[14] In any case, it is not as if the author himself presents his setting as an everyday, oft-repeated situation. In the Prologue of the dialogue, Decius Albinus asks Postumianus about the details of one such renowned *convivium*, an especially successful occasion

---

[13] Cameron's intention in Ch. 10, "The real circle of Symmachus", is mainly to refute the historical authenticity of such an idealized circle of learned and intellectual noblemen, something traditionally taken for granted by many scholars, and mostly based on Macrobius' own picture. L. Cracco Ruggini (1986: 97-98) had already mentioned the more than likely idealized nature of this group (a brilliant contrivance of the writer), although she also offered some interesting insights (based on Symmachus' own extant correspondence) regarding the complementary relationship between this kind of learned *otium* and the elite *negotia* which could weaken some of Cameron's arguments: "Il tipo di *otium* che Simmaco sembra auspicare e preferire è pertanto quello *litteratum*, che nulla ha a che vedere con la *iners desuetudo* da ogni impegno che offusca l'intelligenza come acqua fangosa, con gli *otia muta* che isolano dal contatto sociale e politico, con la *quies* piacevole ma inoperosa e quindi *virtuti infructuosa*" (1986: 99). In a previous paper (1985: 295), she also states, "the *otium litterarum* consecrated to *libris veterum ruminandis* in reality was felt to be both complementary and qualitatively akin to the activity carried out in *publica negotia*. It was not an alternative."

[14] "Neque enim recte institutos animus requiescere aut utilius aut honestius usquam potest, quam in aliqua opportunitate docte ac liberaliter colloquendi, interrogandique et respondendi comitate"; "A well-bred spirit can find no more useful or more honorable form of relaxation than in the sort of learned conversation worthy of free men, and in the gracious exchange of questions and answers," (*Sat.* I.2.4.).

(*Sat.* I.2.2-5, my italics: "De *hoc ipso* quaesitum venio"; "The *very one* I'm here to ask about!"). On the other hand, nothing seems to happen in a particularly spontaneous way in the *Saturnalia*, least of all the appearance of erudite or technical topics in the midst of the amiable conversation. On the contrary, the encyclopaedic result of the dialogue is explicitly planned several times and seemingly over-justified.

The first level of justification is the meta-discursive Prooemium, in which the author himself asserts his pedagogic intentions regarding his dedicatee and son Plotinus Eustathius:

> [s]o that all that I have toiled through–in various books of Greek or Latin, both before and since you were born–might be available to you as a fund of knowledge. If ever the need arises for a piece of history, say, that lurks hidden from the common run of men in a mass of books, or to call to mind some memorable deed or saying, you will be able to find and produce it easily as though from your own private store of culture.[15]

He then speaks about his own method for creating a unified work out of such a variety of sources, as bees make their honey from different flowers:

> We ought to imitate the bees, if I can put it that way: wandering about, sampling the flowers, they arrange whatever they have gathered, distributing it among the honeycomb's cells, and by blending in the peculiar quality of their own spirit they transform the diverse kinds of nectar into a single taste.[16]

---

[15] "et quidquid mihi vel te jam in lucem edito vel antequam nasceris, in diversis Graecae seu Romanae linguae voluminibus elaboratum est, id totum sit tibi scientia supellex, et quasi de quodam litterarum peno, siquando usus venerit aut historiae quae in librorum strue latens clam vulgo est, aut dicti factive memorabilis reminiscendi, facile id tibi inventu atque depromptu sit" (*Sat.* Proo. 2.). The verbs *invenire* and *depromere*, translated by Kaster as "find" and "produce", belong to the rhetorical vocabulary of *invention*, the oratorical task of excogitating and selecting material (particularly arguments) for a discourse. Regarding the rhetorical aspects of the encyclopaedic intentions involved in the *Saturnalia*, see Olmos (forthcoming).

[16] "Apes enim quodam modo debemus imitari, quae vagantur et flores carpunt, deinde quicquid attulere disponunt ac per favos dividunt, et sucum varium in unum saporem mixtura quadam et proprietate spiritus sui mutant" (*Sat.* Proo. 5.). Thus, Macrobius' work explicitly presents itself as a "deuteronomic text", to use R. Netz's expression (1998), something particularly characteristic of late-antique technical literature.

And finally, he makes his own literary ambition explicit: "That is my goal for the present work: it comprises many different disciplines, many precepts,[17] examples drawn from many periods, but brought together into a harmonious whole."[18] Such unity would be attained by means of the dramatic setting itself, the choice of characters, the justification given to the gathering and the literary logic of the plot.

And yet, the author feels it should be justified on another level, and so within the dialogue the characters themselves make explicit a secondary plan for their own meeting: a plan that should yield the expected results. Thus, before the three-day formal gathering begins, there is a previous meeting on the eve of the festival (with a subset of the characters present), at which an erudite issue is already discussed (the origins of the *Saturnalia* festivities). This is brought about by Rufius Albinus' assertion: "Neither we [Rufius, Praetextatus and Avienus] nor these gentlemen [the newly arrived Symmachus, Caecina and Servius] can take more pleasure in any discussion that one involving matters of scholarship,"[19] after which they decide to *arrange* the main meetings, mentioning the proposed guests and their suitability for such an aim.

The funny twist, however (something for which Macrobius' literary genius should be praised), is that in fact, nothing goes as first *apparently* planned. Except for the topics discussed in the first book, selected more or less naturally,[20] almost all the rest–including the sections that have given rise to allegedly all-encompassing descriptions of the *Saturnalia* as "a conversational commentary on Virgil" or "a convivial dialogue"–are only made possible because certain uninvited guests appear who will alter the

---

[17] I diverge here from Kaster's translation, which renders *praecepta* as "lessons", because I see in the *praecepta/exempla* opposition a technical distinction between theoretical precepts (belonging to a certain art or discipline) and historically situated case studies. Such a distinction is related to the polarity between knowledge of general principles vs. particularities, a divide always present in rhetorical discussions about the topics (*quaestiones*) involved in discourse (either general *theseis* or particular *hypotheseis*, in Greek technical terms).

[18] "Tale hoc praesens opus volo: multae in illo artes, multa praecepta sint, multarum aetatum exempla, sed in unum conspirata" (*Sat.* Proo. 10.).

[19] "nec enim ulla alia de re quam doctis quaestionibus colloqui aut nobis aut his potest esse iucundius" (*Sat.* I.2.17.).

[20] Corresponding to the "Morning of the First Day" and including Praetextatus' oft-referenced discourse on "solar theology" which is the basis of the extended Neoplatonic interpretation of Macrobius' *Saturnalia* (Neri and Ramelli, [eds.], 2007: 124). An interesting commentary on other possible alternative interpretations of this fragment, which will not be the focus of our discussion, is to be found in Cameron (2011: 264-268).

initial plans.[21] The so carefully and explicitly planned turn-taking for expositions on Virgil's different proficiencies, which will occupy the mornings of the second and third days, is actually decided upon by the main guests as a reaction against Evangelus' impertinent distrust of Virgil's erudition (*Sat.* I.24.1-25)–mockingly expressed in terms of a similar distrust of Cicero's accomplishments and of learned Greeks:

> Unless perhaps, just like the Greeks, who puff their accomplishments up out of all proportion, we too want to claim that even our poets are philosophers, when Tully himself, who claimed to be a student of philosophy no less than of speech, squanders what glory he accumulated with his oratory thanks to the half-baked account he provides, the moment he speaks of the gods, or on fate or on divination.[22]

On the other hand, although most editors define the *Saturnalia* as a "sympotic" or "convivial" dialogue, and Macrobius himself refers to Plato's *Symposium* in his Prooemium, this is only half true, as it only applies to the conversations taking place in the afternoons and evenings around the table (books II, 2nd part of III, and VII) and explicitly dedicated to suitable lighter topics. In the mornings–"ad meliorem partem diei" (*Sat.* I.24.24.)–the guests will conduct more serious or profound talks and expositions (books I, 1st part of III, IV, V and VI). Moreover, much of what is said during the convivial part is only possible because there is a doctor (Dysarius) among the uninvited guests and much of the higher degree of vivacity and playfulness brought into the conversations around the table is explicitly due to the uninvited Evangelus' remarks and proposals.

Far from being the secondary, obscure character (traditionally and mistakenly interpreted as a Christian) that many scholars seem to take him for, Evangelus is the soul of the *Saturnalia*'s best accomplishments. He embodies the "systematic doubt" that will make us listen to a series of encyclopaedic doctrinal discourses, all of them following the thread of Virgil's works:

---

[21] Although this is also part of the genre's literary tradition, in Macrobius the contrast between the careful plans and their subsequent disruption is somehow more complex: it is precisely the disruption that finally brings about the goals made explicit in the prearranged plan.

[22] "nisi forte, ut Graeci omnia sua in immensum tollunt, nos quoque etiam poetas nostros volumus philosophari, cum ipse Tullius, qui non minus professus est philosophandi studium quam loquendi, quotiens aut de natura deorum aut de fato aut de divinatione disputant gloriam quam oratione conflavit incondite rerum relatione minuat" (*Sat.* I.24.4.).

a) Virgil's philosophical and scientific erudition, especially astronomy (by Eustathius, now lost);
b) Virgil's knowledge of augural law (by Flavianus, lost);
c) Virgil's knowledge of pontifical law (by Praetextatus, III.1-12);
d) Virgil's oratorical skills (by Symmachus, lost);
e) Virgil's emotional or "pathotical" rhetoric (by Eusebius, IV.1-6, V.1);
f) Virgil's imitation and knowledge of Greek poetry (by Eustathius, V.2-22);
g) Virgil's imitation of archaic Latin poetry (by Rufius and Caecina Albinus, VI.1-5);
h) Virgil's figures of speech and puzzles (by Servius, VI.6-9).

It is precisely thus that Evangelus' profound distrust of what he himself identifies as an originally Greek cultural ideal–the harmonious attainment of a profound, all-encompassing knowledge expressed, moreover, through eloquent discourse–being contradicted by the other guests' performances, grants us the privilege of receiving Macrobius' own Latinized version of such an ideal.

The afternoon and evening talks around the table cover topics that are, as we have said, somewhat lighter but no less agreed upon beforehand. The first of the convivial sections (Book II) leads to (following Symmachus' suggestion) a collection of *dicta* that will prove the wit of great men and the guests' memory and erudition in a round of turn-taking. It ends, however, with more substantial philosophical matters, as Eustathius defends the ideas of Plato (II.8.5-9) and Aristotle (II.8.10-15)–also mentioning Socrates and Hippocrates (II.8.16.) before the final *lacuna*–on the appropriateness of certain pleasures. This is undoubtedly a more technical and less iterative topic, which is again brought about by one of Evangelus' awkward remarks (II.8.4.). In the second convivial section (Book III.13-20), it seems that it is Horus (in a *lacuna*)–now taking his turn as the banquet's Cynic–who makes the comments that result in the Albini's discourses on ancient luxury (III.13-17). This second day ends with Servius' speech on the variety of names of fruits and nuts requested by Symmachus.

However, Evangelus' role as the dialogue's "engine", so far rather significant, reaches its peak in the last convivial section (Book VII). Things start out in a rather low-key mode, with the highest-ranking guests, Praetextatus and Symmachus, discussing the possible place of philosophy among the *quaestiones conviviales*. They ask Eustathius to settle the dispute, which he does (VII.1-3),[23] talking about the advantages of philosophy and its conversational pedagogy by means of *reprehensio*, *fabulae* and *interrogatio* (VII.1.24.) within a convivial situation. At this point, Praetextatus continues to lead the conversation, hinting that they should move on to certain characteristically *convivial* topics (food, drink and bodily matters) and suggesting that they take advantage of the presence of a doctor to pose a round of questions:

> [e]specially given that our friend Dysarius is here, whose learning and skill will be able to contribute a very great deal to questions of this sort. If it's agreeable, let's draw lots so that each one of us in order might put forward a question that he deems worth answering.[24]

There is such round then (although not determined by lot, but following the usual rank order), in which at one point Dysarius talks about the dangers of a varied diet (VII.4.32.). And then comes Evangelus' moment of (however infamous) glory:

> This is really beyond bearing–to have our ears held captive by this Greek and be forced to agree with his orotund pronouncements, fooled by the rapid flow of speech that tyrannizes its listeners to exert their belief. Since we confess that we can't keep up with this labyrinthine discourse, come Vettius, let's urge Eustathius to take up the other side and share with us what can be said on behalf of a varied diet, so that Dysarius, with his linguistic torrent, might be hoisted with his own petard–one Greek snatching the applause away from another, like one crow poking out another's eyes.[25]

---

[23] Much of the content of this part is taken from Plutarch's *Moralia*, 612F-614D (*Symposiakōn*, I.1), but, as Kaster says, "with some rearrangements and supplements in M.'s usual fashion" ([ed.], 2011: v. III, p.143, n.2). I will look at this kind of "deuteronomic" procedure in other parts of the dialogue that are more significant for my purposes.

[24] "praesente maxime Disario nostro, cuius plurimum ad hoc genus quaestionum poterit ars et doctrina conferre, sortiamurque, si videtur, ut per ordinem unus quisque proponat quam solvendam aestimat quaestionem" (*Sat.*VII.4.1.).

[25] "nihil tam indignum toleratu quam quod aures nostras Graeca lingua captivas tenet et verborum rotunditati adsentiri cogimur circumventi volubilitate sermonis, qui ad extorquendam fidem agit in audientes tyrannum. Et quia his loquendi labyrinthis impares nos fatemur, age, Vetti, hortemur Eustathium ut recepta contraria disputatione quicquid pro vario cibo dici potest velit communicare nobiscum, ut suis telis lingua violenta succumbat, et Graecus Graeco eripiat hunc plausum, tamquam cornix cornici oculos effodiat" (*Sat.* VII.5.1-2).

Such an impertinent demand[26] is surprisingly welcomed (with just a slight criticism of its wording) by no other than Symmachus, who takes advantage of the moment to relate the dispute to rhetorical practice and restate the Greek character of such confrontations:

> You've made a pleasant proposal, Evangelus, in quite an unpleasant way: an attack on so substantial and shapely a theory is a thing that could be both useful and enjoyable, but we ought not to make the attempt as though we were preparing ambushes in a contest of wits, out of mere envy for star turns. Nor do I deny that I too could, as it were, have sung a palinode: it's one of the preliminary exercises in the schools of rhetoric to treat general topics by arguing both sides in alternation. But since an account by other Greeks will perhaps more readily respond to the discoveries of Greeks, we all ask you, Eustathius, to refute Dysarius' positions with contrary arguments and thereby reinstate the gaiety of our banquet that was dishonourably dismissed.[27]

After Eustathius' long reply, defending the varied diet (VII.5.5-32), Dysarius is already rather piqued and points out: "You have been speaking as a dialectician, Eustathius, I, as a doctor. To determine what regime to follow one should examine one's normal practice, and experience will teach a man what serves best to keep him healthy."[28] Nevertheless, the

---

[26] Cf.: Parker (2009: 205), commenting on the customary reading of poetry as an entertainment in Roman private gatherings and banquets, mentions that the usual way this was done was to have *lectores* or *anagnostae*. A guest, whether a poet or not, would never be expected to stand up and recite poetry for the others. He makes it clear, however, that: "Here, of course, we must distinguish between formal performance of texts as dinner entertainment and the informal exchange of poetry between learned persons, in other words, poetry as part of the conversation and *filologa* that Cicero praises." Evangelus' proposal stands somewhat midways between the two, but leans towards the wrong side by asking some guests to act as entertainers for the others. The image of the crow fight is clear enough in this sense.

[27] "Rem iucundam, Evangele, amarius postulasti. Audere enim contra tam copiose et eleganter inventa res est quae habeat utilem voluptatem, sed non tamquam ingeniis insidiantes et gloriosis tractatibus invidentes hoc debemus expetere. Nec abnego potuisse me quoque tamquam palinodiam canere. Est enim rhetorica praelusio communes locos in utramvis partem inventorum alternatione tractare. Sed quia facilius Graecorum inventionibus a Graecis forte aliis relata respondent, te, Eustathi, oramus omnes ut, sensa et inventa Dysarii contrariis repellendo, in integrum restituas exauctoratum conviviorum leporem" (*Sat.* VII.5.3-4).

[28] "Obsecutus est, inquit, Eustathi, dialecticae, ego medicinae. Qui volet eligere sequenda usum consulat, et quid sit utilius sanitati experiencia docebit" (*Sat.* VII.5.33.). Here I use Davies's translation (1969), which I find more technical or

round of questions continues. But it is clear that nothing is settled between these two and without further encouragement by Evangelus, they will engage in more serious and at times rather bitter disputes (VII.14-15.), on which I will comment later. Things go so far that, at a certain point, Evangelus feels he has lost his role, and tries to recover it. He does this first by posing a new, half-mocking question directed at Dysarius about the chicken/egg priority problem, which the doctor tries to answer rather seriously (VII.16.1-14.). Then, after a second question on the conservation of meat has also been answered by Dysarius (VII.16.15-18.), Evangelus addresses Eustathius rather provokingly: "Evangelus looked at Eustathius and said: 'If you agree with this account, you should indicate your assent, or if something leaves you unsettled, don't be slow to raise it, since the forceful speech of you Greeks has made me a brilliant listener.'"[29] Although we have the beginning of Eustathius' rather balanced response (he will only dispute a minor point and clarify some details, VII.16.20-34), the final *lacuna* deprives us of Dysarius' reaction, and more importantly, of the way this prolonged Greek *agon* would be concluded and probably somehow appraised by one of the main characters (Praetextatus, most likely) at the end of the work.

So, as we see, it will also be through the good yet impertinent offices of Evangelus that we will finally receive not only the above-mentioned harmonious ideal of a broad and eloquent culture, but also some notion of the theoretical limits of such harmony, all included within just one dialogue/work. The staging of certain unresolved and conflicting approaches to different problems shown by the confrontations between Eustathius and Dysarius in Book VII, in addition to creating some of the livelier atmosphere in the convivial sections (as opposed to the erudite

---

methodological and contextually appropriate than Kaster's ("You've followed the precepts of dialectic, Eustathius, I of medicine: the person who wants to choose a course to follow should give thought to what's proved useful, and experience will show what is more useful to health"). This dispute and many (though not all) of the arguments used in it are taken from Plutarch's *Moralia* 660D-664A (*Symposiakōn*, IV.1). However, apart from the use of different sources, I find that Macrobius' most interesting contributions are the meta-discursive and methodological remarks: the way the dispute is introduced in the dialogue and encouraged by the guests; the opposition, mentioned by Dysarius, between *examples* and *rational explanations* as different types of arguments (suitable for different audiences, *Sat.* VII.4.4-8) or Dysarius' final comprehensive evaluation of the dispute as one between dialectical and medical approaches. *None* of this is in Plutarch.

[29] "Et dictis Evangelus Eustathium intuens 'si rationi dictae adsentiris', ait, 'annuas oportet: aut si est quod moveat, proferre non pigeat, quia vis vestri sermonis obtinuit ne invita aure vos audiam'" (*Sat.* VII.16.19.).

morning speeches), brings to the fore the polemic and competitive aspects of the Greek scientific tradition as something valuable *per se*. Both opponents fare rather competently in the parts we have and I would even suggest that, although the ascendancy of philosophy is somewhat emphasized, there is no definite winner in these dialectical exchanges.

The technical content of the *Saturnalia* is thus brought into the conversation through different twists and turns of the plot, some expected from the beginning and some more unexpected, but all of them carefully justified by the main author's literary construction. The fact that he mines the sympotic tradition for some of his ideas (most of the topics in Book VII are, for example, taken from Plutarch)[30] and characters (uninvited guests, presence of the Cynic elements, etc.) does not mean that he is not somehow innovating with his particular mix of genres and explicit encyclopaedic intentions.

Throughout the dialogue, Evangelus will compete with the more serious and higher-ranking personages in trying to lead the conversations towards his own ends, mostly carrying the day and bringing in some humour with his outlandish lack of respect for the others' most cherished and allegedly self-evident convictions. In the end, the result is to grant them the opportunity for an effective, explicit and enriching defence. The Greek compound term *spoudogeloion* (useful and enjoyable) so extensively used by scholars to characterize the encyclopaedic work *De nuptiis Philologiae et Mercurii*, by Macrobius' contemporary Martianus Capella[31] and related to the Menippean tradition is, I think, likewise due here and with similar pedagogic intentions. In fact, the character of Symmachus mentions something rather like it in *Sat.* VII.5.3., responding positively to Evangelus' proposal for a dispute: "res est quae habeat utilem voluptatem".

It is quite evident that Evangelus embodies the satirical element in the *Saturnalia*. He in fact makes his first contribution by quoting Varro's Menippean *Satires* (just after joining the company, *Sat.* I.7.12.) in a rather self-mocking manner–suggesting he and his companions should rather leave and making Praetextatus formally invite them to stay, after they have come uninvited. The name of the particular Varronian satire mentioned here ("Nescis quid vesper vehat", *You Know Not What The Evening Will Bring*), speaks for itself with regard to Evangelus' intentions to break with initial expectations, just to bring about better results for the work as a

[30] See table of contents and sources in Kaster ([ed.], 2011: il-liii).

[31] E.g. by Shanzer (1986: 41). In *De nuptiis,* Satire is the Muse that recounts her learned tale to Martianus, including the expositions of the seven liberal arts (see Olmos, 2011; forthcoming).

whole. That Evangelus should be especially impertinent and demanding with the Greeks might therefore tell us something about the importance assigned to *their* contributions by Macrobius.

The Performance of the Greeks: The Systematic Nature of their Main Speeches

Let us now take a closer look at the individual presentations, doctrinal discourses and formal and informal contributions of the Greek scholars of the *Saturnalia*.

Eustathius, the philosopher, is by far the most respected character of the three. He is the only one that is officially invited on the eve of the festivities (*Sat.* I.5.13-16), when his accomplishments are praised by Symmachus:

> But our friend Eustathius, who has studied the doctrines of every school–choosing to follow those that are most susceptible of proof–and combines in himself all the styles of speech that those Greeks[32] displayed, is so competent to act among us as his own interpreter that it would be impossible to say in which tongue he would discourse with the greater ease and elegance–in Greek or Latin.[33]

---

[32] i.e. the Academic Carneades, the Stoic Diogenes and the Peripatetic Critolaus, mentioned in *Sat.* I.5.15 as the three ambassadors sent to Rome by Athens in 155 BCE and whose very diverse modes of eloquence are described by Symmachus thus: "Fuit, ut relatum est, facundia Carneades violenta et rapida, scita et tereti Critolaus, modesta Diogenes et sobria", "As report has it, Carneades' eloquence was forceful and swift, Critolaus' clever and well turned, Diogenes' balanced and sober" (*Sat.* I.5.16.).

[33] "at hic noster cum sectas omnes adsecutus sed probabiliorem secutus sit omniaque haec inter Graecos genera dicendi solus impleat, inter nos tamen ita sui locuples interpres est ut nescias qua lingua facilius vel ornatius expleat operam disserendi" (*Sat.* I.5.16.). In this case I have also used Davies's translation (1969), which I prefer to Kaster's less anaphoric version (2011): *Cf.* "But our friend Eustathius, having attended all the philosophical schools, has aligned himself with the one he most approved, and he fully employs, by himself, all the styles of speaking known to the Greeks; yet at Rome he is so much his own reliable interpreter that you would not know in which language he speaks more fluently or more resourcefully."

He is also presented as a close friend of the noble Flavianus.[34] Moreover, he is given priority by Praetextatus when deciding on Virgil's speeches at the end of Book I:

> Philosophy–that singular gift of the gods and the discipline of disciplines–should have the honour of being treated first: accordingly, let Eustathius keep in mind that all the other topics are giving way and granting him the opportunity to speak first.[35]

It is all the more regrettable, therefore, that his main discourse on Virgil's "knowledge of astrology and of all philosophy" (as announced in I.24.18.)[36] should have been lost. However, we do have his contribution about his *secondary* interest in Virgil's poetry: his borrowings from Greek poets, particularly Homer (V.2-22).[37]

Here we should emphasize the way he is explicitly and repeatedly asked to be systematic. Although he starts by quoting some scattered

---

[34] "Flavianus et Eustathius, par insigne amicitiae, ac minimo post Eusebius ingressi"; "Flavianus and Eustathius entered –a distinguished pair of friends–followed quickly by Eusebius" (*Sat.* I.6.4.).

[35] "Philosophia, quod unicum est munus deorum et disciplina disciplinarum, honoranda est anteloquio, unde meminerit Eustathius primum sibi locum ad disserendum omni alia professione cedente concessum" (*Sat.* I.24.21.).

[36] "Eustathius deinde, 'maxime', inquit, 'praedicarem quanta de Graecis cautus et tamquam aliud agens modo artifici dissimulatione, modo professa imitatione trastulerit, ni me maior admiratio de astrologia totaque philosophia teneret, quam parcus et sobrius operi suo nusquam reprehendendus aspersit'"; "Then Eustathius said: 'I'd make special mention of his attentiveness to matters Greek, and how he made his borrowings now artfully disguised-as though intent on some other purpose-now expressly imitative, did I not admire still more his knowledge of astrology and of all philosophy, which he sprinkled sparingly and soberly over his work, nowhere leaving himself open to criticism'" (*Sat.* I.24.18.).

[37] Incidentally, a change in plans that is brought about by Evangelus' remark about Virgil's ignorance of the Greek letters, contradicted by Eustathius thus: "cave, Evangele, Graecorum quemquam vel de summis auctoribus tantam Graecae doctrinae hausisse copiam credas, quantam sollertia Maronis vel adsecuta est vel in suo opere digessit. Nam praeter philosophiae et astronomiae amplam illam copiam de qua supra disseruimus, non parva sunt alia quae traxit a Graecis et carmini suo tamquam illic nata conseruit"; "Don't suppose, Evangelus, that any of the Greeks–even among the greatest authors–drank as deeply of Greek learning as the skilful Maro, or incorporated as much in his own work. For beyond the abundant knowledge of philosophy and astronomy that we discussed before, he derived no small amount of other material from the Greeks and implanted it in his poetry as though it were native growth."

parallels between the *Aeneid* and the Homeric epics, when he pauses, Avienus demands an exhaustive survey so that he may tell "*all* that our poet borrowed from yours" ("*quaecumque* vestro noster poeta mutuatus est", V.3.16.). Eustathius then asks for a copy of the book (a *codex* already), tries to open it at random, but is once again forced by Avienus to proceed "ab initio per ordinem" (V.4.1.). Eustathius will moreover make a certain refinement to this mandate, by further classifying the borrowings according to the relative degree of excellence of the original vs. the imitation, or vice versa.

It is also interesting to note how at the end he extends the subject by talking about borrowings and information drawn by Virgil from other Greek authors in a search for truly obscure pieces of erudition. This reminds us of Macrobius' comments on his programmatic Prooemium about what "lurks hidden from the common run of men in a mass of books". Eustathius says here "I come now to things fetched from the inner sanctum of Greek literature, things no Roman knows, save those who have earnestly drunk full draught of Greek learning."[38] And this comes after he had already said at the beginning of his speech that he would not even mention what is "well-known ground" ("quae vulgo nota sunt", V.2.4.) and would pass over "these items and others like them that comprise the schoolboy's catechism" ("sed et haec et talia pueris decantata praetero", V.2.6.). There is, therefore, an increase in difficulty as we proceed, for which a Greek scholar is ultimately required.

I have already talked a little about Eustathius' speeches in the convivial sections and his final confrontations with Dysarius will be discussed further. However, Eustathius himself may also give us some clue about the somewhat disappointing role of Horus, the Egyptian. Unless his contributions in the *lacunae* made him a more interesting character,[39] his remarks in Book I do little more than help Praetextatus slightly with his allusions to Egyptian religion. When he is asked to make a more substantial contribution, assessing Praetextatus' account of the calendar, because he is "a native of the land that holds dominion over numbers" ("gentis accola numerorum potentis", I.16.37.), it is Eustathius that answers in his stead (I.16.38-44), displaying greater learning which includes Egyptian, Latin and Greek elements. Furthermore, Horus does not act as the real Cynic in the symposium, a role sufficiently represented by Evangelus. He has, moreover, no Virgilian theme–which Evangelus,

[38] "Ad illa venio quae de Graecarum litterarum penetralibus eruta nullis cognita sunt, nisi qui Graecam doctrinam diligenter hauserunt" (*Sat.*V.18.1.).

[39] He is supposed to have been talking previous to what we have of the second day's afternoon conversations (*lacuna* between III.12 and III.13.).

Dysarius and Avienus do not have either. His best contribution in the whole dialogue is probably his quoting of a nice couplet by Plato in the *dicta* section (II.2.15.).[40]

Finally, we may also take a look at his attempt to answer Avienus' enquiry about the convention of wearing rings on the "medicinal finger" of the left hand. After Horus tries to complete Dysarius' answer with some "numerological" symbolism (VII.13.10.), Caecina intervenes, offering instead the *Etruscan* alternative explanation, which is totally practical, earth-bound, and entirely satisfactory (VII.13.11-16). Caecina finishes with what appears to me as a rather ironic remark: "Let each person pursue either the Etruscan opinion or the Egyptian, as he pleases."[41] Horus' role in the *Saturnalia* cannot be compared to that of the other foreigners, the Greek scholars that are the focus of this paper. He rounds off the number of guests, also completes the group of three uninvited characters–following another pattern in the work where the guests are always introduced in groups of three–and obviously has some lines throughout the dialogue, but his personal capacities do not seem to be particularly required for Macrobius' purposes.

As already stated, Eusebius, the rhetorician, comes as a substitute guest. Nevertheless, he is introduced in very positive terms by Praetextatus: "since he [Postumianus] is declined, I've invited Eusebius to take his place, a rhetorician distinguished by his Greek learning and eloquence."[42] And he will be responsible for conveying the conversation to Postumianus (as reported in the Prologue, I.2.6-15), and ultimately to us all, thanks precisely to his memory and discursive capacity as a rhetorician. He is not, however, in so confident a position as Eustathius and does not seem to contribute so spontaneously on any subject as the latter does. In fact, he does not pick his own Virgilian topic as the rest do, but has to accept what Symmachus will leave him of the connections between Virgil and rhetoric:

> And lest I [Symmachus] seem to claim all topics for myself, I undertake to make plain the rhetorical skills of Virgil's oeuvre–its most forceful conceits and expressions–though I do not want to deprive Eusebius, the

---

[40] *Anth. Gr.* 5.78.

[41] "unus quisque ut volet vel Etruscam vel Aegyptiacam opinionem sequatur" (*Sat.* VII.13.16.).

[42] "in cuius abnuentis locum Eusebium Graia et doctrina et facundia clarum rhetorem subrogavi" (*Sat.* I.6.2.).

> most eloquent of speakers, of the chance to treat Maro's oratorical skill, since he will be better suited by both his scholarship and his teaching.[43]

As Symmachus' Virgilian discourse is completely lost, as is the beginning of Eusebius' (or perhaps most of it with, in any case, what would probably have been a significant transition between them), we do not know exactly what the difference between the two was. However, I would say that Symmachus would most likely have chosen to comment upon some favourite poetical passages and choice examples of Virgil's genius, while Eusebius' enterprise is rather more methodical and technical in nature. In his review of Virgil's expression of *pathos* (which we have in IV.1.-6), he in fact follows a systematic list of *topoi*, explicitly asserting that his survey starts with certain definitions found in rhetorical textbooks, notwithstanding their legal resonances:

> Now let's see how the tenor of one's speech makes one's emotions clear, starting with the textbook line on such speech: emotional speech must be aimed at producing either indignation or pity (in Greek, deinōsis and oiktos, respectively), the former relied on by the prosecution, the latter by the defendant.[44]

He then proceeds to give examples of Virgilian expressions which precisely seek to communicate indignation or pity, giving technical names to the resources used by the poet (*ecphonesis* or exclamation, hyperbole, irony, *argumentum a minore*, *argumentum a simili*), after which he gets even more systematic, going through a kind of list of "pathotic topoi":

a) Those related to the core elements of any action (usually called *in re*, although the expression is not mentioned here), including those based on a person's condition (*ex habitu*), geographical location (*a loco*), the timing of an action (*ex tempore*), an action's cause (*a causa*), manner (*a modo*), or means (*a materia*), plus those based on a combination of different sources (*duobus aut plurimus locis coniunctis*);

---

[43] "et ne videar velle omnia unus amplecti, spondeo violentissima inventa vel sensa rhetoricae in Vergiliano me opere demonstraturum, Eusebio autem, oratorum eloquentissimo, non praeripio de oratoria apud Marinem arte tractatum, quem et doctrina et docendi usu melius exsequetu" (*Sat.* I.24.14.).

[44] "Nunc videamus pathos quo tenore orationis exprimitur. Ac primum quaeramus quid de tali oratione rhetorica arte praecipitur. Oportet enim ut oratio pathetica aut indignationem aut ad misericordiam dirigatur, quae a Graecis oiktos kai deinōsis appellantur. Horum altereum accusatory necessarium est, alterum reo" (*Sat.* IV.2.1.).

b) Those related to external features (*circa rem*), including arguments from likeness (*a simili*, in Greek *paradeigma*, *parabolē* or *eikōn*), arguments from what is lesser (*a minore*), arguments from what is greater (*a maiore*), arguments from what is "contrary to expectation" (*prater spem*), homoeopathy, addresses to things either inanimate or mute (*sermo dirigitur vel ad animalia vel ad muta*), uncertainty (*aporesis*), and testimony (*attestatio rei visae*).

In all these cases, he uses the expressions in parentheses in the formulaic way characteristic of school parlance–as things surely well known to his educated interlocutors–that the rather explanatory translations offered here cannot evoke. That he is, in fact, following a particular textbook tradition (one among the different rhetorical trends found in Late Antiquity) was demonstrated by Bestul (1975) in an article in which Eusebius' discourse was compared with the chapter "On emotions" (*De adfectibus*) from Julius Severianus' *Praecepta Artis Rhetoricae* (Halm 364-369), which has a similar (although not identical) list of "pathotic topoi".

Eusebius' systematic survey occupies the whole of Book IV. Book V will bring Eustathius' discourse on Virgil's poetic borrowings. But before that, its first chapter (V.1) is still dominated by Eusebius, creating a kind of coda to what was surely a much longer rhetorical study of Virgil, attributing to him the role of the complete orator (according to Cicero's *Orator*), one that masters all the styles: "the Mantuan's eloquence is many-sided and diverse, embracing every style"; "he combined the divergent styles of the ten orators who flourished in the Athens of Attica."[45] Eusebius is almost totally silent in the convivial sections. He will do no more than just take his appointed turns to offer a funny saying (from Demosthenes, as could be expected, II.2.11.) and to put a brief question to Dysarius that serves only as a detail in his characterization as an old man (VII.10.1-14).

Our third Greek, Dysarius, coming uninvited, does not pick a subject of Virgilian erudition to expound upon. His role is restricted almost entirely to Book VII, where all the guests are encouraged to ask him about health issues (considered to be *quaestiones conviviales*). Apart from that, he makes only a brief and almost completely lost appearance in Book II (II.2.14.). He is presented, however, as one "who then was reputed to be

[45] "facundia Mantuani multiplex et multiformis est et dicendi genus omne complectitur" (*Sat.* V.1.4.); "decem rhetorum qui apud Athenas Atticas floruerunt stilos inter se diversos hunc unum permiscuisse" (*Sat.* V.1.20.).

the foremost physician in Rome."[46] No other social situation could have been proper for Praetextatus' household. Yet, his status is not so highly regarded as Eustathius' and his knowledge is not deemed or presented as so systematic in its content. However, he is the one that brings about almost all the remarks regarding certain methodological issues concerning either scientific research or the proper exposition of scientific topics, which seems to reveal a certain need for self-justification in front of other sorts of experts and a context of polemical negotiation of authority.

For example, even in the first question received, the one presented by Praetextatus about the plain or varied diet (VII.4.3.)–which will ultimately end in the first Eustathius/Dysarius confrontation, via Evangelus–he seems to prepare himself for polemics, asserting from the beginning the methodological parameters of his exposition. If he were talking to "someone from the uninformed commons" ("aliquis ex plebe imperitorum"), *exempla* would be the best mode of persuasion. And he does in fact offer some particular illustrations, notwithstanding his questioner's status (about healthy animals with a simple diet, doctors' usual prescriptions for sick people and the wine/nourishment analogy). But then he goes on, changing his approach, for when dealing with one who has "perfect command of all fields of learning [Praetextatus], I should use not examples but rational analysis."[47] However, by "rational analysis", he means a search for internal causes, in medical/physiological terms (the four humours and four processes of digestion), which also allows for the role of "practice" and "experience" as sources (*usu docente*, VII.4.10.) in scientific survey. That is why he feels somewhat justified in confronting Eustathius' alternative account in terms of the dialectics/medicine polarity, vindicating the role of practice ("usum consulat") and experience ("experientia docebit") in medical matters.[48]

The role of one's experience in physiological topics, however, opens the field to non-experts, and this is shown in the way many of Dysarius' answers are not only contested by Eustathius, in a critical discussion format, but also commented on by the other participants. A good example is the brief debate after Dysarius' response to Flavianus' question on the atypical nature of women's drunkenness (VII.6), in which another example of negotiating authority is staged. Symmachus comments on the

[46] "qui tunc Romae praestare videbatur ceteris medendi artem professis" (*Sat.* I.7.1.).

[47] "cui soli perfectionem omnium disciplinarum contigit obtinere, non tam exemplis quam ratione tractandum est" (*Sat.* VII.4.8.).

[48] Again, in his answer to Caecina's question: "You know from your own experience, Albinus"; "usu tibi, Albine, compertum est" (*Sat.* VII.8.10.).

"cold" nature of women's bodies.[49] Horus replies that women are in fact "warm". And then Symmachus, before actually refuting this assertion, tries an *ad hominem* rejoinder ("'Our friend Horus', replied, 'is making a stout effort to abandon his Cynic philosophy and to appear as a Sophist instead'")[50] only to finish his reassertion of women's cool nature with a conciliatory or perhaps ironic "But let each person judge these matters as he wishes" ("Sed de his singuli ut volent iudicent", VII.7.13.).

Later, I will examine some rather significant statements within the Eustathius/Dysarius *agonistic* performance. But first let us return to Evangelus and his characteristic treatment of the Greeks.

Scorn and Opportunity in Evangelus' Remarks

Although it is not focused on the particular polarity between Evangelus and Greek culture as embodied by the foreign scholars in the *Saturnalia*, let us first look at Kaster's ([ed.], 2011: xxxviii-xxxix) commentary on Evangelus' attack on Servius (II.2.12.), in which he seeks to make him take part in the round of famous sayings:

> You declare us all shameless, schoolteacher, if you want to appear to keep silent as a way of safeguarding your own sense of shame: neither your haughtiness nor that of Dysarius or Horus will escape being branded as arrogance if you do not choose to imitate Praetextatus and the rest of us.[51]

For Kaster, this is a significant fragment that shows Macrobius' "attention to characterization". Both Evangelus' "mean and sardonic cast of mind" and Servius' modesty (*verecundia*)[52] are believed to be clearly exposed here by their ways and words. Kaster further claims that a key point is Evangelus' derogatory use of the vocative "grammatice" and compares it with other "terms of abuse" used by him in the dialogue, such as the previously mentioned "Graeculus" (VII.9.26.). Kaster concentrates on these cases as "breaches of etiquette" characteristic of Evangelus, but does not delve into their role in moving the dialogue forward. In fact,

---

[49] On the traditional medical ideas on women's temperature, see the essay by King, in this same volume.

[50] "Bene, inquit, Horus noster temptat videri orator ex Cynico" (*Sat.*VII.7.8.).

[51] "'omnes nos', inquit Evangelus, 'impudentes, grammatice, pronuntias, si tacere talia vis videri tuitione pudoris, unde neque tuum nec Dysarii aut Hori supercilium liberum erit a superbiae nota, ni Praetextatum et nos velitis imitari'" (*Sat.* II.2.12.).

[52] Remember that the grammarian's required *verecundia* is one of Kaster's most cherished topics (1980, 1988).

despite all his harshness, Evangelus is basically right in encouraging Servius (and then Dysarius and Horus) to participate, which they do indeed after his outburst, and everybody rejoices ("These anecdotes gave rise to much merriment"; "Orta ex his laetitia", VII.2.16.). Evangelus is in fact providing an excuse, one rather "levelling" in nature, for the three lowest-ranking guests to join the others in telling jokes (that is why he mentions the other two when he addresses Servius) and his satiric mood has a positive outcome (and nobody complains).

In the case of the term "Graeculus", directed to Dysarius after answering Evangelus' medical question, it in fact marks a slight (although somewhat temporary) change in Evangelus' attitude, from absolute negativity to a certain positive slant. Let us look at the whole paragraph (I have underlined the elements I will comment on):

> "Hats off to *our* little Greek!", Evangelus said, "for revealing to us things that nature has kept hidden *so clearly* that we seem to see every detail he described. But now *I yield* to Eustathius, *whose turn I took* in asking my question."[53]

This is the first time in the dialogue that Evangelus uses the friendly "noster",[54] however patronizingly or ironically, to apply to anything Greek. So far, Greek things have always been for him "vester" or "suum", in contrast with what is Latin: "nisi forte, ut Graeci omnia *sua* in immensum tollunt, nos quoque etiam poetas *nostros* volumus philosophari" (I.24.4.); "quod aures *nostras* Graia lingua captivas" (VII.5.1.); "vel *vestri* oris nota volubilitas" (VII.9.9.).[55] Indeed, it is also the first time that he says anything positive about Greek eloquence ("tam plane nobis ostendit") so far quite poorly rated by him ("volubilitate sermonis", "loquendi labyrinthis", VII.5.1.). He will, no doubt, revert to some derogatory terms in VII.16.1 ("in ostentationem loquacitatis"), but

---

[53] "'*Euge*! Graeculus noster', Evangelus ait, 'tam plane nobis ostendit res opertu naturae tectas ut quidquid sermone descripsit oculis videre videamur. Sed Eustathio iam cedo, cui praeripui consulendi locum'" (*Sat.* VII.9.26-27).

[54] Already twice used by Praetextatus, referring to Eustathius, "at hic noster" (I.5.16.), and Dysarius, "praesente maxime Dysario nostro" (VII.4.1.), and once by Symmachus in a more patronizing way: Bene, inquit, "Horus noster" (VII.7.8.).

[55] The contrast is also used by different guests in the dialogue: e.g. with apparent deference towards Virgil and Cicero by the Greeks–"Cicerone vestro" (Eusebius, V.1.4.), "Vergilium vestrum" (Eustathius, V.20.17.)–although not always so admiringly–"quem litteratores vestri nec obscurum putant" (Eustathius, V.19.31.)–revealing a certain Greek/Latin tension.

will also retain some admiring notes in VII.16.19 ("quia vis vestri sermonis obtinuit ne invita aure vos audiam").

Lastly, he acknowledges he has stepped in and taken Eustathius' turn, something seen by Kaster as a sign of his inability to "assume his proper place" as an uninvited guest, however noble and Latin. Kaster ([ed.], 2011: xlii) compares the two rounds of turn-taking in the dialogue (II.2.; VII.4-13), assuming that the first one is in theory correct, while there are two significant changes in the second: Avienus' wise and mature downgrading of himself and Evangelus' stubborn and impertinent upgrading. But there are surely more nuances here, as nobody discusses Evangelus' precedence with regard to Eusebius and Servius (both invited), already on the first list. It seems that what is in any case remarkable is Eustathius' high position on it, also acknowledged by Praetextatus in the case of the Virgilian speeches. And that is exactly what Evangelus assumes again and emphasizes in this paragraph. Moreover, Kaster offers the following lists and comments on the underlined displacements:

1. *Sat.* II.2., round of famous *dicta*: Praetextatus, Flavianus, Symmachus, Caecina Albinus, Rufius Albinus, Eustathius, *Avienus*, *Evangelus*, Eusebius, Servius, Dysarius, Horus.
2. *Sat.* VII.4-13, round of questions addressed to Dysarius: Praetextatus, Flavianus, Symmachus, Rufius Albinus, Caecina Albinus, *Evangelus*, Eustathius, Eusebius, Servius, *Avienus*, Horus.

However, what in fact happens in the dialogue is that in (2), Eustathius actually refuses to take his turn after Evangelus has spoken and yields instead to Eusebius and the rest, finally asking his question after Horus himself. He *wants* to be last ("nos postea liberiore otio congredimur" VII.9.27.) and bring some more serious controversies into the debate. I think that *this* is what Macrobius wants to emphasize, (i.e. Eustathius' significant disposition for the *agon*), and he again uses Evangelus' somewhat impertinent offices to make it explicit. I do not think the real aim of the anecdote is to characterize Evangelus, but rather to highlight Eustathius' intention to go last. Furthermore, there is also a third list of turn-taking in the *Saturnalia* worth mentioning, the one regarding Virgilian discourses. This one has a different "thematic" order and, as we have said, the uninvited guests are not on it:

3. a) Eustathius I, Flavianus, Praetextatus, b) Symmachus, Eusebius, c) Eustathius II, the Albini, Servius.

We have an initial block (a) about "profound" and "extra-linguistic" knowledge in Virgil, including philosophy, science (astronomy) and Latin religious laws, a second one (b) about Virgil's rhetorical skills and a third one (c) about grammatical topics, i.e. Virgil's poetic and literary wisdom. However, while in the second block the nobles lead and Symmachus precedes Eusebius, in the other two, Eustathius is the first to take part, preceding most noble and Latin guests. Eustathius (and with him philosophical knowledge) is a very important and highly valued character in the *Saturnalia*.[56] And, as we have seen, it is often thanks to Evangelus that we see him act to the full extent of his ability, even though introduced by a satirical remark.

The spark of dissensus between Dysarius and Eustathius (and between the medical and the philosophical traditions) ignited by Evangelus in VII.5.1 will explode in VII.13-15. Here Eustathius, incidentally after contradicting Dysarius' response to Horus (criticising the pseudo-Aristotelian *Problemata*), finally takes his turn and poses a question that he knows will result in polemic.

The Greek *Agon* at the Dialogue's Climax

When Eustathius asks Dysarius about why things "appear larger" in a watery medium (VII.14.1.), saying that all he has heard about it so far does not seem to him either "true or probable" ("nec vera nec veri similia"), he is very conscious of which traditions answering his question will call for.[57] Dysarius will duly invoke Democritus' and Epicurus' theories of perception (*eidola* emanating from objects, VII.14.3-4) and Eustathius will be so glad at this that before he starts to refute his remarks, his face will light up with a smile: "Ad haec *renidens* Eustathius..."

Eustathius then rejects Epicurus' analogy with the other senses and finally turns to Plato's theory of ocular emanation and rational recognition (*Tim.* 45b-d), engaging in a discussion about reason and the senses based

[56] Another reference to philosophy's high status above the arts is Symmachus' remark: "Or am I to suppose that philosophy ought to be brought into a setting where rhetoric, wholly a profession of the people, blushed to come?" (Kaster); "Or am I to suppose that philosophy should show herself where even the practical and popular profession of the rhetorician was ashamed to be seen?" (Davies); "An ego censeam producendam philosophiam quo rhetorica venire ars et professio popularis erubuit?" (*Sat.* VII.1.4.).

[57] This discussion on perception is not taken from Plutarch, but from a very wide variety of sources. See Kaster ([ed.], 2011: v.III, 275-285).

on Cicero's *Academica* and *De natura deorum*, aimed at contradicting Academic Skepticism as well:

> Reason also makes sense of all the things that gave the Academics an opportunity to condemn the senses, though the senses ought to be considered among the more reliable things when they're accompanied by reason, which sometimes finds a single sense insufficient for identifying a specific object.[58]

All the guests admire the *solidity* of his exposition ("dictorum soliditatem") and even Evangelus seems to retain the good mood of VII.9.26 ("even Evangelus himself was not loath to add his testimony [assent]"; "ut attestari vel ipsum Evangelum non pigeret"). But Dysarius' response will be rather bitter. He leaves aside the topic discussed, does not offer any counter-arguments, and merely ridicules philosophers in general by giving a completely different example of their supposed lack of skill in certain matters:[59]

> It's applause of that sort that leads philosophy on to claim to deal with matters alien to its expertise, with the result that it often plunges into obvious errors. For example, while your Plato didn't avoid dealing with anatomy–the proper sphere of medicine–he handed on to posterity an opportunity to mock him.[60]

This seems to me a more serious breach, if not of etiquette, at least of the dialogue's harmony, than Evangelus' remarks. Nonetheless, if we adopt a positive stance towards this paragraph–also taking into account that it parallels Evangelus' previous commentary on Cicero's lack of skill in philosophy (I.24.4.)–it is a breach that is as productive in terms of

---

[58] "et omnia illa [reason] discernit quae Academiacis damnandorum sensuum occasionem dederunt; cum sensus inter certissimas res habendi sint comitante ratione, cui non numquam ad discernendam speciem non sufficit sensus unus" (*Sat.* VII.14.21.).

[59] The anatomical discussion is again taken from Plutarch, *Moralia* 698A-700B (*Symposiakōn*, VII.1), where Erasystratus is opposed to Plato in this particular topic. Plutarch tries (not much convinced himself, cf. Frazier & Sirinelli ([eds.], 1996: 5-6)) to defend Plato's position, and advances the idea of some opposition between medical and philosophical approaches, but not in such sharp terms as Macrobius.

[60] "Isti plausus sunt qui provocant philosophiam ad vindicandos sibi de aliena arte tractatus, unde saepe manifestos incurrit errores. Ut Plato vester, dum nec anatomica quae medicinae propria est abstinet, risum de se posteris tradidit," (*Sat.* VII.15.1.).

dialogue as Evangelus' own trangressions. Dysarius' target in what follows will be the Platonic anatomical doctrine of the *Timaeus* (70c-91a), against which he will hold up Erasystratus' medical account of the different *tubes* in the throat and the functions of the lungs (VII.15.2-11). He will again finish with a further appeal to *experientia* ("We also know from experience that…"; "est et hoc de experientia notum quod…", VII.15.12.) and concludes with another personal attack on Plato: "You see that it would have been better for the prince of all philosophers to avoid matters alien to him than to pronounce on things unfamiliar to him."[61]

Eustathius' reaction to all this is not particularly gracious either. He responds, in Macrobius' own words, "with a bit more heat than usual" ("paulo commotior", VII.15.14.). And in fact he begins with another *ad hominem* attack, only to move on to a resolute assault on medicine itself and its supposed lack of rational scientific method (mere "guesswork", *coniectura*), asserting philosophy's pre-eminence over it, due to its both being a more dignified topic and having a more accurate method:[62]

> I counted you no less a philosopher than a physician, Dysarius, but now you seem to consign to oblivion a fact repeated and believed as a matter of universal consensus, that philosophy is the very essence of all arts and disciplines. Now, with the brazenness of a parricide, medicine launches an attack upon it […] But medicine is the lowest dregs of the physical side of things, a system concerned with bodies of clay and earth. But why did I use the word "system", when medicine is more the realm of guesswork than system? So an art that works by conjecture on filthy flesh has the nerve to mount a charge against philosophy, which deals with incorporeal and truly divine matters according to a reliable system.[63]

Eustathius will ultimately defend Plato's account in terms that are, for us, not indeed very different from Dysarius' ways of reasoning (conjectured, mostly wrong, physiological explanations and the testimony

---

[61] "vides satius fuisse philosophorum omnium principi alienis abstinere quam minus nota proferre" (*Sat.* VII.15.13.).

[62] Cf. Arist., *De anima* I.1, 402a.

[63] "Non minus, inquit, Dysari, philosophis quam medicis inserebam, sed modo videris mihi rem consensu generis humani decantatam et creditam oblivioni dare: philosophiam artem esse artium et disciplinam disciplinarum; et nunc in ipsam invehitur parracidali ausu medicina, […] Medicina autem physicae partis extrema faex est, cui ratio est cum testeis terrenisque corporibus. Sed quid rationem nominavi, cum magis apud ipsam regnet coniectura quam ratio? Quae ergo conicit de carne lutulenta audet inequitare philosophiae de incorporeis et vere divinis certa ratione tractanti" (*Sat.* VII.15.14-15).

of the poets), using only the group of traditional arguments associated with his stance, instead of the opposing ones. In any case, his conclusion goes beyond the particular point at issue and characterizes those who disagree with Plato as "insane": "Since, then, both a rational understanding of the body's make-up and the noble authority of witnesses support Plato, isn't anyone who hold an opposing opinion just insane?"[64]

The disciplinary debate between the Greeks thus escalates to an insoluble and rather bitter confrontation, in which philosophy might in the end be relatively favoured by Macrobius, but does not really win out in any definitive way. Thank heavens Evangelus is there–however jealous and mocking: "gloriae Graecorum invidens et inludens" (VII.16.1.)–to cool the atmosphere with a lighter topic, putting the egg/chicken priority problem to Dysarius as a way to save him from Eustathius' invective. The cooling down of the controversy will be apparent both in Dysarius' confident taking up of the challenge in a resolute but non-partisan way (actually offering arguments for both solutions and giving the questioner the option of making his own choice, VII.16.14.) and in Evangelus' cynical rejoinder, which will again serve as the excuse to encourage further and much friendlier dialogue and debate between the two Greek scholars: "Since your aggressive chatter turns even jokes into sober discussions, I'd like the two of you to solve this puzzle for me, which I've really pondered for a while."[65] When Evangelus asks then his last question (about the preservation of meat),[66] Dysarius offers an answer and Eustathius points out some further details, these two will be chatting away again in a civilized and truly convivial atmosphere up to the final *lacuna* that deprives us of any concluding remarks.

Conclusion: Macrobius and the Greek World

We know that Macrobius was not Roman (Proo. 1.1.), but speculation about his possible Greek or Eastern origins was silenced long ago (Kaster [ed.], 2011: xi). Furthermore, although he may have some minor misunderstandings regarding Greek expressions, as Wissowa had already pointed out (1880), he nevertheless presents to us a gathering in which all

[64] "Cum igitur et ratio corporae fabricae et testium nobilis auctoritas (poetas y otros) adstipuletur Platoni, nonne quisquis contra sentit insanit?" (*Sat.* VII.15.24.).

[65] "Quia ex iocis seria facit violentia loquendi, hoc mihi absolvatis volo, cuius me exercuit vera deliberatio" (*Sat.* VII.16.15.).

[66] Again a topic taken from Plutarch, *Moralia* 657F-659D (*Symposiakōn*, III.10), but wisely adapted to its role in the *Saturnalia*.

Greek poetry and philosophy is quoted in Greek and in which–apart from the initial praise of Eustathius for his fluency in Latin (I.5.16.)–everybody seems to be at least functionally bilingual enough to take part in such conversations without the least complaint. Leaving aside the possible romanticism of this picture, rather than a reliable representation of the historical people portrayed in it, it is undoubtedly part of Macrobius' cultural ideal to offer a wide-ranging compendium of learning in which both Latin and Greek contributions would create a harmonious whole.

His search for unity is, nevertheless, not blind to certain differences and tensions between both its Latin and Greek sides and different approaches within Greek culture. Throughout his monumental *Saturnalia* (of which we have little more than half), there is opportunity to reveal certain aspects of such tensions, mostly thanks to the dialogue format and especially to the role of certain characters in it. The Greeks in the dialogue are professionals, while the Latins (except for Servius) are elite connoisseurs. The Greeks are (and are explicitly required to be) more systematic than the Latins. The Greeks are asked by others to answer indiscriminate rounds of questions or to engage in (more or less bitter) debates, while the Latins enjoy and take advantage of their answers and controversies. But the Greeks are also endowed with knowledge of the most secret pieces of information and obscure lore and that is something that Macrobius values (Proo. 2; V.22.15.; VII.9.26.). It is, moreover, one of the Greeks who represents philosophy, acknowledged as the supreme standard and explicitly given priority. And lastly, Greek polemic ways and agonistic culture are also presented as an integral part of the overall picture and as things worthwhile in themselves.

Yet there are other tensions that are not fully resolved and on which Macrobius himself seems to take an ambiguous stance. Greek vivid eloquence in scientific matters is one of them. Evangelus attacks the Greeks on several occasions for their aggressive way of speaking and their "volubility", and is never directly contradicted about it. On the other hand, Praetextatus praises Eustathius for his mastery of all Greek styles of expression (I.5.16.), although this might also be considered a somewhat artificial, professionalized skill unsuitable for Macrobius' Latin audience and intended pupils (his son and others like him).[67]

As is well known, during the Roman Saturnalia festivities some ritualized transgressions of the norms of interaction between owners and slaves were accepted. While the extent of this was more symbolic than

[67] See Olmos (forthcoming). Within the dialogue itself, Avienus is the model young elite Roman that is there to learn and be instructed by the others' performance.

anything else, it seems slaves could, at least for these days, drop some of the usual formalities, their precedence at meals being the most evident of their temporary privileges. Although this is not a particularly prominent theme in Macrobius' *Saturnalia*, there are a couple of passages where the slave servants come to the fore. We have first the initial discussion about the origins of the Saturnalia itself (one among the *antiquitates* discussed in the dialogue). Here, the custom is justified by a reference to the mythical "age of Saturn", when there were neither slaves nor differences of rank (I.7.26., cf. I.10.22.). In order to establish compliance with such old rules, Macrobius takes care to mention, at the appropriate time (I.24.22.) and trying to give a realistic touch to the dramatic setting, that the guests in his own story will also dine *after* the servants have finished their dinner. But more important is the discussion between Praetextatus and Evangelus regarding the appropriateness of these and any other signs of deference towards slaves (I.11.1-46). In one of his first impertinent (dialogue-creating) remarks, Evangelus takes aim at the levelling principle of the Saturnalia, saying it has no religious import, but is merely a folk custom. Praetextatus defends the slaves against Evangelus' scorn,[68] talking about their many virtues and their loyalty, and finally mentioning their personal participation in the search for knowledge and philosophy (I.11.41.: "Furthermore, the wits of slaves have not been unsuited or unequal to philosophy"; "Sed nec ad philosophandum ineptum vel impar servile ingenium fuit"). The Greeks Phaedo, Menippus and Epictetus are then mentioned as suitable examples of slave-philosophers.[69]

The Greek scholars who act as guests in the *Saturnalia* are themselves neither slaves nor servants and nowhere is there any insinuation of the kind. But what seems rather significant is the choice of such a particularly equalizing festivity to celebrate a rather harmonious gathering of people definitely belonging to different ranks, but all of them playing a distinct role in building the best possible compendium of learning to be passed on to Macrobius' own son and other similar young elite Romans. With all the limitations and distinctions we have thus far tried to identify, the final score between the Greek and Latin contributions to the *Saturnalia*'s ideal of elite "further education"[70] is rather balanced. The systematic nature of the former and the social and cultural significance of the latter make up a

---

[68] "iocone an serio putas esse hominum genus quod di immortales nec cura sua nec providentia dignentur?"; "are you joking, or do you seriously believe that there is a category of human being for which the immortal gods deign neither to care nor to provide?" (*Sat.* I.11.2.).

[69] Taken from Gellius, *NA* 2.18.

[70] See Olmos (forthcoming).

clever and well-contrived mix, while Macrobius' own literary achievement in putting them realistically together stands out as something worthy in itself, moving it far beyond a mere compiler's account.

References

Bestul, T.H. (1975). "The *Saturnalia* of Macrobius and the *Praecepta Artis Rhetoricae* of Julius Severianus", *The Classical Journal* 70/3: 10-16.

Cameron, A. (1966). "The Date and Identity of Macrobius", *Journal of Roman Studies* 56: 25-38.

—. (2011). *The last Pagans of Rome*, Oxford: Oxford University Press.

Cracco Ruggini, L. (1985). "Conservatism and Innovation in the Culture of the Fourth/Fifth Century", *Quaderni ticinesi di Numismatica e Antichità classiche* XIV: 287-303.

—. (1986). "Simmaco: *otia et negotia* di classe, fra conservazione e rinnovamento". In F. Paschoud *et al.* (eds.), *Colloque Genevois sur Symmaque* (pp. 97-118), Paris: Les Belles Lettres.

Davies, P.V., ed. (1969). Macrobius. *The Saturnalia*, New York/London: Columbia University Press (Records of civilization, sources and studies).

Flamant, J. (1977). *Macrobe et le Néo-Platonisme latin, à la fin du IVe siècle*. Leiden: Brill (Études préliminaires aux religions orientales dans l'Empire Romain, 58)

Frazier, F. & J. Sirinelli, eds. (1996). Plutarque. *Oeuvres morales*. Tome IX, 3ème partie: Traité 46. "Propos de Table" (Livres VII-IX). Paris: Les Belles Lettres.

Kaster, R.A. (1980). "Macrobius and Servius: Verecundia and the Grammarian's Function", *Harvard Studies in Classical Philology* 84: 219-262.

Kaster, R.A. (1988). *Guardians of language: the grammarian and society in Late Antiquity*, Berkeley: University of California Press.

—. ed. (2011). Macrobio. *Saturnalia* (3 vols.), Cambridge, Mass./London: Harvard University Press.

Keulen, W. (2009). *Gellius the Satirist. Roman Cultural Authority in Attic Nights*. Leiden/Boston MS: Brill.

Marinone, N., ed. (1967). Macrobio. *I Saturnali*, Torino: UTET.

Navarro Antolín, F., ed. (2010). Macrobio. *Saturnales*, Madrid: Gredos.

Neri, M. & I. Ramelli, eds. (2007). Macrobio, *Commento al Sogno di Scipione*, Milano: Bompiani.

Netz, R. (1998). "Deuteronomic Texts: Late Antiquity and the History of Mathematics", *Revue d'Histoire des Mathématiques* 4: 261-288.

Olmos, P. (2011). "Rhetoric, Satire and Encyclopaedia in Martianus Capella". In D. Hernández de la Fuente (ed.), *New Perspectives on Late Antiquity* (pp. 327-351), Newcastle: Cambridge Scholars Publishing.

—. (forthcoming). "Two Literary Encyclopaedias from Late Antiquity", *Studies in History and Philosophy of Science* (Sp. Issue "Structures and strategies in ancient technical writing", editd by L. Taub, L. Jardine and A. Doody)

Parker, H.N. (2009). "Books and Reading Latin Poetry". In W. A. Johnson y H. N. Parker (eds.), *Ancient literacies. The Culture of Reading in Greece and Rome* (pp. 186-229), Oxford: OUP.

Scarpa, L., ed. (1981). Macrobii Ambrosii Theodosii *Commentariorum in somnium Scipionis libri duo*, Padua: Liviana Editrice.

Shanzer, D. (1986). *A Philosophical and Literary Commentary on Martianus Capella's De Nuptiis Philologiae et Mercurii. Book 1*, Berkeley: University of California Press (Classical Studies, Vol. 32)

GIORGIO VALLA, EDITOR AND TRANSLATOR OF ANCIENT SCIENTIFIC TEXTS*

AMEDEO ALESSANDRO RASCHIERI

Through his teaching and editorial activities, the Italian humanist Giorgio Valla (1447-1500) contributed to sparking an interest in the rediscovery of ancient Greek scientific texts among his contemporaries. His remarkable intellectual education started near Piacenza, in Vigoleno, the small village where he was born. After studying Greek with Costantino Lascaris in Milan, Valla became a student of the mathematician and physiologist Giovanni Marliani in Pavia. There he also deepened his knowledge of Greek under the guidance of Andronicus Callistus.[1] He taught in Pavia, Milan and Genoa, for several years (1467-1484). On the invitation of Ermolao Barbaro,[2] Valla moved to Venice in 1485 and worked as a teacher for the "Scuola di San Marco." This institute had been founded by the Republic in 1446 to supply the chancery clerks with a solid training.[3] In 1496, he was imprisoned and tried for treason under the accusation of supporting Charles VIII of France. He was eventually

* English translation by Paul Warren.

[1] On Giorgio Valla see also Heiberg (1896), Branca (1980: 161-166), Gardenal (1981), Lowry (1984: 239-241), Bietenholz (1987: 371), King (1989: 650-652), Maillard (2003: 416), Tucci (2008). Some recent contributions on Valla's approach to Greek are found in Tessier (2003) and Ventura (2006). Regarding Valla's relationship with Lascaris see Martínez Manzano (1998: 21-25). In this respect the scholar writes: "la historia de Láscaris y Giorgio Valla [...] es, en realidad, la historia de una larga amistad que sobrepasa los límites de la habitual relación profesor-alumno" (22). The funeral oration for Giorgio Valla, written by Bartolomeo Zamberti, was studied and edited by Rose (1976).

[2] On the importance of Ermolao Barbaro for the Venetian culture at the end of the 14th Century see Branca (1980) (especially pages 162-163 for the relationship between Barbaro and Valla).

[3] On the School of San Marco see Branca (1980: 125-127), Ortalli (1996: 32-33). In contrast to the School of San Marco, which had a rhetorical and grammatical approach, the School of Rialto (founded in 1408) gave prominence to philosophical and scientific matters. On the School of Rialto see Lepori (1980).

acquitted of all charges, but such a terrible experience left him deeply scarred and he died only a few years later. He left his adopted son, Giovanni Pietro Cademosto, as his only heir.

Besides being a teacher, Giorgio Valla was also a great collector of ancient manuscripts, a translator from Greek, a commentator, a publisher and an encyclopaedist.[4] Because of his acclaimed reputation as a teacher, he was called to work for the most important families in Venice such as the Pisani, to whom he became particularly close. Amongst the ancient texts he commentated, it is worth remembering Iuvenalis for literature,[5] Cicero for philosophy[6] and Plinius and Ptolemy for science.[7] Documentary evidence shows that Valla gave lectures on Vitruvius, Archimedes and on the history of Greek poetry; there still remain codes by Dioscoride, Teocritus, and Sophocles that are accompanied by his own teaching notes. He had access to a wide range of Greek and Latin manuscripts. Among these, some were particularly precious and sought after by humanists, such as the volume containing Archimedes' complete works. After his death most of his Greek manuscripts eventually ended up in the collection of Alfonso II of Este, Duke of Ferrara. Today this collection is preserved at the Estense Library in Modena.[8]

His great interest in publishing was supported by the extraordinary development printing technology was undergoing in those days in Venice. In his preface to the *Astronomica* (Venice 1488), one of Valla's pupils, Vittore Pisani, recollects a printer (*impressor quidam*) coming to visit his master during one of his lectures. The printer (whose name, *Antonius de Strata*, appears in the colophon)[9] asked the scholar to give him a work to

---

[4] On Valla's activity as a collector of manuscripts see Cataldi Palau (1994). Concerning Giorgio Valla both as a translator and editor see Cortesi-Fiaschi (2008: XXVIII).

[5] Valla's comment on *Satire* by Iuvenalis, edited for the first time in 1486 and printed by Antonio da Strada, was reprinted several times in the following years (Flodr, 1973: 200-201). See also CTC (1970: 223-224).

[6] A comment by Valla on *De fato*, *Topica* and *Timaeus* was printed in Venice in 1485 by Antonio da Strada and subsequently reprinted in 1492 by Bonetus Locatellus, and edited by Octavianus Scotus (Flodr, 1973: 99, 114).

[7] On Valla's comment on Plinius see CTC (1980: 350-351).

[8] On Valla's library see Heiberg (1896), Mercati (1938: 58-74, 203-265, 271), Tucci (2008: 47-54, 505-517). On the Estense Library of Modena see Milano, (1987) (for the manuscripts belonging to Valla see p. 28).

[9] *Hoc opus impressum Venetiis arte et ingenio Antonii de Strata Cremonensis Anno salutis Mccccl xxxviii octavo calendas novembres*. "This work was printed in Venice thanks to the technical skills and ingenious of Antonio da Strada of

print that would be "useful to many and advantageous to him." Valla gave him a codex containing Avienus' complete works,[10] although it appears that what mostly interested both of them were the *Aratea*. Justification was also sought for the preference given to the translation by Avienus over those by Cicero or Germanicus, which were considered as being "largely abridged and incomplete." The *Liber medicinalis*, by Serenus Sammonicus, was added to this collection, not so much for its content as for being, according to Pisani, "composed of splendid verses."[11] Moreover on that occasion, Valla informed Antonius de Strata of his intention of giving some public lectures on Avienus' *Aratea*.[12]

Thanks to his vast knowledge, Valla was able to compile *De expetendis et fugiendis rebus*. This is an encyclopaedia that covers the entire scientific and philosophical knowledge of the time. Such a monumental work was published posthumously by Valla's adopted son, Giovanni Pietro, through the printer and humanist Aldus Manutius in 1501.[13] It is the largest volume ever published by Aldus Manutius. The deep friendship shared by the two humanists is accounted for in a passage from Manutius' Latin grammar published in 1493 (it does not appear in later versions): "quod apud Plautum ostendit mihi Georgius Valla Placentinus vir et graece et latine apprime doctus" ("which is showed to me in Plautus by Giorgio Valla of Piacenza, an extraordinary learned man both in Latin and Greek").[14] Valla's translations from Greek are also of primary importance, especially Aristotle's *Magna Moralia* and *Politics*.

Cremona in the year of salvation 1488, on the eighth day before the November calends (Saturday 25th October)".

[10] Billanovich (1974: 58) assumed that also this code came from Milan: "ritengo che invece quest'altro libro sia emerso per il Valla da un vecchio deposito Milanese."

[11] The insertion of *liber medicinalis* must have caused considerable worry to both the editor and the printer, as at the end of the volume Vittore Pisani readdresses Paolo Pisani ("Victor Pisanus ad Paulum Pisanum in postremo opere commonitio") not only by demonstrating the affinity between Astronomy and Medicine ("propter maximam inter ipsas artes affinitatem"), but also their contiguity ("astronomiam medicine partem esse non mediocrem"). The work by Quintus Serenus is also recalled as it contains the first documentary evidence of the magic word *abracadabra* (v. 935), used during the manufacture of an amulet against malaria (*febre terzana*).

[12] Raschieri (2010).

[13] On Valla's encyclopaedia see Tucci (2008).

[14] Branca (1980: 163-164), Dionisotti (1995: 58, 99). Lowry (1984: 252) points out that the first collaborators and editors of Aldus Manutius were Valla's colleagues and pupils.

However, of no lesser significance are those of many scientific and medical works (by Galen, Alexander of Aphrodisias, Aristarchus of Samos, John Philoponus, Nicephorus and Hypsicles).

As his encyclopaedic collection has already been broadly and meticulously analysed, particularly in recent times by Roberta Tucci,[15] it seems more appropriate to focus on Valla's activity as a translator. In this regard, specific attention will be paid to the collection of works he published in Venice through the printer Simone Bevilacqua in 1498.[16] It is the last of the scholar's works to be published during his lifetime; as opposed to his posthumous encyclopaedia, this appears to be finished and refined.[17] Given the breadth and diversity of this miscellany, it will not be possible to analyse each single translation. Thus, after a general introduction to the volume, time will be spent on examining the prefatory letters that accompany the writings.[18] It may be worth mentioning that there do not seem to be any other in-depth studies in modern times on this collection by Giorgio Valla.

The incunabulum consists of the Latin translations of twenty-three Greek works of different length and content. In addition to these, there is the *Libellus de argumentis*, or *Ratio argumentandi*, that is a work written by Valla himself, which gained popularity only a few decades later. The themes developed are considerably heterogeneous; they regard subjects such as logic, rhetoric, mathematics, astronomy, music, theology, ethics,

---

[15] Tucci (2008). Also see http://www.dm.unipi.it/~tucci/index.html.

[16] For an accurate description of the volume see Coates (2005: 1839-1840) (N-017). The colophon reads: *Impressum Venetiis per Simonem Papiensem dictum Beuilaquam. 1498. Die ultimo Septembris.* The last page of the incunabulus also bears the trademark: "Albero secco sul quale è appollaiato un uccello rapace, due scudi, caricati ciascuno di una corona, pendono dalla pianta. Sul cartiglio: Simon Bivilaqua" (EDIT16, CNCM 511). On Early Modern typography trademarks see Zappella (2001: 559-622). "Simone Gabi [or Bevilacqua], tipografo [worked from 1485 to 1514], figlio di Pietro, nacque a Pavia poco dopo il 1450. Stampò a Vicenza, Venezia, Saluzzo, Novi Ligure, Savona, Cuneo e Lione. A Venezia lavorò anche insieme ai suoi fratelli; a Cuneo fu in società con l'editore Vioto Dolce. Morì a Lione probabilmente nel 1518" (EDIT16, CNCT 690).

[17] "Gli studi hanno preso in considerazione questo incunabolo soprattutto per l'opera valliana *De expetendis rebus et fugiendis*, mentre l'interessantissimo *corpus* delle traduzioni in esso contenute non ha ancora ricevuto l'attenzione che merita" (Cortesi-Fiaschi, 2008: XVIII note 23).

[18] On the dedicatory letters see: the project "I margini del libro" (www.margini.unibas.ch), directed by Maria Antonietta Terzoli, the rich bibliography collected on the website and in the magazine *Margini. Giornale della dedica e altro*; Terzoli (2004), Santoro-Tavoni (2005).

and an ample section concerning medicine. A similar variety is shared by the number of authors that are translated: Nicephorus, Hypsicles, Ps. Euclid (in reality Hypsicles), Nicephorus Gregoras, Proclus, Aristarchus of Samos, Ps. Timaeus of Locri, Cleonides, Ps. Eusebius of Caesarea, Cleomedes, Athenagoras, Ps. Michael Psellus (in reality Theophanes Chrysobalantes), Ps. Alexander of Aphrodisias, Rhazes (a Greek translation of Muhammad ibn Zakariyā Rāzī). In addition, large space is given to works by Aristotle (*De caelo et mundo*, *Ars poetica* and *Magna moralia*) and by Galen (*De bono corporis habitu*, *De inaequali distemperantia*, *De praesagitura* or *De febribus*, *Praesagium experientia confirmatum Galeni*, *De succidaneis*).

The above list clearly epitomises Valla's interest in a wide selection of texts, which ranged from literature and philology to philosophy (with a strict Aristotelian connotation), and from theoretical to applied science (medicine *in primis*). Also contained in this volume are his earlier published works, for instance, the *Magna moralia* by Aristotle (Venice 1496, *per Gregorium de Gregoriis expensis Benedicti Fontanae*)[19] and *De musica* (or *Harmonicum introductorium*) by Cleonides (printed by Simone Bevilacqua in 1497).[20] Some translations, such as Aristotle's *Ars Poetica*,[21] were subsequently published several times in the following decades. Others were used to compose *De expetendis et fugiendis rebus*, which took the scholar several decades to complete. In fact evidence of his encyclopaedic intent can also be found in his 1498 miscellanea, where explicit reference is made to the encyclopaedia which was probably close to completion. As he stated in the dedicatory letter to Guidobaldo of Montefeltro:

> It will certainly be possible for you to examine such information, whatever it may be, which has long been circulating under other people's names, till when, if the giving god will allow me, it will soon be reunited under our

[19] Also see the 1496 Paris edition: *Parisiis, apud Ioannem Higman et Wolfgangum Hopilium, 1496; cur. Jacques Lefèvre d'Étaples* (Cortesi-Fiaschi 2008: 272). The translation by Giorgio Valla was reprinted in Paris in 1497 by Johann Higman and Wolfgang Hopyl (Flodr, 1973: 27).

[20] For the 1497 edition see Flodr (1973: 131). Besides the dedicatory letter to Vittore Pisani, the incunabulus contained *De architectura* by Vitruvius, *Panepistemon* and *Lamia. Praelectio in Priora Aristotelis analytica* by Angelo Poliziano, *De aquis urbis Romae* by Frontinus.

[21] Cortesi-Fiaschi (2008: 248).

> name in forty-nine volumes on that which is to be searched and to be escaped.[22]

Moreover, the 1498 miscellanea bears evidence of collaboration with Giovanni Pietro Cademosto (the editor of *De expetendis et fugiendis rebus*), as amongst Valla's translations there is one by his adoptive son, which is the version of *De succidaneis* by Galen. It was dedicated to his master Ludovico Bonacioli (Ferrara, 1475-1536), the eminent doctor and gynaecologist of Lucrezia Borgia.[23]

The incunabulum appears to be well laid out. There is a frontispiece that presents on *recto* the works contained in the volume,[24] together with the relative authors and the general attribution of the translation of the works to Giorgio Valla (*Georgio Valla Placentino Interprete*). However, there appear to be some inconsistencies in the list,[25] as at times the order is upset. In one case, for instance, reference is made to the *Galeni introductorium*, which is a text that is not included in this collection but had already been printed in Milan by Leonhard Pachel and Ulrich Scinzenzeler in 1482/83 (or 1483/84).[26] The presence of such a page may be explained through the analysis of the *verso* that contains the prefatory letter to the first translated work (*Logica* by Nicephorus) and not to the entire collection, as had been mistakenly believed by the printer. This is explained by a later handwritten note saying: "Such a preface, which, being mistakenly omitted by the printers, was placed here at the beginning, precedes Nicephorus' *Logica*."[27] The graphics are rich in xylographic adornments that very beautifully decorate initial letters, which usually appear in the frontispiece of most works. The same technique is also used

---

[22] "Verum haec, tantisper sub alieno, quoad propediem sub nomine nostro expetendorum ac fugiendorum quadraginta et nouem uolumina, si deus dator concesserit, tibi licebit qualiacumque sunt inspicere."

[23] Giovanni Pietro Cademosto was also co-author, together with Bernardino Saraceno, of a commentary of Plautus' comedies, printed in Venice in the years 1489, 1498 and 1499 (the latter edition by Simone Bevilacqua); see Flodr (1973: 242-243).

[24] On the evolution of the frontispiece in the history of printing see Zappella (2001: 421-499).

[25] Other slight imprecisions may be found among the small headings.

[26] Also see Cortesi-Fiaschi (2008: 534), where the translation is still indicated as included in the 1498 collection. For the following reprints of the Milanese edition see Flodr (1973: 32).

[27] "Haec praefatio prima errore imprimentium quae omissa hic posita est logicam praecedit Nicephori."

to illustrate subject matter, such as mathematics, music and astronomy, by means of geometrical drawings.

In order to understand the true aim of the scholar and identify the readership he wished to address, it seems important to examine the dedicatory epistles, bearing in mind that they need to be stripped of all the rhetorical and panegyric tones which are typical of this literary genre. Valla addresses his patrons who were mostly celebrated internationally, such as Guidobaldo of Montefeltro (1472-1508), Duke of Urbino, to whom he dedicates the whole collection; Agostino Barbarigo (1419-1501), doge from 1486 to 1501; or Alberto Pio (1475-1531), Prince of Carpi, who would later become of invaluable importance for preserving his Greek library. The other referents were prominent members of the Venetian patriciate,[28] who at times became, either directly or indirectly, his own disciples: Girolamo Zorzi, Giovanni and Sebastiano Badoer,[29] Pietro and Antonio Loredan,[30] Vittore and Paolo Pisani,[31] Paolo Trevisan, Costantino Priuli, Giorgio Corner. He also refers to public figures from other cities, such as Giovanni Valla (*iurisconsultus ducis Ferrariae*),[32] Polidoro Tiberti and Niccolò Masini (from Cesena), Giusto de' Giusti (from Verona), Antonio Visconti (*legatus ducis Mediolani*).

The general usefulness of rhetoric

A particular case concerns his Genoese disciple Bernardo Salvatico, to whom he dedicated his work on argumentation (*De expedita ratione argumentandi*)[33]:

---

[28] On the Venetian patriciate see Raines (2006).

[29] On Sebastiano's relationship with G. Valla see King (1989: 452-455).

[30] On the connection between the Venetian Loredan family and Giorgio Valla see Vendruscolo (1995).

[31] On Paolo Pisani and his relationship with G. Valla see King (1989: 616-617).

[32] Valla also dedicated to him his Latin translation of *De motu circulari* by Cleomedes (CTC, 1992: 7-8).

[33] In which, according to Vasoli (1981: 85): "Valla [andò], via via, raccogliendo, ordinando e sistemando elementi, esempi e, addirittura, interi passi forniti da testi assai celebri e che costituivano il fondamento della didattica umanistica, nell'ambito dell'insegnamento non solo della retorica, ma anche della teoria generale dell'argomentazione." See Vasoli (1981: 69-78, on his European fortune and 78-92 about his 1498 edition). As Vasoli says: "Giorgio Valla, dopo il Trapezunzio e prima dell'Agricola, o, perlomeno, negli stessi tempi in cui fu composto il *De inventione dialectica*, operò una decisa inserzione di procedimenti tipici della dottrina retorica nell'ambito dell'insegnamento logico, attribuendo

> Dearest Bernardo Salvatico, you have often asked me to put down in writing, so that it could be better remembered, that which, being free of any impediment, I had exposed to you about dialecticts, when you were my disciple, starting from the daily practice. You surely asked me because you had not found anyone else who could carry out this task in such a friendly way. You most certainly believed that it would not be difficult for me to do so, as I have dealt with all that concerns the oratory art in many other volumes, pertaining both to rhetoric and to other disciplines.[34]

As Valla himself explains, the compilation of this work had been spurred on by Bernardo, who confided in the doctrine, to the validity of which his master had already given evidence in other writings ("pluribus aliis fueram uoluminibus amplexus"). According to the humanist scholar such an argumentation is not out of place, as the oratoric ability ("dialecticen facultatem") does not only cocern rhetoric, but all disciplines ("caeteras ... disciplinas").

The themes concerning the general usefulness of the art of speech (*artis disserendi munera*) and of discipleship by Paolo Zorzi, son of the referent of the work (the Venetian patrician Girolamo Zorzi), are also found in the introduction to the Latin translation of the "short work on the art of speech" ("de arte disserendi opusculum") by Nicephorus. In this case the writing assumes a double function, by allowing both a close approach to "Aristotelian theories" ("ad aristotelicam doctrinam") and also an effective propaedeutics to any other form of learning ("ad alia enucleanda"):

> For your son, Paolo, a student of such lively intelligence, I considered important to render into Latin this short work on the art of speech, both because the author [Nicephorus], who is such an erudite scholar, embraced in a clear and concise manner the notions which seem convenient to those approaching Aristotelian theories, and, in a similarly correct and acute manner, also supplied useful information to those who are learning the

---

un'importanza essenziale alla teoria dell'argomentazione, degli argomenti e dei *loci*" (91-92).

[34] "Saepenumero a me, optime Bernarde Saluatice, poposcisti ut, quae ab interuentionibus uacuus tibi ad me uentitanti de ratione disserendi ex quottidiano usu lingua prodideram, quo tenaciori haererent memoriae, scriptis concrederem, quandoquidem quemquam alium non satis nactus esses, qui aut id posset, aut qui tam amice id praestare moliretur. Rem certe mihi non arduam fore censuisti, quando totam dialecticen facultatem, siue ad rethoricam, siue ad caeteras tendentem disciplinas pluribus aliis fueram uoluminibus amplexus". See Heiberg (1968: 15).

practice of the oratoric art, and because through these teachings it will be much easier to learn some other.[35]

Mathematic and Astronomic Sciences

At times, a compact group of works is dedicated to a single referent. This is the case for the mathematical syllogism dedicated to Alberto Pio di Carpi (1475-1550),[36] which Valla introduces with the narrative of an anecdote on the shipwreck of the philosopher Aristippo (from the *incipit*, book VI of Vitruvius' *De architectura*),[37] where the importance of exact science in the definition of human being is highlighted.

> His Excellence Alberto Prince of Carpi, Aristippo, the philosopher renowned for his Socratic knowledge, when deprived of his wealth by a shipwreck, being stricken by a disastrous tempest, arrived on the coast of Rhodes. There it is said that, on finding geometric diagrams on the ground, he turned to his fellows and exclaimed that they should raise their hopes, as he could see traces of human beings.[38]

In the closure of the epistle, the humanist explains the reason why he grouped together the Latin translations of Euclid, Hypsicles, Nicephorus and Proclus of Licia. This was to satisfy the prince's desire of dealing with

---

[35] "Filio tuo Paulo, peracuti ingenii nostro discipulo, hoc de arte disserendi opusculum latinum faciundum esse censuimus, tum quod ad aristotelicam doctrinam aspirantibus breuiter et dilucide nec minus commode et acute artis disserendi munera initiantibus quae opportuna uidebantur amplexus sit eruditus hic imprimis autor, tum quod per haec ad alia enucleanda longe ei facilior futurus sit accessus."

[36] Prince of Carpi, he was a friend of Aldus Manutius and Ludovico Ariosto. See Marogna, 2005.

[37] "Aristippus philsosophus Socraticus, naufragio cum eiectus ad Rhodiensium litus animadvertisset geometrica schemata descripta, exclamavisse ad comites ita dicitur: bene speremus! hominum enim vestigia video". "The philosopher Aristippus, a follower of Socrates, was shipwrecked on the coast at Rhodes, and observing geometrical diagrams drawn upon the sand, he is said to have shouted to his companions: There are good hopes for us; for I see human footsteps!" Trans. F. Granger.

[38] "Aristippus, Socratica insignis sapientia philosophus, Alberte Carpensis princeps inclyte, naufragio fortunis dispoliatus suis cum rhodiorum littus a scaeua tempestate iactatus tenuisset, ibi geometricas in terra descriptas figuras offendens ad co[m]mites conuersus fertur exclamasse omnibus confidendum ac bene sperandum esse quod humana obtu<e>retur uestigia."

astronomy, by starting from a propaedeutical study of mathematics in order to be able to move on to acquiring the notions necessary for the building and use of an astrolabius:

> Here is the fourteenth book of *Elements* by Euclid, of which us Latins were without, that I have just now translated into Latin to be delivered to you, together with the addition of the comment by the eminent philosopher Hypsicles. So that you can also dedicate yourself to the study of astrology, to which geometry allows secure access, I thought it may be useful to send you some booklets by Nicephorus and Proclus of Licia, which I translated, on the building and use of an astrolabius.[39]

Hence the four translations that follow are grouped together because of their common content regarding geometry and astronomy.[40] These are: *Interpretatio libri decimi Elementorum Euclidis* (or *Disputatio de dodecaedro et icosaedro*) by Hypsicles;[41] *Elementorum quartus decimus liber* by Ps. Euclid (Hypsicles);[42] *De astrolabio* by Nicephorus Gregoras;[43] *De astrolabio* (or *De fabrica usuque astrolabii*) by Proclus.[44]

In addition, the following short treatise is also dedicated to science. This is the Latin translation of *De magnitudinis et distantiis solis et lunae* by Aristarchus of Samos (4th-3rd cent. BCE),[45] dedicated to the Venitian nobleman Giovanni Badoer.

> Although within the mathematical science, many and extremely interesting arguments have been expounded by numerous and excellent scholars, Aristarchus of Samos' treatise on the size of the sun and the moon is by far the greatest work of all. After having translated it, I believed it to be appropriate to dedicate it to you as a symbol of our long-lasting friendship; I am sure that it will be well accepted by you, as, being an expert philosopher, you also dedicate yourself to mathematical disciplines, both

[39] "Iure tibi Euclidis quartumdecimum elementorum uolumen, quo Latini carebamus, quod nuper latinum fecimus tibi destinandum, interpretemque ei ad[i]iciendo Hypsiclem eximium philosophum. Quo astrologiam quoque, ad quam a geometria certissimus est accessus, attingas, latinos a nobis factos Nicephori et Procli Lycii de fabrica [us]usuque astrolabi libellos mittendos esse putauimus".

[40] On ancient Greek mathematics see Acerbi (2010).

[41] Thorndike (1963: 174).

[42] *Ibid*: 35.

[43] *Ibid*: 1442.

[44] *Ibid*: 279.

[45] Heath (1959), Thorndike (1963: 839), CTC (1960: 165), Acerbi (2010: 378). Aristarchus was translated into Latin again by Federico Commandini (Pesaro 1572). See CTC (1960: 165-166).

because of the nobility of the problems and because of the great depths of the enquiry.[46]

In this case mathematical and astronomical argumentations almost assume an aesthetic value, as Valla stated he had chosen, among the Greek scientific books available to him, the texts that, besides being of remarkable interest thanks to the depth of the author's enquiry ("philosophus peritissimus"), also dealt with fascinating topics ("longe omnium pulcherrima"), such as the size of the sun and the moon ("de solis lunaeque magnitudine").

Next comes *De mundo* (or *De universitatis natura*), the Latin translation of περὶ φύσιος κόσμω καὶ ψυχᾶς, a brief treatise in Doric vernacular, attributed since ancient times (for instance by Iamblicus and Diogene Laertius) to the Pythagorean Timaeus of Locri. For a long time, this had been erroneously considered the model of Plato's *Timaeus*.[47] Modern scholars consider this work as a fake however, and many doubts remain as to when it was made. In most recent interpretations it is placed between the 3rd century BCE and the end of the 2nd century CE; a double layering of the text is also supposed: a school teacher's exegetic notes seem to have been added to a compendium of Plato's *Timaeus*.

The short work consists of a synthetic and systematic treatise on cosmology and anthropology. It deals with the following topics: the fundamental principles and original causes of the world, the idea, the *hyle*, the sensitive world; the world and its determinations; the soul of the world; astronomy; the earth and its elements; the theory of perceptions; nutrition and breathing; pathology of the body and soul; pedagogy and the path to happiness. With its Latin tradition, dedicated to the Venetian patrician Pietro Loredan, Valla moves from the study of particular astronomic problems to considering the universe as a whole, so that the reader can understand the fundamental rules inherent to all scientific disciplines. The latter, according to the list compiled by the scholar, are

---

[46] "Cum sint in mathematicis res multae atque perpulchrae, Iohannes Baduare, pluribus ac eximiis autoribus explicatae demonstrataeque uoluminibus, longe omnium pulcherrima de solis lunaeque magnitudine Samii Aristarchi traditio est. Eam, ob conciliatam pridem inter nos amicitiam, latinam a nobis factam tibi destinandam esse censuimus, meae erga te beniuolentiae pignus immortale, quod iucundum et pergratum fore habemus exploratum, tum quod, philosophus peritissimus, mathematicarum quoque sis studiosus disciplinarum, tum ob quaestionum nobilitatem altissimamque indaginem."

[47] Baltes (1972), Marg (1972), Centrone (1982), Tobin (1985). For the Latin translation see Thorndike (1963: 1573).

considered to be Theology, Arithmetic, Music, Geometry, Astrology, Physiology and Ethics.[48]

> Pietro Loredan, I have decided that the booklet by Timaeus of Locri on the nature of the universe, which I have just finished translating from Latin, should be especially dedicated to you. [...] In honour and consideration for your family and your genius, it was not possible for me to open my soul to you with a more deserving token of my affection than by dedicating to you this sanctuary of the liberal arts by the outstanding philosopher Timaeus, as even Plato was not embarrassed to admire, imitate and follow a philosopher of such quality. Within his work, you will be able to observe the reason of the creation of the world, which, together with Theology, Arithmetic, Music, Geometry, Astrology, Physiology and Ethics, he wrote for the human kind that: it strengthens the mind by bringing it from remote dark places to the light of the truth of reality, which assigned the creation of the world to the optimum and maximum god and eliminated the sophisticated and cunning tricks of some bad philosophers.[49]

Valla's deeply-rooted Aristotelianism is therefore combined with Platonism in an effort to overcome a particularist and sectorial vision of science and, more importantly, to avoid the doctrinary catches concealed in the oversophisticated reasoning of some authors ("sophisticas sycophantias strophasque quorundam philosophantium").

Universal harmony can find a human correspondence in the ability to harmonise mundane reality and administrate in a balanced manner, which is proper of political men such as the Venetian patrician Vittore Pisani, to

---

[48] The Greek text translated by Valla was probably the one contained in a manuscript which is now preserved in the Estense Library of Modena, see Puntoni (1896: 387-388, n. 10 = II A 10); Mercati (1938: 219, 224, 241). This miscellaneous code contains at ff. 1-333 the *Epicteti Dissertationes ab Arriano digestae* and at ff. 336-362 the short work by Ps. Timaeus. This was written in the 15th century by the learned Bizantine Matthaeus Camariota, and contained Valla's *ex libris*, later deleted (f. 362v): Γεωργίου τοῦ βάλλα ἔστι τὸ βιβλίον.

[49] "Timaei Locri de uniuersitatis natura opusculum, quod nuper latinum fecimus, Petre Lauretane, tibi potissimum destinandum mancipandumque esse existimaui. [...] Tua domo tuo ingenio non fuit ut pignore digniore meum erga te animum aperirem quam ut Timaei peritissimi philosophi hoc tibi liberalium disciplinarum dicarem sacrarium, quod ne Platonem quidem tantum, inquam, philosophum admirari imitari et sequi puduit. In cuius sinu tibi licebit mundanae fabricae intueri rationes, quas Theologia Arismetica Musica Geometrica Astrologia Physiologia Ethicaque humano generi exarauit: mentem confirmauit, ex auiis tene[ri]bris in ueritatis lucem perduxit, mundi ortum optimo maximo deo deputando et sophisticas sycophantias strophasque quorundam philosophantium eliminando."

whom Valla dedicates *Cleonidae harmonicum introductorium* (or *De musica*).[50] Moreover, as in the previously analysed case of Aristippo, the dignity expressed by a philosophical-scientific concept finds confirmation within the ancient Pythagorean and Platonic knowledge:

> According to a Pythagorean and Platonic maxim, in all things, in those that are generated as in those to be generated, nature has nothing more ancient than harmony. [...] Hence, illustrious Vittore Pisani, a natural inclination towards all this is evident in you, I was not mistaken in rendering for you into Latin such a highly erudite and acute short work on harmony by Cleonides.[51]

The centrality of Aristotelian reflection

Evidence of Giorgio Valla's belief in the doctrinal centrality of Aristotelian teaching (along the same lines as Ermolao Barbaro's new Aristotelianism)[52] is given by the fact that the central position in the collection is occupied by the Latin translations of three large works by the Stagirite: *De caelo et mundo*,[53] *Ars poetica*[54] and *De ethica* (*Magna Moralia*). Such a huge admiration for the Greek humanist is already clearly perceivable in the preface of *De caelo*, dedicated to the Venetian nobleman Paolo Trevisan, where the authorship alone makes the work interesting.

> Paolo Trevisan, I know with great certainty that the extent to which Aristotle is held in high esteem by the most erudite of men in all the liberal disciplines, is widely renowned. All the argumentations by the Peripatetics confirmed how elaborate was that which he expounded, discussed and stated, with a nectar-like eloquence, as if he were writing in the sky. Thus

[50] For his previous edition see 20. As stated by Gallo (1981: 311), with this work Valla was "il primo a dare alle stampe la traduzione latina di un trattato musicale greco." The scholar also recollects that the humanist "nella sua ricca biblioteca veneziana […] possedeva [...] l'attuale codice gr. III C 2 della Biblioteca Nazionale di Napoli, contenente i testi di Gaudenzio, Cleonide, Euclide, Aristosseno". See also Thorndike (1963: 596).

[51] "Pythagoreorum platonicorumque sententia in cunctis tam productis quam producendis rebus natura nihil harmonia habet antiquius. [...] Quorum omnium cum in te, inclyte Victor Pisane, iam indoles administrandorum cernatur perspicua, non iniuria censui Cleonidis de harmonia opusculum, eruditione et acumine refertissimum, a me tibi latinum esse faciendum."

[52] On Barbaro's Aristotelism see Branca (1980: 129-133).

[53] On Valla's translation see Thorndike (1963: 1182).

[54] The Greek work is in M 100.

> we dedicate this book to you, which we have translated from Latin; we know that it will be cherished by you both because the discourse is entirely by Aristotle and because it deals with the sky, by which circle all is covered, in which all things have a place and of which no other work is more sublime.[55]

An interesting problem tackled by Valla, concerning Aristotle's works, is that of the other Latin translations which circulated in his day (in particular, in this passage he refers to the *Magna moralia*, which he dedicated to the Veronese nobleman Giusto de' Giusti).[56] These appear completely unsatisfactory to the eyes of the humanist scholar, as they reveal the translator's scarce knowledge both of the original language (Greek) and of the target language (Latin). This was worsened by the use of a word-by-word translation, which is typically mediaeval, and by an excessive use of loan-translations. Consequently, the content results distorted by rough and coarse expressions ("barbaris rusticisque dictionibus") that discourage the reader.[57]

> When recently asked to explain these writings, I felt a slight hesitation as I did not know who had already tried to translate them into Latin. All I knew was that whoever it had been did not have a good knowledge either of the Greek language or of ours, and that he made the mistake of translating word by word, which inevitably shed darkness over the Aristotelian concepts. In addition, as he was not aware of which words in our language corresponded with those belonging to the foreign language, by giving

[55] "Quantus cunctorum astipulatu doctissimorum in liberalibus omnibus disciplinis, Paule Triuisane, sit Aristoteles existimatus, satis receptum esse habeo exploratissimum. Quam elaborata quoque omnia de caelo scribens suo illo eloquentiae nectare expresserit disseruerit confirmarit, peripateticorum omnes scholae comprobarunt. Hunc igitur librum, quem latina expressimus oratione, tibi destinamus, quem gratum fore certo scimus tum quia Aristotelis, tum quia de caelo, cuius ambitu cuncta conteguntur, in quo quicquid usquam est et quo nullum opus sublimius, tota futura est oratio."

[56] There is a letter by Giorgio Valla to Giusto de' Giusti in Heiberg (1896: 83).

[57] On humanist translations in general see: Cortesi (2007a); Berti (2007). Also see: *ENTG. Edizione Nazionale delle traduzioni dei testi greci in età umanistica e rinascimentale* (http://www-3.unipv.it/entg/index.html), a project directed by Mariarosa Cortesi. On the translation problem in Giorgio Valla see Gardenal (1981: 44-53); for other translations see CTC (1970: 130-132) (Alexander Aphrodisiensis, *Problemata*); CTC (1980: 151-152) (Paulus Aegineta, *De re medica*); CTC (1986: 19-20) (Horapollo, *Hieroglyphica),* CTC (1986: 55-57) (Nemesius, *De natura hominis*); CTC (1992: 7-8) (Cleomedes, *De motu circulari*); CTC (1992: 304-305) (Gregorius Nyssenus, *De anima ad Tatianum*).

improper translations, he cast a further considerable cloud over the meaning of the work that he maimed to the point of rendering it full of coarse and rough language; more than communicating to us Aristotle's ideas, he seems to have discouraged anyone from reading them.[58]

The usefulness of Medicine

The last group of translations is dedicated to Medicine, a discipline of which Valla first of all underlines the prime usefulness in keeping people healthy and prolonging their life[59] in the prefatory epistle to *De salubri uictu* (or *De uictus ratione*) by Ps. Michael Psellus (in reality Theophanes Chrysobalantes, who dedicated his work to Constantine VII Porphyrogenitus)[60] addressed to the doge Agostino Barbarigo,

> Agostino Barbarigo, *serenissimo* Prince, I have regarded it to be important to translate for you from Latin *The healthy nourishment to the Emperor Constantine* by Psellus, philosopher and extraordinary medic, as according to an ancient tradition it does not seem good to come in visit to such a sublime prince without a gift, I have thought of a present that is suitable for me as a donor and to you as a receiver. Hence, as nothing is more preferable to me than a wise prince, and neither for you anything in this lifetime should be more dear than your health and longevity in such a prosperous condition, I felt it most convenient to offer you this book as a reminder of this, rendering you less worried about your health.[61]

---

[58] "Haec nimirum cum nuper explicari a me poscerentur, haesitauimus parum modo quod compertum haberemus tentasse nescio quem ipsa iampridem latina facere, eum cum exploratum habuissemus, nec graeca, nec nostra nouisse, mali interpretis more uerbum uerbo reddere studuisse, et sensus aristotelicos, quod fieri necesse fuit, multa obruisse caligine, cumque nostra peregrinis respondentia uerba ignorasse, graecis abutendo aliam adhuc legentibus nubem non exiguam offudisse, denique barbaris rusticisque dictionibus opus totum confudisse, ut non tam tradidisse nobis Aristotelem quam plane iugulasse uideatur adeo ut omnes ab eius auerterit lectione."

[59] On medicine in Valla's encyclopaedic work see Landucci Ruffo (1981).

[60] Sonderkamp (1987: 1-4). The translation was prepared according to the *Mutinensis* α.U.9.4 (gr. 61) Greek code, which at f. 99 contains the *ex libris* by Valla (Sonderkamp, 1987: 126-127).

[61] "Pselli philosophi singularisque medici de salubri uictu ad constantinum imperatorem tibi, Augustine Barbadice princeps serenissime, latinum faciendum esse censuimus, quod prisco ritu tam sublimis princeps sine munere adeundus non uideretur, eiusque generis esse munus oportere, ut mihi quidem exhibere, tibi uero conueniret admittere. Proinde cum nobis sapiente principe nihil optabilius, nec tibi in hac uita carius sanitate et uitae longitudine in tam eximia foelicitate esse debeat,

However, a dedication can strike even more personal chords when it is Niccolò Masino of Cesena, a medic himself, grieving over the loss of his daughter; the study of a work by Galen (*De praesagitura* or *De febribus*)[62] can spur on the urge to continue to practice the medic art in order to overcome one's grief.

> While I was considering what I could offer you, through which you could recognise that I requite your love, it came to my mind that by translating into Latin the short work, of considerable usefulness, on the signs of ilnesses, written by Galen, an admirable physician and a remarkably erudite commentator of Hippocrates, I could replenish your soul by means of taking care of the sick, not only by not distancing you from her, but by moving you towards her and, which may relieve the pain in your soul caused by the recent death of your daughter and if not eliminate and clear completely your internal sadness, at least it could render this milder, by bringing you back from your mortal to your immortal thoughts, thanks to the acknowledgement of reality.[63]

As to a medic, besides a deep doctrine, there is the need to have some illustrious example to imitate, Giorgio Valla proposes Alexander of Aphrodisias as a model, whom he assumes to be the author of *De causis febrium*[64] and that, according to him, was able to conjugate in an effective way theoretical preparation, practical ability and prognostic skills ("doctum et tam experientia quam iuditio sollertem"):[65]

> Why should we not devote ourselves to the study of the causes and nature of fevers, in order to remedy them more easily and to give aid to the

---

iure hunc librum protulimus qui haec te commoneat et nos de tua salute minus habeat sollicitos." The Latin text of the letter is found in Sonderkamp (1987: 2).

[62] Thorndike, 1963: 386.

[63] "Mecum ergo reputanti, quidnam praestare possim, unde meum esse mutuum erga te amorem cognosceres, succurrit Galeni [ac] admirandi medici <ac> Hippocratis interpraetis doctissimi morborum praesagiturae opusculum, utilitatis non contemnendae, latinum tibi faciendo, tuum posse animum oblectare ab aegrotantium cura, non modo non distinendo, sed ad ipsam potius impellendo, quodque tuum possit animum morte filiae proxima exulceratum mulcere et inuectum moerorem, si non penitus leuare ac abstergere, saltem quiret reddere leniorem, te a moribundis cogitationibus ad immortales rerum scientia reuocando."

[64] See Tassinari (1994); on Valla's translation see vii-viii; 90-115 (reproduction of the Latin text). The Latin translation is dedicated to the Venetian patrician Giorgio Corner. Also see CTC (1970: 125-126).

[65] On this preface see Cranz (1958: 540).

> suffering or succumbing nature of the body and to its virtue? In fact it is essential that the medic be learned and experienced as much in the practice as in his ability of judgement and that his intelligence be acute; so it was without any doubt Alexander of Aphrodisias (besides being a famous philosopher), whose book on the causes of fevers I thought I had to translate for you from his language to ours.[66]

However, Valla is able to evaluate his Greek sources in a critical manner, including the most prestigious such as Galen, of whom he exposes the inadequate coverage given to the subject of the plague, which is mentioned only marginally in his writings. For such a topic, which he regards as important and always present, he resorts to an Arabic source, *De pestilentia* by Muhammad ibn Zakariyā Rāzī, already translated into Greek ("de syriaca lingua in graecam") and here presented in Latin with a dedication to Paolo Pisani.

> Paolo Pisani, I believed it would be fitting to dedicate to you the translation into Latin of this short work on the plague by Rhazes, which was translated for some emperor from Syriac into Greek, both because this teaching is considered to be written by a distinguished author and because such knowledge is considered by everyone as absolutely necessary. Moreover Galen, who is the principal author of medical texts, has not left any detailed writing on the plague, which is a subject that should concern the medical art. In truth, he only refers to it in a hasty manner, thus giving the impression of having paid little attention to the treatment of the plague.[67]

[66] "Cur non imprimis cognoscendis febrium causis studuerimus ac naturis, quo procliuius ipsis possimus occurrere et laboranti aut succumbenti naturae corporisque uirtuti ferre suppetias? Talem namque, doctum et tam experientia quam iuditio sollertem, medicum esse oportet, cor cuius sapiat, qualis sine controuersia fuit Alexander Aphrodisieus, nedum philosophus nominis celeberrimi, cuius de febrium causis librum ex ipsius lingua in nostram tibi conciliandum existimaui."

[67] "De syriaca lingua in graecam Rhazae de pestilentia opusculum imperatori cuidam conuersum latinum, Paule Pisane, tibi faciendum esse existimauimus, tum quod ab insigni autore haec traditio emanare certo perspicitur, tum quod perquam necessaria haec prorsus omnibus habita cognitio est eoque magis quod Galenus, medicinae autor praecipuus, de pestilentia, quod ad artem medicinae pertineret, nihil accuratius scriptum reliquit; cursim uero eius rei meminit dumtaxat, ut eius intentio nulla ad pestilentiae curam habita uideatur."

Valla translator and editor

Valla's translations enjoyed a widely spread popularity during the humanist age, which is proven by their innumerable reprints. Nevertheless, there were some negative judgements on his works, such as that by Robertus Balforeus (Robert Balfour) who, at Bordeaux in 1605, published his own commentated Latin version of *De motu circulari corporum caelestium* by Cleomedes, which had already been published by Carolus Valgulius (Brescia 1497) and, as mentioned above, by Giorgio Valla (Venice 1498). In fact, in his preface[68] he writes that:

> Giorgio Valla, who was the first to spread it in Latin lands and to make it familiar to the Latin people, frankly speaking, made inaccurate assumptions throughout the work and, because of the bad translation, he attributed to it so many false and useless interpretations. It was as if he had wanted to destroy it rather than translate it, and not so much bring Cleomedes to Latium as drive him away and condemn him to an eternal exile.[69]

Modern scholars express more positive opinions, starting from Branca's[70] considerations on Valla's activity in the last period of the flourishing of humanist translations and its importance in forming a Scientific Latin lexicon. This was later developed by Gardenal,[71] who states:

> Despite some indecision and uncertainties, in his translations [he] displays a scientific aim that is particularly noticeable in his search for a meticulous closeness to the original text. This is carried out by means of using the most semantically appropriate word to express the reality of the concepts and *res* […], especially by dropping periphrases, which were so dear to Mediaeval authors and first-generation humanists.[72]

---

[68] CTC (1992: 9).

[69] "Georgius Valla Placentinus, qui primus illud in Latias sedes transferre et Latinae consuetudini tradere aggressus, libere dicam, tam pueriliter toto opere ineptit, et interpretatione prava tam multa saepe vana, saepe falsa, illi affingit, ut non tam illud vertere quam pervertere, nec tam Cleomedem in Latium ducere quam inde eiicere, et aeterni exilii infortunio mulcare voluisse videatur."

[70] Branca (1980: 165-166).

[71] Gardenal (1981: 44-53, cited on p.52).

[72] "nonostante alcune indecisioni e incertezze, [egli] rivela nelle sue traduzioni un proposito scientifico che si manifesta soprattutto nella ricerca di una continua fedeltà al testo tradotto e che si verifica mediante l'utilizzazione della parola di più univoca pertinenza semantica alla realtà dei concetti e delle *res* […] e soprattutto

In addition, as the result of an attentive comparison between the Greek original and the Latin translation, Tassinari's opinion on *De febribus*, by Ps. Alexander of Aphrodisias, seems also significant:

> This version [by Valla] is characterised by an exact closeness to the Greek text, which if on the one hand it may appear awkward because of conforming excessively to the darkness of the original, on the other hand it appears remarkable for its effort in creating a clear scientific language, which well reflects the Greek *concinnitas* and is free of the circumlocutions of Mediaeval translators.[73]

Besides revealing his particular attention for the Greek language and ancient scientific doctrines, Valla's translations are also significant to the understanding of his editorial sensitivity and of his effort made in spreading such knowledge by means of printing. In addition, thanks to the examples analysed, it is possible to understand the breadth of Valla's culture. He rejected all forms of dogmatism or excessive specialization when compiling his general encyclopaedia of science. Moreover, as appears from a brief examination of his 1498 collection, his kind of Humanism did not involve a simple juxtaposition of literary-philological knowledge, philosophico-theological reflections and scientific disciplines; he regarded all forms of knowledge as coming together in a systemic and complex manner so as to include and reflect universal harmony.

References

Acerbi, F. (2010). *Il silenzio delle sirene. La matematica greca antica*, Roma: Carocci.

Baltes, M. (ed.) (1972). Timaios Lokros, *Über die Natur des Kosmos und der Seele*, Leiden: Brill.

Bietenholz, P.G. and Th.B. Deutscher, eds. (1987). *Contemporaries of Erasmus. A Biographical Register of the Renaissance and Reformation*, III, Toronto; Buffalo; London: University of Toronto Press.

Berti, E. (2007). "La traduzione umanistica." In Cortesi (ed.) 2007b: 3-15.

con l'abbandono delle perifrasi, care agli autori del Medio Evo e agli umanisti della prima generazione."

[73] "la versione [di Valla] è caratterizzata da un'esatta fedeltà al testo greco, che se da un lato può risultare talvolta ostica per il troppo assecondare l'oscurità dell'originale, dall'altra è notevole per lo sforzo di creazione di una limpida lingua scientifica, ben rispondente alla *concinnitas* greca e libera dalle circonvoluzioni dei traduttori medievali." (Tassinari, 1994: 90).

Branca, V. (1980). "L'umanesimo veneziano alla fine del Quattrocento. Ermolao Barbaro e il suo circolo." In G. Arnaldi and M. Pastore Stocchi (eds.), *Storia della cultura veneta*, vol. 3/1, *Dal primo Quattrocento al Concilio di Trento* (pp. 123-175). Vicenza: Neri Pozza. [Now in V. Branca, *La sapienza civile. Studi sull'Umanesimo a Venezia* (pp. 59-127), Firenze: Olschki, 1998].

—. ed. (1981). *Giorgio Valla tra scienza e sapienza*, Firenze: Olschki.

Cataldi Palau, A. (1994). "Su alcuni Umanisti possessori di manoscritti greci, I. Alcuni manoscritti appartenuti a Giorgio Valla, II. Un nuovo manoscritto appartenuto a Marco Musuro", *Studi umanistici piceni* 14: 141-155.

Centrone, B. (1982). "La cosmologia di pseudo Timeo di Locri e il *Timeo* di Platone", *Elenchos* 3: 293-324.

Coates, A. *et alii* (2005). *A Catalogue of Books Printed in the Fifteenth Century now in the Bodleian Library, Oxford*, 6 voll., Oxford: Oxford University Press.

Cortesi, M. (2007a). "Vitalità della traduzione umanistica." In Cortesi (ed.), 2007b: IX-XVI.

—. ed. (2007b). *Tradurre dal greco in età umanistica. Metodi e strumenti.* Atti del Seminario di studio, Firenze, Certosa del Galluzzo, 9 settembre 2005, Firenze: SISMEL – Edizioni del Galluzzo.

Cortesi, M. and S. Fiaschi, eds. (2008). *Repertorio delle traduzioni umanistiche a stampa. Secoli XV-XVI*, 2 voll., Firenze: SISMEL – Edizioni del Galluzzo.

Cranz, F.E. (1958). "The Prefaces to the Greek Editions and Latin Translations of Alexander of Aphrodisias, 1450 to 1575", *Proceedings of the American Philosophical Society* 102: 510-546.

CTC (1960-2003). *Catalogus translationum et commentariorum. Mediaeval and Renaissance Latin translations and commentaries: annotated lists and guides*, 8 voll., Washington: The Catholic University of America press.

EDIT16. Censimento nazionale delle edizioni italiane del XVI secolo, Roma: Istituto Centrale per il Catalogo Unico delle biblioteche italiane e per le informazioni bibliografiche (ICCU), http://edit16.iccu.sbn.it.

Flodr, M. (1973). *Incunabula classicorum. Wiegendrucke der griechischen und römischen Literatur*, Amsterdam: Hakkert.

Gallo, F.A. (1981). "La trattatistica musicale." In G. Arnaldi and M. Pastore Stocchi (eds.), *Storia della cultura veneta. Dal primo Quattrocento al Concilio di Trento*, vol. 3/3 (pp. 297-314). Vicenza: Neri Pozza.

Gardenal, G. (1981). "Giorgio Valla e le scienze esatte." In Branca (ed.), 1981: 9-54.

Heath, Th., ed. (1959). Aristarchus of Samos, *The ancient Copernicus: a history of Greek astronomy to Aristarchus together with Aristarchus's Treatise on the sizes and distances of the sun and moon*, Oxford: Clarendon press.

Heiberg, J.L. (1896). *Beiträge zur Geschichte Georg Valla's und seiner Bibliothek*, Leipzig: O. Harrassowitz.

King, M.L. (1989). *Umanismo e patriziato a Venezia nel Quattrocento*, 2 voll., Roma: Il Veltro. [*Venetian Humanism in an Age of Patrician Dominance*, Princeton: Princeton University Press, 1986.]

Landucci Ruffo, P. (1981). "Le fonti della «medicina» nell'enciclopedia di Giorgio Valla." In Branca (ed.), 1981: 55-68.

Lepori, F. (1980), "La scuola di Rialto dalla fondazione alla metà del Cinquecento." In G. Arnaldi and M. Pastore Stocchi (eds.), *Storia della cultura veneta. Dal primo Quattrocento al Concilio di Trento*, vol. 3/2 (pp. 539-605). Vicenza: Neri Pozza.

Lowry, M. (1984). *Il mondo di Aldo Manuzio. Affari e cultura nella Venezia del Rinascimento*, Roma: Il Veltro. [*The world of Aldus Manutius. Business and Scholarship in Renaissance Venice*, Oxford: Basil Blackwell, 1979.]

Marg, W., ed. (1972). Timaeus Locrus. *De natura mundi et animae*, Leiden: Brill.

Marogna, M.A., ed. (2005). *Alberto Pio da Carpi contro Erasmo da Rotterdam nell'età della Riforma*, Pisa: ETS.

Martínez Manzano, T. (1998). *Constantino Láscaris: semblanza de un humanista bizantino*, Madrid: Consejo superior de investigaciones científicas.

Mercati, G. (1938). *Codici latini Pico Grimani Pio e di altra biblioteca ignota del secolo XVI esistenti nell'Ottoboniana e i codici greci Pio di Modena con una digressione per la storia dei codici di S. Pietro in Vaticano*, Città del Vaticano: Biblioteca Apostolica Vaticana.

Milano, E. (1987). "Profilo storico della Biblioteca Estense". In *Biblioteca Estense. Modena* (pp. 11-47). Firenze: Nardini.

Ortalli, G. (1996). *Scuole e maestri tra Medioevo e Rinascimento. Il caso veneziano*, Bologna: Il Mulino.

Puntoni, V. (1896). "Indice dei codici greci della biblioteca Estense di Modena", *Studi Italiani di Filologia Classica* 4: 379-536.

Raines, D. (2006). *L'invention du mythe aristocratique. L'image de soi du patriciat vénitien au temps de la Sérénissime*, 2 voll., Venezia: Istituto Veneto di Scienze, Lettere ed Arti.

Raschieri, A.A. (2010). *L'orbis terrae di Avieno*, Acireale; Roma: Bonanno.

Rose, P.L. (1976). "Bartolomeo Zamberti's funeral oration for the humanist encyclopaedist Giorgio Valla." In C.H. Clough (ed.), *Cultural aspects of the Italian Renaissance. Essays in honour of Paul Oskar Kristeller* (pp. 299-310). Manchester: Manchester University Press; New York: A.F. Zambelli.

Santoro, M. and M.G. Tavoni, eds. (2005). *I dintorni del testo. Approcci alle periferie del libro.* Atti del Convegno Internazionale, Roma, 15-17 novembre 2004 – Bologna, 18-19 novembre 2004, Roma: Edizioni dell'Ateneo.

Sonderkamp, J.A.M. (1987). *Untersuchungen zur Überlieferung der Schriften des Theophanes Chrysobalantes (sog. Theophanes Nonnos)*, Bonn: Habelt.

Tassinari, P., ed. (1994). Ps. Alessandro d'Afrodisia, *Trattato sulla febbre*, Alessandria: Edizioni dell'Orso.

Terzoli, M.A., ed. (2004). *I margini del libro. Indagine teorica e storica sui testi di dedica.* Atti del Convegno Internazionale Basilea, 21-23 novembre 2002, Roma; Padova: Antenore.

Tessier, A. (2003). "Un corso veneziano su Sofocle di Giorgio Valla (con un piccolo addendum euripideo)", *Italia medioevale e umanistica* 44: 189-198.

Thorndike, L. and P. Kibre (1963), *A catalogue of incipits of mediaeval scientific writings in Latin*, London; Cambridge, Mass.: The Mediaeval Academy of America.

Tobin, Th.H., ed. (1985). Timaios of Locri, *On the Nature of the World and the Soul*, Chico (California): Scholars Press.

Tucci, R. (2008). *Giorgio Valla e i libri matematici del* De expetendis et fugiendis rebus*: contenuto, fonti, fortuna*, PhD Thesis, Università degli studi di Pisa.

Vasoli, C. (1981). "Nota sul *De expedita ratione argumentandi* di Giorgio Valla e sulla sua fortuna." In Branca (ed.), 1981: 69-92.

Vendruscolo, F. (1995). "Lorenzo Loredan/Λαυρέντιος Λαυρετάνος 'copista' e possessore di codici greci", *Italia Medioevale e Umanistica* 38: 337-363.

Ventura, I. (2006). "Aristoteles fuit causa huius libri: On the Reception of Pseudo-Aristotle's Problemata in late Medieval Encyclopaedic Culture." In P. De Leemans, M. Goyens (eds.), *Aristotle's Problemata in Different Times ad Tongues* (pp. 113-144), Leuven: Leuven University Press.

Zappella, G. (2001). *Il libro antico a stampa: struttura, tecniche, tipologie, evoluzione*, vol. 1, Milano: Bibliografica.

PART II

ARTS, SCIENCES AND TECHNOLOGIES OF QUANTIFICATION

THE MORE IT CHANGES … REFLECTIONS ON THE WORLD HISTORICAL ROLE OF GREEK MATHEMATICS

REVIEL NETZ

Let us consider the longue durée of Greek science: that is, consider Greek science as part of such wider phenomena as *reception* and *the classical tradition*. What was Greek science–in particular, Greek mathematics–for its history not of individual scientists producing this or that work, but rather *through two thousand years*?

This question is not merely interesting for the historian of ancient science. It is also of significance for, let us say, world history. It seems plausible to suggest that one of the defining features of world history is the growth of knowledge leading to the explosive moment of the Scientific Revolution (a term that may be misleading but remains useful); and it seems plausible to suggest that this explosive moment could be understood, among other things, by putting it in the context of the longue durée of Greek science. In short, the question at hand is the place of Greek science in world history. A big question. I shall offer a position, employing the broad brush which my theme demands.[1] For reasons which will become apparent, I will concentrate on the Greek exact sciences. One caveat is yet in order. I will unfold a thesis. It is not meant as "the explanation of the scientific revolution". I think it is no more than one factor among many contributing to the rise, and shape, of the scientific revolution. What it does is to provide one of the key ways in which the particular history of the Greek exact sciences helps explain the Scientific Revolution and through it the place of Greek science in world history.

[1] As a consequence, I will also forego the fine chisel of the footnote: this–I openly admit–is an essay, not a paper.

The Ancient Exact Sciences: Periods and their Diversity

Here are two basic intuitions about mathematics: that it is always the same; that it changes according to culture. It is either math or maths, singular or plural. We historians usually concentrate our efforts (at least for the last generation or so) in debunking the singular image of mathematics and in promoting the plural one. It is easy for us, historians, to make fun at essentialist accounts that lump together the variety of diverse cultural forms. Yet maybe we should not rush to mock the unifiers, and the very temptation to lump together may be significant. Intuitions should never be ignored: they hold keys to understanding. And so I try to account for both intuitions, to show both the diversity as well as the unity of the Greek exact sciences.

I begin with what may be the easier part for an historian: the diversity. Indeed it is possible, I would argue, to identify several periods in the history of Greek mathematics, and to see how, at the different periods, what we call "Greek mathematics" was a completely different project serving completely different purposes. So let me start with such a brief survey, based in part on my two books Netz (2004) and Netz (2009), extending their account in some ways. The following, then, is a tentative division of Greek mathematics into five periods. I am aware of course that such periodization in and of itself tends to lump together diverse forms. Science does not move in a lockstep and the diversity of scientific practices in any one given time would not necessarily be less than that between different periods. However, changes across the time axis are important, and for our purposes it would be useful to consider such ideal-type simplifications, as an example of the way in which Greek mathematics did change.

Pre-Socratic Mathematics: The Turn of the Fifth Century BCE

Here the evidence is minimal. A heavily mediated report of a result by Hippocrates of Chios (which however seems to hold a kernel of the original text); a report, somewhat less mediated perhaps, of a result by Archytas (which however clearly is in a form different from that of the original). It is likely that the most important author in the exact sciences of the period could have been a different one altogether, namely Eudoxus; but too little survives from him to form any clear sense of how his mathematical work might have looked like. Less survives from other authors such as Oenopides or Theatetus. Most of our evidence comes from much later sources, mostly philosophical, mostly from Late Antiquity. Their presuppositions concerning the nature and function of mathematics would colour their reports. Thus, should we believe Proclus who seems to say that works such as Euclid's *Elements* were already written (several times!) by authors preceding Euclid, starting with Hippocrates of Chios? (I do not: I think Proclus imagines that the form of the *Elements* is so essential to mathematics that its origin must be as early as that of the science itself).

One's impressions are necessarily speculative. Here are some observations.

First, mathematics appears to emerge rather suddenly, and forcefully, during the second half of the fifth century BCE. It is an irruption correlated with an overall explosion of new creative forms in an Athenian dominated world that also sees Phidias, Euripides, Anaxagoras.

Second, the results we can firmly associate with this period are showpieces. They are difficult to achieve; obviously important in and of themselves (and not as tools for some other result); they are radically original (one recalls in this context the discovery of irrationals, which can not be firmly associated with any particular author but which no doubt forms a key achievement of this mathematical period); an emphasis seems to be placed on the virtuosity of a construction, underlining the sense of a tour-de-force; a showpiece.

I use the term "Pre-Socratic". Chronologically and perhaps also intellectually, a more precise connotation can be "Sophistic". It stands to reason, then, to suggest that Greek mathematics emerged in the world of, say, Gorgias. By which I do not imply that early Greek mathematical arguments were merely sophistic (neither were those of the sophists). Rather, I suggest that Greek mathematics could have emerged in response to a similar effort of gaining prestige through the power of the showpiece argument.

Philosophical Mathematics: the Fourth Century BCE

This then would be in sharp opposition to another, generally speaking later form of mathematics, the one now firmly attested–with the authors Autolycus and especially, of course, Euclid. (The extant works of Aristoxenus are a useful corroboration of the overall trends in the exact science and its related fields). In all three authors, the most remarkable thing is how they seem to *eschew* the remarkable. The negative evidence is not trivial: among the many hundreds of results in Euclid's oeuvre (as well as those of Autolycus and Aristoxenus) the tour-de-force is very rare:

results are relatively unspectacular, and their proofs do not call for very complicated constructions. There are some important results encrusted within the text of the *Elements*, some striking–the infinity of prime numbers, the volume of the cone–but they are slowly and painstaking developed and when finally reached they are as it were almost an afterthought. Euclid seems to wish to understate his moments of the spectacular. It is telling how the most directly engaging result in Book X (otherwise a very understated work, repetitive to the point of being nearly algorithmic and so devoid of any sense of authorial design)–the proof that the ratio of diagonal to side in a square is irrational–is what appears like a later supplement to fill in what was felt as a gap in the treatment. Euclid provided a systematic and unspectacular classification of irrationals, and later readers found it important to reinsert the "Pre-Socratic" proof for the existence of irrationality which Euclid avoided altogether.

The extant works by Euclid (mostly, but not exclusively in geometry, though of course the authorial status of his works is always in doubt), Autolycus (astronomy) and Aristoxenus (music) are elementary. To some extent this is a bias of survival. Aristoxenus, for instance, certainly wrote more than survives, and in a more sophisticated vein; but perhaps his more sophisticated works would not reflect a similar quasi-mathematical form of writing but be more akin to Aristotle's philosophical writings). There does seem to be a unique interest, in this period, in the production of elementary results: it is after all the period from which the *Elements* derive. "Elements" are not merely elementary, in our modern sense. They are also systematic (perhaps part of the meaning of the ancient term *stoicheia*). The one striking feature of Euclid's *Elements*–and the one which would forever make him a key figure for philosophers eager to learn about Greek mathematics–is his axiomatic method. More than any other Greek mathematician, Euclid is invested in the explicit demarcation between the provable and the unprovable, in the mapping of the paths leading to results and not just in results themselves. This methodological concern is of course also very evident in Aristoxenus, and for a good reason: he was, after all, primarily a philosopher. The fourth century is when Greek philosophy was most directly engaged in reflection upon scientific methodology, in Plato's work and in the work of Aristotle (mostly extant to us) and Epicurus (mostly fragmentary now, and negative in its attitude towards science). It is plausible to suggest that the mathematics of the time would reflect a similar methodological interest. The philosophers are right, to care especially for Euclid; it was he who cared especially for them. This would then mark his kind of mathematics as completely different from that of Hippocrates of Chios–perhaps, as different as Aristotle was from Gorgias.

Alexandrian Mathematics: Third to Second Centuries BCE

I now reach the theme of Netz (2009). The title of the book was *Ludic Proof* and I argued for the centrality of *play* in the mathematics of the period (which of course does not mean mathematicians did not take their writing seriously). The play is about the mathematics: the choice to concentrate on complicated results which essentially bring together highly distinct fields; it sets itself seemingly inconceivable tasks, whose origins often lie outside mathematics in other cultural fields (Archimedes' *Sand-Reckoner*–counting the grains of sand in the universe, following a Homeric cue–would be a good example). The play is about the audience: the writing is meant to elicit maximum suspense and surprise, a complex encounter of raising expectations so as to quash them.

The number of extant writings from the period is substantial. One of the defining features of the period, indeed, is the (relatively) central position of science in the literary system of Alexandria. Authors range in date from Aristarchus (near the beginning of the third century BCE) to Hypsicles (the middle of the second century BCE). At the center of the period stands the ideal Hellenistic mathematician, Archimedes, whose works mostly are extant. He also fits the bill of the ludic mathematician most perfectly (Apollonius, the other great mathematician of the period, is perhaps less of a ludic author–though still clearly a creature of the same culture). I make a choice, in excluding Euclid from my survey and identifying him with an earlier period. If we do follow Euclid's traditional dating, my choice then translates into a break not with the end of the fourth century and the foundation of Alexandria but a bit later: at some point during the first half of the third century.

This dating is meaningful in the terms of Alexandrian cultural history. This is the age of Callimachus, Theocritus, Aratus and Apollonius of Rhodes: a period during which a group of self-confident authors, employing a largely original stylistic approach, re-appropriate the canonical forms of the past and manage–the first to achieve this for a long time–a canonical position for themselves, even while writing in the most canonical genres of poetry. (Previous major authors, for many generations, achieved canonical status only via prose or drama but never through the most canonical, and "Archaic" form of non-dramatic poetry). And indeed, the "ludic" themes of surprise, tension and overall generic experimentation are central to their work. Hellenistic mathematics would seem to respond to wider cultural sensibilities of its time–nothing surprising in that–especially visible in the poetry of Alexandria.

This Hellenistic mathematics would be closer to the Pre-Socratic than to the Philosophical one. Archimedes would be more like Archytas, less like Euclid. Perhaps, in the need to create larger poetic structures that build gradual surprise and that combine together various alien strands, the tendency of Hellenistic mathematics would be to look for a larger-scale achievement than that of Pre-Socratic Mathematics, which was perhaps more focused on the single, isolated tour-de-force. There is a difference between Hippocrates' Lunules and Archimedes Spirals, which is perhaps rather like the difference between Gorgias' *Helen* and Theocritus' *Idyls*. Be it so: such speculations are inherently plausible but are misleading, as we compare the relatively secure understanding of a mostly extant Archimedes, with the fragmentary, mediated evidence for Hippocrates. What is definitely clear is that the scientific personality emerging from Euclid's *Elements* is very different from that emerging from Archimedes' works. Perhaps this is an accident about isolated personalities; perhaps it simply attests to the particular project Euclid was engaged in; and perhaps my (traditional) early dating of Euclid is wrong. Nothing is certain; and yet it remains true that the most simple reading of the evidence is that, between the late fourth and the middle third century BCE, Greek mathematics went through a sharp rupture.

The Roman Synthesis: First century BCE to second century CE

I said that Hellenistic mathematics could have responded to wider cultural sensibilities of its time. But which? According to my account, those were those of Alexandria. Hellenistic mathematics did not belong to the world of Athens. And indeed, why should it have? No one in Athens seems to have been interested in science, either–while the actual space of mathematics (where mathematicians were, with whom they corresponded) was firmly anchored in Alexandria itself.

I mentioned the close interest of fourth century philosophy in the details of scientific practice. Such interest sharply declined as the Hellenistic schools crystallized at the turn of the century. A new practice of philosophy emerged, powerfully dominated by a handful of schools in Athens (which now became distinguished above all–for the rest of antiquity!-as a center of philosophical study), and the new spatial and institutional realities went hand in hand with a new intellectual emphasis.

Epistemology remains a central concern, but it was now more generally "methodological" and did not involve considerations of actual scientific practice of which Athens, it appears, now knew little.

Nor did Alexandria direct its interests closely towards Athens. The one form of cultural expression largely neglected by Alexandria was philosophy (so that not only were very few philosophers active in Alexandria, it also appears that, throughout the Hellenistic era, very few intellectuals moved from Alexandria to Athens–otherwise a pan-Mediterranean magnet–to pursue studies in philosophy). Alexandrian and Athenian cultures were exclusive and complementary: philosophy, perceived as a repository of a certain Hellenic past tradition, in Athens; new generic forms that engage more experimentally with tradition, in particular scientific ones, in Alexandria.

So, the Hellenistic world. The spatial and political relations underlying this duality came apart as Rome came to assume a more direct rule over the Mediterranean. Athens got broken; Alexandria lost its primacy. The mutually defining pair was eroded at both hands. And so one finds, at around the middle of the first century BCE, a new phenomenon: philosophy in Alexandria. (Why not? Philosophy was no longer Athenian; Alexandria was no longer the anti-Athens). Above all, one finds a new arrangement of the disciplines, first of all visible in Rome's rapid domestication of Greek knowledge in the first century BCE but then replicated across the Greek Mediterranean through following centuries down to the Antonine heyday of the empire. Roman poets recover the ground achieved by their Alexandrian predecessors, but with a much closer attention to Athenian philosophy (Lucretius being but the most obvious example of this Athenian and Alexandrian mélange). Philosophers recover their faith in science and the philosophical traditions most intent on scientific practice–dogmatic Platonism as well as Aristotelianism–are finally recovered after a long period neglect; Stoicism itself becomes Platonized. And science, finally, comes to be much more focused on its potential significance within a wider philosophical project.

The mathematics of Euclid was "philosophical" as a display piece of a methodological concern. This would be different in the Roman synthesis, which reacts after all not directly to the epistemological debates of Plato and of Aristotle (however much these were revived) but to the accumulated philosophical traditions of several centuries since. For mathematics to be of interest to a philosophically educated public, it would have above all to tell it something about the world and man's place in it. It is typical that whereas much of the evidence we have covered so far was mostly from pure geometry, much of the evidence we have from this period is in astronomy, the most obviously "philosophical" of all mathematical disciplines. This is the era of Ptolemy (but also of Geminus and likely Cleomedes), as well as of Hero and Vitruvius. The exact sciences are brought into connection with philosophy and in doing so are brought into connection with the world.

Late Ancient Synthesis: Fourth to Sixth Centuries CE

The Roman world went through a political and economic crisis in the third century CE. Such cultural activity as did take place during that period was invested in the growing sophistication of new religious forms, above all Christianity (but also in the rise of, say, Neo-Platonism). Local elites in Oxyrrhynchus and, apparently, everywhere else, went on studying their Homer, Demosthenes, Euripides, but this continuous transmission of pagan cultural monuments was not accompanied by a continuous practice of new contributions to that tradition.

In the fourth century CE accommodation was achieved: Christianity became a stable church, a state one at that; at the same time, the pagan monuments came back to attract a more direct, creative engagement. The rupture was real enough and it also involved some deep material transformations. The world of the papyrus roll was fast disappearing, replaced by a new world of the parchment codex. Less common and more heavily invested (and now sharing in the reflected glory of the book form in the culture of scriptural religion), the book came to be the dominant arena of intellectual life.

From now on, intellectual life would take mostly the form of re-engaging with past texts, in the effort to remake them textually: by collecting, excerpting and above all attaching commentary. In the exact sciences, the most interesting and creative author–Pappus–dedicates himself to a reworking as well as excerpting of past achievements, occasionally in the explicit form of commentary. Eutocius writes commentaries to Archimedes and Apollonius; Proclus and others–to Euclid.

This then is a culture of second-order texts, a "deuteronomic" culture as I have characterized it in Netz (2004). There, I emphasized the ways in which such an avowedly conservative project–seeking to maintain the study of a canonical group of texts–ends up defining a new form of science. Conservatism is a force for change. Archimedes, after all, was not interested in creating seamless wholes. To the contrary, his works reveled in the jarring juxtaposition of diverse elements. The effort to fit the heritage of Greek mathematics into a single systematic format would have to transform such sources. The net result of Late Ancient mathematics, then, is a certain "Euclideanized" mathematics, foregrounding more elementary, methodical and well-structured works. (It may well be, indeed, that the central role of Euclid's *Elements* as in some sense emblematic of Greek mathematics as a whole dates to this period and not before). But the concern with method and structure does not answer primarily to philosophical motivations. The motivation is much more direct, in method and structure as such, textual strategies valued in and of themselves. This is not a period of axiomatic; it is a period of the cross-reference. A long way from Hippocrates of Chios.

The Ancient Exact Sciences: Form and its Unity

Hippocrates, Euclid, Archimedes, Ptolemy, Pappus: the differences seem to overwhelm the similarities. True, we may broadly divide the periods into two groups, "surprise" and "structure" (mathematics is about structured surprise; sometimes it emphasizes surprise and sometimes structure). Pre-Socratic and Hellenistic mathematics were more about *surprise*; Philosophical, Roman and Late Ancient Mathematics were more about *structure*. But this broad pattern hides the many differences. While we cannot say much with any certainty about Pre-Socratic mathematics (so that it is not safe to try and demarcate it from Hellenistic mathematics in any specific way, beyond the presumption that, had we known more of it, more would be known of the historical specificity of Pre-Socratic mathematics, hence of its difference from the Hellenistic), it is clear that Euclid was no Ptolemy, who was no Pappus. All three, perhaps, reacted to philosophy: Euclid, by presenting his field as a model of method; Ptolemy, by engaging with the physical world; Pappus, by demarcating his own field, mathematics, against that of the philosophers, and insisting upon its own internal well-ordered nature. The projects are different; the things you look at are different; the end-products are different. These are different *mathematicses*.

One might be tempted to say that we call those various things *one* merely for external, misleading reasons. Perhaps we call all of these "mathematics" for no reason other than the fact that all of them share a family resemblance with a certain practice which today, within our own institutions, has a form locus–the mathematics curriculum. We have an institution, hence a discipline; but the ancient did not have the institutional framework for the exact sciences, hence they did not share our own division into disciplines or indeed our disciplines tout court. There was no such thing as ancient mathematics, then: just a plurality of things we call such.

Tempting, and wrong. It is clear that ancient mathematics was a reality–that this was a well defined field, clearly understood as such by its practitioners. It is also clear that it was not supported by institutions. Indeed, if we need to divest ourselves of an anachronistic modern assumption, then this is not our notion of a well defined discipline, but rather our presupposition that disciplines are defined by institutions. In antiquity, disciplines were real enough, and they were defined by genres. And it is such–as a literary genre–that the ancient identity of mathematics is undeniable.

The way of writing associated with the ancient exact sciences is easy to characterize (this formed the subject of Netz, 1999). Verbal text is punctuated by visual diagrams marked by letters of the alphabet standing on isolated lines or on points where lines intersect. The text itself refers to the diagram (that follows it) via the letters, calling out the objects via their diagrammatic labels but also through a system of abbreviations based on them: thus e.g. "the by the ABC" may mean (depending on the gender and number of the article, both marked in Ancient Greek) "the rectangular area contained by the straight lines AB, BC". Most terms are elided, resulting in a highly concise, regimented use of language. This formulaic aspect extends to the language as a whole so that there are essentially no terms of natural Greek: with hardly an exception, every word emerges out of a fixed stock of phrases (perhaps several hundreds for the entirety of Greek mathematics and sometimes no more than several dozen for an individual treatise). Both aspects–reliance upon diagrams as well as the formulaic language–mark out the text of the Greek exact sciences sharply against those of any other ancient genre. The very language is different; as indeed is the very form of textuality (Greek writing, anywhere else, is free of visual illustration and is most typically conceived as a performative text in natural, spoken Greek).

Not all texts in the ancient exact sciences are composed of diagrammatic-formulaic texts. Euclid comes closest, with the barest discursive introductions in the form of definitions and other axiomatic statements alone; in Hellenistic mathematics, discursive introductions are sometimes longer and may also combine with the main text in richer ways. In later authors, one very often finds more discursive passages, whether in the more philosophical and physical reflections of Ptolemy or in the general statements of Pappus. Such passages hardly change the character of the diagrammatic-formulaic passages which remain mostly self-enclosed (more discursive claims are usually made preceding or following the technical passages). The formulaic expressions remain the same across the thousand years of Greek mathematics, or indeed for two thousand years, if we wish to extend our view to later Byzantine sources as well as to the translated forms in Arabic, Hebrew or Latin. The characteristics of the diagram do not essentially change throughout this period. In the west, Greek mathematics is the most stable genre.

As mentioned above, it is rare to find treatises consisting entirely of diagrammatic-formulaic passages. Such passages form the great bulk of pure geometrical works, especially preceding the Roman synthesis. They are more closely balanced with the rest of the text in later works. We can therefore characterize the Greek exact sciences as follows. They are the Greek writings where a particular, highly marked form of text is encrusted among other textual practices. Greek mathematics, in other words, is always the same in terms of its atomic building blocks–the individual diagrammatic-formulaic passage or, indeed, simply the individual proposition; it changes, however, and quite dramatically so, in terms of the context in which such propositions appear: their discursive neighbourhood, and the overall purpose for which they are marshaled. The duality of unity and diversity, therefore, can be summed up neatly. On the one hand, unchanging atoms–propositions. On the other hand, dramatically changing arrangements within which such atoms may appear.

Hippocrates of Chios produced proofs of results concerning the quadrature of lunules, perhaps in the complete form we are familiar with from later Greek mathematics (or perhaps in a simpler, earlier form). The proof was there for a purpose, which might have been *simply to impress*. And if so, it was very different from the purpose to which Euclid has marshaled his own proofs–clearly, in some sense, this was so as to arrange and elaborate an axiomatic order. The proofs, as such, would be similar in character–the same diagram (perhaps not yet lettered in Hippocrates' case), the same formulaic language. And yet the function would have been very different. The context within the treatise (perhaps minimal in the case of Hippocrates, consisting of a very tight structure in the case of Euclid) would imply a context outside of the treatise: Hippocrates would belong in a world of competing masters of wisdom; Euclid, in a world of emerging philosophical schools.

Different again would have been the function of the propositions in Archimedes, Ptolemy or Pappus. Archimedes arranges his proofs in complex, winding structures of surprise. Outside the treatise, this puts him in the context of Alexandrian literature. Ptolemy most directly positions his proofs together with philosophical and physical reflections, putting himself in the context of contemporary philosophical or perhaps encyclopedic scholarship. Pappus, finally, with his second-order comments on the correctness of procedures and their place within wider systems, turns individual proofs into participants in a larger project of systematization. This, finally, puts him in the context of Late Ancient commentary culture. And yet take out of context an individual proof, from the lunules of Hipocrates down to Pappus' domain of analysis, and they are essentially the same.

Are we to decide on a balance? To say that Greek mathematics was rather more "the same always" or rather more "not the same always", finding the mean between the two observations made so far–historical variety as well as formal uniformity? But this would be to miss the point: historical variety is *underwritten* by formal uniformity. We can say that Greek mathematics changes, precisely because there is a thing that undergoes change and remains the same. Remove the formal continuity of the Greek mathematical proposition–and would we even be justified in considering Archimedes' measurements of curved objects as well as Ptolemy studies of stars, rays and harmonies, as belonging to the same enterprise? Perhaps we might but, then, such a judgment would indeed reflect mostly our anachronistic lumping together of various separate ancient projects that just happen to belong to a single modern one. But there is no doubt that the two did belong together, historically, and this is because of their shared formal properties. And thus one could have the same thing, "the ancient exact sciences" (by which we mean writing incorporating the diagrammatic-formulaic genre), belonging to very different projects.

The issue is almost a matter of logical necessity. For the same thing to be different, there has to be a clear sense in which it is the same. Thus formal sameness can underwrite functional difference. And this logical necessity has an important historical corollary: the strict formal invariance of the Greek exact sciences allowed them to display a great diversity in terms of context and function, without losing any of their sense of unity. Greek mathematics could be maximally diverse, because it was maximally unified.

Consequences of Diversity and Unity

It is quite a generalization, to consider the entirety of Greek mathematics; but let us move on to contextualize this further, bringing in the entirety of Greek knowledge. This may be usefully considered in three main compartments–besides the exact sciences, there is Greek medicine, and then there is Greek philosophy.

In both those other fields of knowledge, we will also find diversity and unity. Indeed we can use to a large extant the same periodization and notice the difference between a Pre-Socratic Hippocrates of Cos, or Democritus–and a Roman Galen, or Alexander of Aphrodisias. The projects are extraordinarily different. And yet there is also no question that they are identified as such–unique traditions encompassing, within each, a major diversity.

So much is similar. Beyond that, medicine and philosophy would appear to be more similar to each other, and more distinct from the exact sciences. First of all, their identity is not provided by an underlying stable form. There are only very weak generic requirements from an ancient philosophical or medical text: the works in the Hippocratic corpus range from speeches such as *On the Art* to case histories such as in the *Epidemics*; philosophy could have been turned into didactic epic (by Lucretius) or even satire (by Menippus). The identity would instead be formed through an intertextual relation with canonical authors: the constructed author "Hippocrates" in the case of medicine; a set of canonical authors, typically one each per Hellenistic philosophical school; the identity of those canonical authors perhaps best understood in terms of their standing in an intertextual relation with another constructed author, Socrates. (The difference then would be that the authorial presence of Hippocrates of Cos was constructed through the attribution of works, while Socrates was conveniently left as the oral author at the center of the philosophical written tradition).

So here is one observation: medicine and philosophy unite differently from mathematics. Mathematics unites through its formal style; medicine and philosophy unite through their intertextuality with canonical authors. (For this reason, mathematics made less of the need of a canonical author; it only came late to define Euclid, effectively, in such a way–who was by no means a founding figure. The true founding figures of Greek mathematics–perhaps Eudoxus above all?–have simply disappeared).

Another observation is that medicine and philosophy divide differently from mathematics. I did note systematic differences in the way mathematics was pursued, locating them chronologically: at different periods, mathematics was part of different projects. But while one can say the same about medicine and philosophy, this would not be the most natural way to divide them. The ancients themselves divided both medicine as well as philosophy vertically rather than horizontally, as it were, across periods and not according to periods: there were different flavors of medicine (Galen liked to talk about Rationalists, Methodists and Empiricists) as well as of philosophy (the four major Athenian schools, Academy, Peripatos,

Stoa and Garden were a genuine institutional divide and represented genuine intellectual boundaries). The philosophical division is more clear-cut and is more obviously independent of periodization, but it remains clear that at any given point, no one was in antiquity merely a "philosopher", merely a "theoretical medical author". One was instead a philosopher or a physician within a particular tradition: a Cynic, a Pnematicist, a New-Platonist or a follower of Galen. At least from the fourth century BCE onwards, there was no other way in which to be an author in the fields of philosophy or of medicine.

The ways in which philosophy and medicine, on the one hand, and mathematics, on the other hand, united and divided, are of course interrelated. Philosophy and medicine united relative to a certain body of contents and so quite naturally were defined intertextually (towards a shared source of contents), and divided by their contents (the typical ways in which they added their own content to that shared source). Mathematics united through its form and for this reason did not need a canonical author and did not divide according to the different kinds of content it provided. Indeed, from the point of view of the practitioners themselves, it did not really divide. It is my interpretation, to note differences according to periods; but from the point of view of ancient practitioners, ancient mathematicians were just that: they were "just mathematicians". And this is exactly the kind of thing ancient philosophers or medical authors were not. And so, there were no mathematicians "of a certain school": there were no mathematical schools. As it were, the identity based on formal genre did not merely substitute identity based on institution: it prevented the formation of any strong identity based on institutional relations.

Now, when identity is based on doctrinal school divisions, and when this is manifest to the practitioners themselves, it becomes much harder to form continuities across internal divisions. This is most apparent inside schools, in the cases where school doctrine has changed rapidly enough and visibly enough for the practitioners to notice. The most extreme case of that was with the Platonic school, whose doctrine was skeptical during much of the Hellenistic period. As the school reverted back into Platonic dogma, from the first century BCE onwards, the skeptical phase was erased from its continuous tradition. There was no sense in which being a Platonist in the third century CE involved any reference to what was being a Platonist in the third century BCE. The skeptical authors of the academy were simply forgotten by Neo-platonists and, quite simply, were no longer circulated among Platonists. This indeed is a more general phenomenon: as schools went out of fashion, their works were not absorbed into other schools but instead were generally forgotten. Not only the Skeptics, but also to a large extant the Epicureans and indeed also the Stoics were gradually dropped from circulation; what survived as the Greek philosophical tradition was a particular synthesis of Platonism which also included a component of Aristotlianism: so Plato, Aristotle and (above all) their Late Ancient commentators. Nearly everything we know about the rest of Greek philosophy has to be pieced together from testimonies from within this tradition, as well as stray works that happen to survive through other interests, perhaps literary or historical.

A similar process happened with Greek medicine, with the difference that schools were mostly short-lasting and local. One would have the followers of Erasistratus or of Herophilus, but the school identity such authors formed would not last for more than a few generations and as a consequence their works would be replaced by the works of other, more recent schools. In a few centers, one could still probably find (some) works by medical authors of the distant past, but such works did not form such a major part of the contemporary medical tradition, always dominated by more recent schools. Galen's division into three schools gives us a snapshot of how he viewed his contemporary medical scene; what he called "rationalism" was his own private synthesis. Following the Late Ancient decision by Oribasius and perhaps others, to elevate Galen's version of "rationalism" as the canonical form of Greek medicine, virtually any other Greek medicine was lost. What we have from antiquity is essentially Hippocrates, Galen and their Late Ancient reception. The rest are fairly random survivals, and the most important authors are known through testimony alone.

Both philosophy and medicine erased their redundant past–a past that came became redundant because, identity being formed by doctrine, extra-doctrinal works could not serve in one's own professional project. Most significantly, the redundant past, in both cases, ended up encompassing the totality of the achievement of the Hellenistic era. The vast production of the philosophical works by the Hellenistic schools of Athens, as well as the scholarship of Alexandrian doctors, was all essentially lost from memory. What replaced those was a Roman synthesis (in the case of medicine) and a Late Ancient synthesis (in the case of philosophy). The most interested, creative period was no longer transmitted. The difference from mathematics is obvious: in mathematics, the major Hellenistic authors were preserved and indeed the most significant body of literature to survive from the Hellenistic era, in any field, is that of mathematics. And, overall, the continuous tradition of Greek mathematics ended up encompassing major works from four of the five periods I identify. Greek mathematics was at the same time Euclid, Archimedes, Ptolemy, Pappus.

Late antiquity may have foregrounded Euclid; but all four would have been equally available for later reception and equally seen as part of the ancient mathematical tradition.

We end up with a paradoxical-sounding result: because Greek mathematics was more obviously "one", it could survive in a more diverse form. A result which may appear paradoxical but is in effect quite straightforward: the identity of Greek mathematics was formal and for this reason it could be bent and extended in more ways, its identity unbroken.

Let us finally pause and consider the consequences of this result for the place of Greek mathematics in the scientific revolution.

Greek mathematics was diverse. And this was helpful for the scientific revolution in at least two ways. First, it provided the authors of the scientific revolution with a rich array of tools available from within the tradition: there was something useful about the fact that one could summon up simultaneously Euclid, Archimedes, Ptolemy and Pappus. Second, it provided the authors of the scientific revolution with a certain latitude: there was something useful about the fact that the definition of their field was sufficiently wide to allow further variations.

Mathematics, in the scientific revolution, participated in the discussion concerning method; it was useful to have Euclid as a reference point to that. It participated in a literary, rhetorical culture of invention and genius; it was useful to have Archimedes as a reference point to that. It participated in the study of nature; it was useful to have Ptolemy as a reference point to that. And it participated in an effort to bring structure into the fields of knowledge; it was useful to have Pappus as a reference point to that. All were Archimedean: Galileo maybe an Archimedean revisionist of Ptolemy; Descartes maybe an Archimedean revisionist of Pappus; Newton, finally, a Euclidean-Archimedean author of a new grand synthesis.

Qua mathematicians, they were unconstrained. True, the Ptolemaic model became enshrined as doctrinal–but this was true within the *philosophical* tradition. Mathematicians felt free to act as mathematicians–true to the tradition inherited from antiquity–while invoking new doctrines for, after all, they kept to what mathematics most fully demanded, namely the ancient mathematical form. And since the tradition encompassed many ways of encasing the ancient mathematical form within new contexts, they felt free to do so as well. Most important was the willingness to encase the geometrical proof within other, algebraical techniques. There were precedents enough for this type of freedom within Archimedes, Ptolemy or Pappus, who often put their geometrical proofs in a context containing other forms of knowledge. An Aristotelian, in the 16th century, would have to be just that–an Aristotelian. But an Archimedean could do whatever pleased him. What chance, then, to the Jesuits, confronted by Galileo?

Standing on the shoulders of giants is no easy feat. The past is not automatically useable. The past may be overbearing and stifling, demanding a degree of conservatism, enforcing a limitation of horizons, so that, instead of serving as a stepping stone to the growth of knowledge, it ends up acting as a barrier to such growth. To make a sweeping claim: such, arguably, was the historical role of the Chinese classics, a relatively monolithic synthesis allowing little space for innovation. The same could perhaps be said for much of the Western tradition. The function of Aristotle, in the west of the 16th century, was perhaps not unlike that of Confucius in the contemporary East. Philosophy, in the West as in the East, was a monolith. The West had another tradition, preserved via its form and for this reason containing, underneath its monolithic cover, a fertile diversity. This mathematical tradition came to define the scientific revolution. And so, in this indirect way, the Greek mathematical style created some of the preconditions for Newton.

References

Netz, R. (1999). *The Shaping of Deduction in Greek Mathematics: a Study in Cognitive History*, Cambridge: Cambridge University Press.

—. (2004). *The Transformation of Early Mediterranean Mathematics: From Problems to Equations*, Cambridge: Cambridge University Press.

—. (2009). *Ludic Proof: Greek Mathematics and the Alexandrian Aesthetic*, Cambridge-New York: Cambridge University Press.

MUSICAL TERMINOLOGY IN PLATO'S DIALOGUES: THE IMAGE OF CONCORD IN THE *REPUBLIC* AND IN THE *TIMAEUS*

MARIANNA TOMASELLO

The studies on Greek cosmology and the link they stress between harmonics and astronomy usually call to mind the image of "harmony of the spheres" and the same link between music, mathematics and cosmology especially working in the Pythagorean thought. What seems to us an overused "metaphor" is, however, only one way by which a more general musical model works. Besides the use of derivatives from ἁρμονία, other terms and other lexical families express, within the same model, different nuances and different perspectives in representing the order of the universe. We can think how Greeks use the same words to mean "song" and to represent the god's order, or how the same terms derived from the names of musical instruments characterize a particular event, to be aware that musical terminology stretches to different areas beyond the strictly technical meaning.[1] Within this semantic framework, I turn my attention to the lexical family of σύμφωνος and to the imagery of concord it concurs to create, not only for the higher productivity of its derivatives, and for their catachrestic use in the sense of "to agree", but also because the use of such terms in cosmology provides a

[1] Cf. Pind. fr. 32 Snell-Mahler; Pind. *Pyth.* I 2-16; Scyth. fr. 1 West; Aesch. *Prom. Vic.* 550-551; Aesch. *Ag.* 158-159, 1186-1187, 1472-1474; Aesch. *Ch.* 468; Aesch. *Eum.* 306-318, 328-333; Soph. *Tr.* 642-643; Eur. *Ph.* 1728-1731. Regarding the effectiveness of the masterful discourse in Archaic Greece I refer to Detienne (1973^2); Cozzo (2001: 93); Carastro (2006, 24-26, 121 ff). The role of music in Aeschylean theatre, especially to characterize characters and events, is analysed in Scott (1984). Moreover, it has to be noticed that *Lyra* is still the name of a constellation, and that the association of the musical scale to the planetary system goes back to the Pythagoreans and has evidence in the Orphic-Pythagorean poem "Lyra": cf. test. 107 Kern=ps.-Lucian. *Astr.* 10; West (1983: 40-44).

complementary perspective to the one in harmony. The aim of my work is therefore to investigate such perspective in its main features, and to consider especially its relevance in two of Plato's dialogues, the *Republic* and the *Timaeus*:[2] the link between macro and micro-cosmos shown by these texts is useful in fact for the knowledge of physical phenomena.

The nature of cosmological science and the effort to give reason to realities which are far away from human perception make necessary, in ancient times as well as nowadays, the use of theoretical models: what I present is therefore only one of the possible images of universe, and, failing to indicate the only and universal way of thought, it has the feature to hold together several aspects of Greek scientific thought, making easier the dialogue between different areas of research. If within the Platonic thought the meeting between harmonics and astronomy, and the mathematical basis on which it stands are a legacy of the Pythagorean tradition, the possibility of identifying a *trait d'union* between celestial events and earthly ones is an effective way for understanding the physical phenomena, which can be definitely enriched by the model of concord.

Most scholars consider the theoretical constructs of the *Republic* and the *Timaeus* the first attempt to give mathematical basis to physics: since the work of Geoffrey Lloyd on the teleological approach in the study of nature, Plato's conception of science and scientific knowledge is much discussed, being subject to the legacy of the Platonism (Lloyd, 1978; cf. Anton, 1980). The point at issue is not only the interest of Plato in the research on nature, but the real effectiveness of such a research and the possibility that it can reach any achievement. The exhortation in book VII of the *Republic* to turn to the real astronomy, as well as the establishment in Plato's thought of an eidetic world which has the ontological features of reality, represent the basis of every argument against the possibility to get at truth throughout the observation of physical phenomena. Neither dialogue, though, shows a real quantitative analysis of physical phenomena, and even within the *cursus studiorum* of the Book VII of the *Republic* mathematics gains only the level of *dianoia*. However, the close link between the study of nature and mathematics in the first dialogue, and the unifying character of the movements of the world soul, which allows the teleological vision of the second one, are sufficient grounds to affirm not only the interest of Plato towards the study of nature, but also the

---

[2] The translations of the passages from the *Republic* and the *Timaeus* are respectively taken from *Plato in Twelve Volumes, Vols. 5*-6, translated by Paul Shorey. Cambridge, MA: Harvard University Press, 1969, and *Plato in Twelve Volumes, Vol. 9*, translated by W.R.M. Lamb, Cambridge, MA: Harvard University Press, 1925.

possibility that this study has a cognitive power (Lloyd, 1978; Sedley, 1997).

In representing sensible phenomena, as well as in describing their knowledge, music plays a central role, both as a discipline of the educational process, and as a model to understand the universe. The most quoted reference to astronomy in Plato's philosophy is *Rep.* 527d-530c, and its description as a subject-matter belonging to a *cursus studiorum* in book VII: after a few lines, in 530e-531c, harmonics appears as well as a subject-matter included in the plan for the educational training of warriors, and, in the same manner of astronomy, it is subjected to a detailed evaluation, which picks out not only the difference between a mathematical and an empiricist approach, but also the role of the former in the achievement of a comprehensive understanding of the universe. This passage has been especially studied with regard to its last statement and to the characterization of different methods in musical science (Meriani, 2003; Barker, 2007: 23 ff). As in the definition of the real astronomy, also in the account of its ἀντίστροφος Socrates points to the need for a higher consideration of formal aspects, leaving aside sensible data and focusing on mathematical rules beyond them. For this reason scholars have seen in Plato the first attempt to give mathematical basis to the study of nature. As harmonics concerns, this particular point of view identifies an opposition between an empirical approach, based on auditory perception and useful to provide a spatial representation of the scale, and a mathematical one, typical of the Pythagoreans and aimed at finding ratios that are able to give a rational representation of scale and intervals.[3] So, if the main purpose of astronomy is to draw a diagram of the movements of stars and, consequently, of their mutual relations, harmonics aims to build a system able to explain the laws of music composition, that is the scale which divides sounds into intervals according to their pitch.

[3] The other mentions of the Pythagoreans occur in *Rep.* 530d and 600d, but they don't provide a detailed account of the Pythagorean thought. A contrary opinion is expressed just with regards to *Rep.* 531c-d by Meriani, which affirms that "questo è l'unico luogo della *Repubblica* nel quale i Pitagorici vengono esplicitamente citati come gruppo di studiosi con una visione unitaria di un problema–si direbbe come 'scuola'" (Meriani, 2003: 566). Regarding the relationship between Plato and the Pythagorean, cf. Barbera (1981). However, *Rep.* 531c-d, rather than the vague reference to a Pythagorean mathematical approach, seems to show the awareness of the research of Archytas of Taranto in harmonics and mathematics. Cf. Barker (1978); Barker (1989: 46 ff); Barker (2007: n. 14, p. 316); Burnyeat (2000: 47 ff). The relationship between Plato and Archytas is also the topic of Lloyd (1990).

Not only the theorists, but also the musicians, described scalar systems by means of relations between the intervals of a tetrachord: the same ratios were as useful for the characterization of different scales in harmonics as for the definition of a practical way of attuning, constituting the so called "method of concordance" (Burkert, 1972: 389 ff; Barker, 1989: 49 ff; Huffman, 1993: 74 ff.; Huffman 2005: 418 ff). Such a relationship between music and mathematics isn't therefore only a legacy of Pythagorean thought, but it is deep-rooted in musical practice, shaping both the terminology and the topics of the two sciences: according to the Pythagoreans, in fact, the most important concords are the ones of fourth, fifth and octave, not only because they represent whole ratios, but also because such ratios are included within the *tetraktys* (Huffmann, 2005; Barker, 2007; Riedweg, 2002: 108 ff).[4] In virtue of such features they are called just συμφωνίαι, and it is because of this that they are taken as an example by Plato. Besides the role of concords in describing the scalar system, the debate between opposed schools is just on the superiority of mathematical laws or sense to determine which intervals were concordant and which were not.

An evidence of Plato's awareness of such a debate is just *Rep.* 530e-531c, especially the criticism towards the empiricists, who "rack the chords" in order to measure audible concords.[5] Nevertheless, Plato's disapproval is directed to those who search for number in these audible concords, making clear that his position is not completely comparable to the Pythagorean one and, in general, to the perspective of those who consider λόγος as an absolute and objective criterion in opposition to the

[4] The correspondence between concordant intervals and multiple and epimoric ratios proves, according to the Pythagoreans, the "beauty" of numerical ratio, and, at the same time, it is the evidence that the reason of concord consists in its numerical properties. Cf. Meriani (2003: 582 ff), concluding that "lo stretto collegamento tra l'eccellenza matematica di questi rapporti e l'eccellenza sensoriale esercitata sull'udito dalle consonanze indusse a ritenere che la prima fosse controparte formale della seconda, e che fosse in grado di spiegarla" (Meriani, 2003: 590).

[5] The Greek term for "rack" is βασανίζειν, that, together with κόλλοψ, alludes to the legal metaphor. The same idea is still in the image of chords' denial (*Rep.* 531b 5-6). According to Barker this metaphor expresses "untrustworthiness of sense-perception" (Barker, 1989: n. 4, p. 56); concerning the insistence on the empirical nature of measurement techniques, cf. Meriani (2003: 593 ff). Moreover, the same passage refers to a spatial representation of sound, displayed also in πυκνώματα, as Barker points out (1989: n. 3, pp. 55-56).

senses.[6] This is the point of the last line of the *Republic*'s account on harmonics and the greatest challenge of Platonic reform. Instead of the empiricist method, the main target of Plato analysis is the Pythagorean research of sound ratios. As noticed by Andrew Barker, according to Plato the Pythagoreans are still too close to sense data, not only since they start from them to identify abstract rules, but also because they return to them, failing a theoretical justification (Barker, 1989: 46 ff; Barker, 1978: 341-342). The significance of the whole passage, but also the essence of the real harmonics, is therefore founded on the understanding of what σύμφωνοι ἀριθμοί means, and in particular on the consideration of such a musical model in representing the universe. For this reason it is necessary to consider how the image of concord works in natural science, deferring at the end of the work a more careful analysis of this passage.

The reference to music in cosmological context comes back in the dialogue, albeit in the form of a myth, and the musical imagery is still used in representing the real composition of *cosmos* and the dynamics of its elements. It is the case of the myth of Er in the *Republic* X, and it will be the first step to demonstrate which is the use of derivatives from σύμφωνος in order to construct an astronomical discourse.

In the last book of the *Republic* Socrates tells about the warrior Er and what he saw in his journey in the world after death. This story is used by Plato to express his cosmological conception, and, in particular, to explain his theory about the fate of the soul after death: the essence of Plato's discourse is therefore ethical, since his aim is to establish the right reward that every soul will receive for its behaviour, and the place it will hold in an afterlife world. But beyond the eschatological reading of this myth, that works together with Plato's political project, its most meaningful part for my purpose is the description of heaven. According to the story told by Plato the afterlife world is located inside the universe, so that, as Franco Repellini explains, "i luoghi del racconto escatologico sono in effetti coordinati con i luoghi della descrizione cosmologica emergente all'epoca", and, he continues, it represents "un aspetto proprio della concezione di Platone, proprio del modo in cui egli comprende la connessione tra anima e cosmo" (Franco Repellini, 2003c: 368, 369).

This is a novelty of Plato's cosmological construction and it carries as natural consequence that the harmony between the parts of universe is the same that rules the fate of souls, and, what isn't obvious, that any

[6] This ancient debate that opposes λόγος to αἴσθησις as a criterion for the description of the musical scale, an opposition overcome by Aristoxenus' point of view, is attested e.g. in Aristot. *Anal. Post.* 78b34-79a6; Aristox. *Elem. Harm.* II 32, 18 *ss.*; Ptolom. *Harm.* I 1-2, 5-6; Porph. *In Ptolom. Harm.* 22, 22-28, 26.

discourse about "the harmony of the spheres" is closely connected to ethical issues and serves as their natural background.[7]

Although such a spatial model, organized according to a high-low scheme, is used in representing the universe even before Plato, the Athenian philosopher picks out further elements, and all of them are contained in the image of the spindle and the shafts. The reference to the structure of a spindle in order to express the frame of the universe is not only useful to the cosmopoietical discourse, but it also makes clear the eschatological values and the intimate connection with the account about justice done in the previous books. So, after describing the composition of the spindle and explaining its arrangement in eight shafts of decreasing size, placed each within the next, Plato describes its circular movement and lays on each orbit a Siren to express the sound produced by such a movement:

> and up above on each of the rims of the circles a Siren stood, borne around in its revolution and uttering one sound (φωνὴν μίαν ἱεῖσαν), one *note* (ἕνα τόνον), and from all the eight there was the concord of a single harmony (ἐκ πασῶν δὲ ὀκτὼ οὐσῶν μίαν ἁρμονίαν συμφωνεῖν). And there were another three who sat round about at equal intervals (δι' ἴσου), each one on her throne, the Fates, daughters of Necessity, clad in white vestments with filleted heads, Lachesis, and Clotho, and Atropos, who sang in unison with the music of the Sirens (ὑμνεῖν πρὸς τὴν τῶν Σειρήνων ἁρμονίαν), Lachesis singing the things that were, Clotho the things that are, and Atropos the things that are to be. And Clotho with the touch of her right hand helped to turn the outer circumference of the spindle, pausing from time to time. Atropos with her left hand in like manner helped to turn the inner circles, and Lachesis alternately with either hand lent a hand to each (*Rep.* 617b-c).

In addition to the image of *Ananke*'s spindle, that we have briefly mentioned and that alludes to *Theogony* and to the representation of Moire as they spin fate for men, the structure of *cosmos* is constructed by another symbolic field, the one of cosmic *symphonia* that I am going to analyze and that denotes the sound produced by the song of the Sirens. The use of both *topoi* is not accidental and creates an interesting *mélange*. It isn't the first time that music is associated with spinning, and more generally with

---

[7] Cf. Franco Repellini (2007), especially p. 397: "nella visione di Platone sembra esserci, come tratto caratteristico, un nesso tra il campo delle scelte e l'ordine cosmico". The scholar also adds that the reference to Necessity persuaded by *nous* in *Ti.* 48a can introduce another level for individual action in the choice of one's fate.

activities related to texture, but such an association appears also in Plato, and lies in the common ground of harmonic connection. Like music, even spinning is based on concepts of connection and unity, so that a syncretic action follows to a diacritic and analytical one, in order to weave together different elements.[8] The same reference to an idea of regular intervals and to a principle of alternation is found as well in the preposition διὰ (δι' ἴσου, διαλείπειν) and in the syntagma ἐν μέρει, which together characterize the intervals of space and time according to which the *Moire* are arranged and according to which they pluck the circles, just as they were chords. Both lexical means have therefore the same purpose to represent not only the periodicity of the cosmological motion, but also the eschatological movement of the Moire, taking part in building the idea of well-ordered connection and harmony. This alternate motion has in fact its opposite in the song of the Sirens, that represents the physical expression of cosmic and eschatological movements: even if each Siren moves with the same circular motion of the ring on which is laid, and just for this reason produces a single and personal voice, anyway the sound produced is concordant, since each voice is composed in a particular harmony.

Several interpreters have noticed an incongruity in the scene represented by the myth, since, if the voices of the Sirens were referred to the eight notes of scale, their simultaneous sound would be cacophonous, and so far from the ideal of beauty and agreement implied by σύμφωνος.[9] I think, however, that just this apparent contradiction makes clear the symbolic feature of the passage and the fact that by the image of the spindle of *Ananke* Plato does not allude to a real melody or to a rule

[8] The image of spinning and its implications in cosmology and ethics are considered by Campese (2007), who underlines its two complementary phases of *diairesis* and *synkresis* and how Plato uses them according to the context (in this regard the scholar refers to *Pol.* 282d ff.); Franco Repellini (2007: 387 ff), after noticing the cosmopoietical function of the complex setting of the myth of Er, explains that the image of the spindle is just the core of the myth, and, recalling the principle of connection proper of spinning, it expresses the global order, both of universe and of soul.

[9] A different opinion about this passages is expressed by Andrew Barker, who translates ἁρμονία as scale and explains that: "though the scalar *harmonia* is indeed sounded, it is not itself the celestial music, but constitutes the permanent framework, the reservoir of elements and relations, on which that music is based. Melody itself is moving, dynamic; the melodies of the Fates are not eternally self-same, but are musical representations of the events in time. The *harmonia*, by contrast, is eternal. It stands to the melodies rather as a preordained syntax, grammar and vocabulary might stand to the sentences of a language" (Barker, 1989: n. 11, p. 58).

quantitatively determined in the cosmic order, but instead he introduces a conceptual framework developed from a musical background. Indeed, as melodies are composed according to intervals and sounds of a given harmony, in the same way both the scheme of celestial motion and the order of universe are organized following a law of regular connection between different elements. Συμφωνεῖν μίαν ἁρμονίαν means therefore to vocalize the rules and the laws that govern the universe, emphasizing their ethical implications and indicating the concordance between the fate of the universe and that of men.[10] Although the cosmic order expressed by the myth of Er is necessary and it is effective as well in heaven as in the afterlife world, and although the movement of the sky is governed by the very harmonic law of Necessity, at the same time the sky is not described in terms of structure, that is by means of a kinematical rule that controls the well-ordered movement of the stars, but, since the main point is the achievement of justice in the world of the sensible, it is characterized by the sound that the real movements produce. So, albeit in a mythical context, Plato allusion to concord and the emphasis on voice that characterizes the whole passage betrays the attention to sensible phenomena and not only constitutes the complementary image of *harmony*, but it is also the *fil rouge* which is able to bring together sensible data and human process of understanding.

The role of concord to represent the well-ordered working of the sensible world is even more evident if we consider that, when the purpose is to describe the structure of the universe and to make clear the law that governs physical phenomena, no derivative from σύμφωνος appears, but harmony expresses the abstract and formal rule behind each execution. It is what happens in the cosmological discourse of *Ti.* 35b-36c, where the World Soul is described by specific mathematical relations. Although

[10] As far as the movement of universe is concerned, it is necessary to pay attention to *Ti.* 40c-d, where, at the end of the account of the generation of gods, Timaeus affirms that "But the choric dances of these same stars and their crossings one of another, and the relative reversals and progressions of their orbits, and which of the gods meet in their conjunctions, and how many are in opposition, and behind which and at what times they severally pass before one another and are hidden from our view, and again re-appearing send upon men unable to calculate alarming portents of the things which shall come to pass hereafter–to describe all this without an inspection of models of these movements would be labour in vain". Whereas the movements of sky seem a concord to hearing, the sense of sight perceives them as dancing: behind both cases there is the same musical model, understood in the broadest sense proper of Greek culture. Indeed, as the music is played, even the rhythm of dance is organized according to particular intervals and measures.

there isn't any term that refers explicitly to music, and although neither ἁρμονία nor anyone of its derivatives occur, the Soul is drawn by particular measures, which, if we pay attention, are nothing but the proportions that define the boundaries of four octaves, two fifths and a tone. I do not surely mean to say anything new about this passage extensively studied by Luc Brisson, I wish instead to emphasize the presence of a musical model, albeit not explicitly shown, and moreover to underline the application of the image of harmony in all those cases where Plato wants to focus on the structures and forms at the base of sensible data.

According to Brisson, the construction of the World Soul involved and the direction of her motion are useful to represent the real movements of the heavenly bodies, especially the main features of regularity and permanence that characterize ancient thinking about stars, despite the observation data. So, as Brisson explains:

> … même si cette âme doit être le principe de l'ensemble des changements ordonnés dans tout l'univers, elle a, entre autres, pour fonction de rendre compte de l'ordre des mouvements des corps célestes; en fait, en décrivant la fabrication d'une sphère armillaire par le démiurge, Platon décrit la constitution d'un modèle mathématique explicant les mouvements astronomiques (Brisson, 1992: 38; cf. Brisson, 1997[3]: 314 ff).

The use of such a mathematical scheme to justify the variety of planetary motions has suggested to scholars the reference to Eudoxus' studies, supposing an influence of the Athenian philosopher on the Academic mathematician (Knorr, 1990: 317 ff; Brisson, 1995[2]: n. 170, p. 234-235, and n. 227, p. 239; Brisson, 1997[3]: 38 ff.; Fronterrotta, 2006[2]: n. 147, p. 225). Whether Plato evidence is linked with Eudoxus or not, the armillary sphere is, however, the outcome of the effort to give geometric expression to astronomical movements, as well as the composition of the World Soul in the *Timaeus* displays the use of mathematics as a representative pattern. Moreover, the Platonic construction reflects the teleological program of the *Timaeus*, and, in virtue of the role of the soul as principle from which everything originates and to whom everything is directed, expresses the preservation of order within the whole *cosmos*.[11]

It is therefore clear that, beyond any possible import, the point at issue in *Ti.* 35b-37b is the building of a representation model. Indeed, at the end

[11] This can be clearly inferred from 34c 4-5. It is because of the very mathematical construction of the World Soul that at the beginning of his account Timaeus addresses to experts, as the only ones able to understand his topics.

of the passage, the Soul is explicitly defined λογισμοῦ μετέχουσα καὶ ἁρμονίας (36e 5-37a 1), referring to the same principles of proportion and relation that will be more precisely recalled later, when in dealing with the generation of sensible world Timaeus says that the four components of the *chora*, first arranged ἀλόγως καὶ ἀμέτρως (53a 8), are afterwards organized εἴδεσί τε καὶ ἀριθμοῖς (53b 4-5). Not only the verb to express such an organization, διασχηματίζεσθαι, refers to a structure that respects particular rules, but the whole passage affirms that a certain homogeneity resides at the base of universe, and that man is able to express it by mathematical laws. While forms refer to geometric figures that represent the movements of the soul and the four elements, numbers are the factors of the ratios that govern such movements (Brisson, 1995^2: n. 386 p. 252; Brisson, 2003: 200-201).[12]

As scholars have effectively noticed, the chance of knowledge for man is founded just on the regularity that at the same time governs the movements of the soul and the one of sensible world, and on the middle position between sensible and intelligible that the soul holds.[13] The stress on homogeneity of structure, and on the possibility of its expression by means of mathematical laws is the point of two works on the interpretation of the *Timaeus*. I refer to *Inventer l'univers. Le probleme de la connaissance et les modeles cosmologiques*, where Brisson and Meyerstein show the premises to scientific knowledge, and to the analysis conducted by Fronterotta in *Il* 'Timeo' *e la matematica* '*embodied*', where the author tries to overcome the widespread belief that the application of mathematics in the *Timaeus* is the origin of the Platonic mathematical idealism (Brisson-Meyerstein, 1991; Fronterrotta, 2007). According to both readings, not only the soul holds an intermediate position between the intelligible world and the sensible bodies, and is thus the intermediary for the transmission and the conservation of mathematical order, but the fact

---

[12] The same term σχήματα appears in the section on astronomy of book VII of the *Republic*, where ἀληθινὰ σχήματα are the diagrams that describe the law at the base of stars motion.

[13] Cf. Brisson (2003: 202-203) makes clear that the use of such a mathematical model explains not only why the soul is considered the primary cause, and, therefore, the reason of the orderly motion, but also the fact that cognitive activities are dependent on her orderly structure. Regarding the coordination between the movements of the soul and the ones of the body, a coordination made possible just by means of the same geometric and rational structure and considered from a teleological perspective, I refer to Johansen (2006^2: 3-4, 139 ff., and 102) where the same mathematical structure, glimpsed in reference to ἴχνη in 53b, is what makes possible the persuasion of *ananke* by means of *nous*.

that the same mathematical laws, that found as well the constitution of the soul and the one of the sensible world, are assumed as the base of the cosmic order, is the way to make accessible to man what would otherwise escape his senses. It is just by the encounter of its movements with the ones of bodies, in fact, that knowledge is given to men, and it is on such an epistemological perspective that the counterpoint between mathematical and harmonic laws in the structure imposed by the demiurge to the Soul is grounded.

If at first glance, the representation of the Soul and of her knowledge seems to be built mainly by mathematical terms, when in the same dialogue Plato wants to express the accomplishment of physiological processes, as well as the process of approximation to divine knowledge, the image of harmony and the laws that it displays give way to the model of concord.

Σύμφωνος and συμφωνία appear in the second part of the *Timaeus*, when, after describing composition and movements of the world soul, Timaeus goes on to analyze sensible bodies, and, within the account of senses, he ascribes hearing to the encounter of elementary particles with the organ of perception. Concord between two sounds is so discussed in 67a-c e 80a-b, and both examples explain sound phenomena from a perceptual point of view. Despite what happens in Pythagorean thought, they aren't defined by the mechanical feature of the sound flux and by the possibility expressed according to mathematical laws, but Plato is underlining just the ways through which the sound flux meets the sense organs. Whereas the first passage briefly alludes to the analysis of concord after describing hearing in purely dynamic terms, the second offers, as Timaeus had promised, a more detailed account, based on the principle of ὁμοίωσις.

The acoustic theory of Archytas, as we are allowed to read it in fr. 1, starts with the statement that "it is not possible that there be sound, if an impact of some things against one another does not occur" (Huffman transl., οὐ δυνατόν ἐστιν εἶμεν ψόφον μὴ γενηθείσας πληγᾶς τινων ποτ' ἄλλαλα). The pitch, given in absolute terms, depends only on the speed of sound movement, and the only mention to human perception is intended to justify all those cases where, although there are movements, there is no sound audible to the human ear. The nature of sound, therefore, is understood in itself, assuming that every body in motion produces sound, whether it is perceived or not. Plato's analysis, by contrast, offers a clear explanation of sound phenomena closely linked with human perception. As Huffman rightly notices about 67b, Plato makes no mention of any impact between bodies, but instead he speaks of the impact of air that moves within our body brain and blood (Huffman, 2005: 130-131, 135).[14] So, it is necessary that the movement of sound meets the one inside human perceptual system to be perceived. In this way a *symphonic* image of hearing is taking shape, and its understanding, in my opinion, is made more clear by the particular stress on the phenomenon of concord within the debate on auditory perception and sound diffusion.

In the first of the two above mentioned passages the tuning between voice and hearing is just suggested between the lines, when hearing is defined as the internal corresponding of voice movement, the topic being the perception of two different sounds as concordant. Once again human perception is assumed as specific criterion, since the question at issue is the sounds "as they appear", whether they are ἀνάρμοστοι, because of the inequality between the movements they produce in our body, or σύμφωνοι by virtue of the similarity. Hence the following description focuses on the changes that those movements, generated inside the human body, are subjected to, after their impact with the sound flux. Concord depends in fact on speed and it consists on the possibility that the faster sound reaches the slower. In the case of two sounds that are linked by a relationship of concord, the movement of the high sound and that of the low one will not interfere with each other, but, since they move in the same direction (ἐληλυθυίαι εἰς ὅμοιον), will merge and produce a single perception. The high-pitched sounds and the low ones are finally concordant when their movement achieves the same speed, striving in the same way to the sense organ, but they are *discordant* when their movements do not achieve this uniform course. The emphasis on the principle of similarity and on the tension towards the same goal, shown by Plato by means of his insistence on ὅμοιος and on the image of *reductio ad unum*, is useful for the description of perceptual phenomena, and it makes clear not only their dynamic nature, but also underlines the approximation between the object perceived and the organ of perception.

The dynamic character of hearing physiology, but more importantly its description in *symphonic* terms, together with the mathematical structure, common to different areas of the sensible world, represents a further basis for the possibility of human knowledge. The same principle of similarity, which seems to be so important for understanding perception, appears again in the description of the assimilation to the divine, and unifies under the same sign the account of both epistemological *and* ethical issues. Timaeus, in lines 80a-b, concludes, in fact, connecting the principle of

[14] Moreover, Huffman underlines that the term used is φωνή which, unlike the more general ψόφος, usually means "voice".

similarity to the process that describes man's participation in the divine, and in lines b 5-8 he notices that, by means of achieving uniformity, the high and low-pitched sounds:

> ...blend in one single sensation, furnishing pleasure (ἡδονή) thereby to the unintelligent (τοῖς ἄφροσιν), and to the intelligent (τοῖς ἔμφροσιν) that intellectual delight (εὐφροσύνη) which is caused by the imitation of the divine harmony manifested in mortal motions (διὰ τὴν τῆς θείας ἁρμονίας μίμησιν ἐν θνηταῖς γενομένην φοραῖς).

The perception of concord that may look like just a sense phenomenon, can therefore lead, for those who have the ears to hear, to the imitation of divine harmony. While in this passage Timaeus refers fast to harmony and does not further specify its nature, the same harmonies are reminded elsewhere in the dialogue, when the debate is focused on the order of universe. Further back in the text and always in connection to the sense of hearing, the reference to harmony returns in lines 47b-e, when Plato introduces the difference between intelligent causes and contributory ones in order to display the work of sight and hearing. Although we apparently see by means of the encounter between the particles of fire outside and those that, together with other elements, constitute our body, the real purpose of sight is to allow us the vision and the understanding of the movements of divine reason. Indeed, the god gives eyes to men in order to look at the circular movements of *nous*, and, because of the similarity between them and the movements of human *dianoesis*, to regulate the latter "après avoir acquis le pouvoir de les calculer correctement en conformité avec ce qui passe dans la nature (47c 1-2 ἐκμαθόντες δὲ καὶ λογισμῶν κατὰ φύσιν ὀρθότης μετασχόντες)" (Brisson, 1992: 144). The same arguments and the same participation in a principle of rationality and in the correctness of reasoning by nature are discussed in regard to hearing, taking into account also the voice in a global view of sound.

> Concerning sound also and hearing (φωνῆς τε δὴ καὶ ἀκοῆς), once more we make the same declaration, that they were bestowed by the gods with the same object and for the same reasons; for it was for these same purposes that speech (λόγος) was ordained (τέτακται), and it makes the greatest contribution thereto; music too, in so far as it uses audible sound (ὅσον τ αὖ μοθσικῆς φωνῇ χρήσιμον), was bestowed for the sake of harmony (πρὸς ἀκοὴν ἕνεκα ἁρμονίας). And harmony, which has motions akin to the revolutions of the Soul within us (συγγενεῖς ἔχουσα φορὰς ταῖς ἐν ἡμῖν τῆς ψυχῆς περιόδοις), was given by the Muses to him who makes intelligent use of the Muses (τῷ μετὰ προσχρωμένῳ), not as an aid to irrational pleasure (ἐφ' ἡδονὴν ἄλογον), as is now supposed, but as

> an auxiliary (σύμμαχος) to the inner revolution of the Soul (ἐπὶ τὴν γεγονυῖαν ἐν ἡμῖν ἀνάρμοστον ψυχῆς περίοδον), when it has lost its harmony, to assist in restoring it to order and concord with itself (εἰς κατακόσμησιν καὶ συμφωνίαν ἑαυτῇ). And because of the unmodulated condition, deficient in grace, which exists in most of us (διὰ τὴν ἄμετρον ἐν ἡμῖν καὶ χαρίτων ἐπιδεᾶ γιγνομένην ἐν τοῖς πλείστοις ἕξιν), rhythm also was bestowed upon us to be our helper (ἐπίκουρος) by the same deities and for the same ends (*Ti.* 47d-e).

The link between human and cosmic perspective is clearly expressed in the text, as well as using the same terminology to describe the physiological movements and the ones of heaven, and in the definition of περίοδοι as συγγενεῖς: so, it is just because of such a dynamic characterization and of the stress on the homogeneity of the movements that affect the physical phenomena, that, in my view, the assimilation to the divine can be better understood by the reference to the image of concord.[15] While in this passage the approximation of the movements of the soul to the divine harmony is only hinted at, and the reference is rather to a principle of order and proportional relationship than to knowledge, the same definition in terms of similarity appears at the end of the dialogue,

---

[15] Not only the heavenly revolutions and the soul ones, but also the movements of sound diffusion are συγγενεῖς. For this reason hearing more clearly represents the way through which balance and order are achieved. Such an opinion is expressed by Brisson (1995^2: n. 322, p. 246), who explains that "les mouvements de notre âme et donc ceux des astres obéissent aux même lois que ceux qui produisent le son, étant donné la structure mathématique de l'âme du monde". Cf. Miller (1957) and Maso (2003: 251 ff), for a general account of the application of the conception of movement to represent the human body and its physiology. The harmony mentioned in the passage is, then, the common direction and the *homoiotes* that concordant sounds achieve, as well as the gradual assimilation between sound and hearing. In this way, also the meaning of ἄλογος ἡδονή becomes clearer, since by means of hearing musical harmony, as noticed Milena Bontempi, man "si dispone, insomma, al riconoscimento e all'ascolto del *lògos* vero che è in definitiva l'opera musicale e l'armonia somma emessa dalle corde della sua anima. [...] L'armonia prodotta dalle corde e la vibrazione che la produce sono, in musica, un unico e medesimo atto: allo stesso modo, si può dire, per l'anima l'attivazione dei moti ad essa propri e il *lògos* che ne risulta si danno assieme. Movimento dell'anima ed esercizio del *lògos* [...] sono uno ed un solo processo" (Bontempi, 2007: 129). As regards the reference to a linguistic principle in the composition of the *cosmos*, I refer to Casertano (2003: 38, 45 ff). Moreover, the connection between λόγος, motion, and musical harmony also occurs in *Ti.* 30a and 53 a, where the adverbial syntagmata "πλημμελῶς καὶ ἀτάκτως" and "ἀλόγως καὶ ἀτάκτως" express the primordial disorder.

when, in reference to diseases of soul and their treatment, the ὁμοίωσις θεῷ and the process of assimilation to the movements of the immortal soul are made the basis not only of understanding but also of achieving the state of *eudaimonia* typical of the divinity itself.[16]

As explicitly stated in 90d, corrupted movements of the individual soul not only are rectified (ἐξορθοῦν) by learning heavenly harmonies and revolutions (διὰ τὸ καταμανθάνειν τὰς τοῦ παντὸς ἁρμονίας καὶ περιφοράς), but, man, "making the part that thinks like unto the object of its thought" (κατανοοθμένῳ τὸ κατανοοῦν ἐξομοιῶσαι), achieves just by virtue of this likeness the better life set before by gods (Sedley, 1997: 330-332).

According to a harmonic relationship–even before that between body and soul–between the parts of the soul itself, the process of likeness to the divine order is achieved by means of understanding the whole harmonic motion. To balance the tripartite soul, in fact, it is necessary that a part dominates the others, so, whereas a soul ruled by its mortal part produces mortal thoughts, the rule of the divine part provides a well-working mind that is able to grasp truth (Joubaud, 1991: p. 234; cf. *Thaeet.* 153b-c). The role of the divine part of the soul and the assimilation of its movements to cosmic harmony point thus towards pedagogical issues, and place music as well as gymnastics within the cosmological context, making clear the interaction of microcosm and macrocosm.[17] Moreover, the representation of such a motion of assimilation is described by the same terms (περιφοραί and περίοδοι) used in the account of sound perception. In this way, the likeness to god not only realizes, as noticed by Casertano, the

[16] Plato explicitly refers to the likeness to god, as well as in the passage quoted, in *Symp.* 207e-209e, *Theaet.* 176b-c and *Rep.* 611d. Although Sedley notices the lack of studies, the same topic has been recently developed, in particular with reference to Hellenistic and Platonic philosophy: cf. Lavecchia (2006); Russel (2004); Mahoney (2005); Thomas (2006); Brown (2008).

[17] Moreover, the reference to *paideia* is useful to describe the rule of the divine part of the soul: in 89 d, in fact, terms as διαπαιδαγωγεῖν, παιδαγωγεῖν and παιδαγωγία are used in order to speak about the workings of the soul. Cfr, Joubaud (1991: 246 ff); Brisson (1995^2: 243), especially n. 280, where the scholar, regarding the common nature of soul and universe movements expressed in 47d, points out the importance of education for the attainment of virtue, and pays attention to 87b and 88b, besides 90d, as further illustrations of the development of this topic. As regards the role of music and gymnastics in the *Timaeus* and in the *Republic*, cf. Johansen (2006^2: 156): "here the aim is not to affect the soul through the body, at least not directly, but to keep a proportion between the motions of the soul and those of the body independently of each other. This assumes a degree of interest in bodily process independently of their relation to the soul."

same cooperation between causes both in human psychology, as in the whole universe, and so stresses the relationship between cosmology and anthropology, physics and ethics, typical of the dialogue (Casertano, 2003: p. 61)[18], but by virtue of such a dynamic image it also underlines the reference to the broader semantic field of concord and to its capability to encompass both the description of acoustic phenomena, and the representation of the working soul and of the processes of knowing.

The reference to ὁμοίωσις θεῷ and the words that describe the likeness to god lead us to ethical issues and so make clearer the hint at concord in book VII of the *Republic*, when Socrates defines the real astronomy and the real harmonics as subjects taught withinthe *cursus studiorum* of his *kallipolis*. The relationship between astronomy and harmonics as it is established in the *Republic* is, in fact, part of the pedagogical program through which warriors must ensure the acquisition of δύναμις τῶν μαθήνατων, that is the capacity to reduce sensible multiplicity to unity. After music and gymnastics, that are the *technai* able to rectify the movements of the soul and the ones of the body, the *cursus* provides the study of mathematics, whose primary purpose is to turn the eyes from sensible events to abstract laws. If music, as sound performance, is relegated to the first level of education, its science, together with astronomy, represents the last degree of knowledge before dialectic. The relation between musical science and astronomy, as noticed in the beginning of this work, is founded on the common connection with mathematics and on the ability to discern figures and proportions in the observation of sound movements on the one hand, and of celestial ones on the other. The fact that the study of planetary motions, as well as the one of sounds, must go beyond the observation of empirical data, and that it must search for the mathematical rules on which the multiplicity of real events is based (*Rep.* 529b ff, 530d ff), however, doesn't lead to a purely mathematical understanding of physical phenomena. As we can read in *Rep.* 530e-531c in regard to harmonics, the criticism of those followers of the empiricist school, which torture the strings to find the smallest audible

[18] Casertano (2003: 47) notices the same interaction of intelligent and contributory causes on the human level in *Phaed.* 97c-99b, and concludes that the introduction of a telelogical explanation of every ποιεῖν involves "l'apertura di un orizzonte per così dire 'etico' che l'uomo deve necessariamente inserire nella sua visione della realtà ai fini del *suo* agire per il meglio." Regarding the relationship between cosmology and human nature, cf. Natali (2003: 231), where the scholar speaks of antropology in the *Timaeus* as a link between cosmology and politics and "perno centrale di tutta l'economia dell'esposizione di Timeo."

interval, doesn't suggest hunting the numbers, but urges to investigate their nature.

> Their method exactly corresponds (ταὐτὸν γὰρ ποιοῦσι) to that of the astronomer; for the numbers they seek (τοὺς ἀριθμοὺς ζητοῦσιν) are those found in these heard concords (ἐν ταύταις ταῖς συμφωνίαις ταῖς ἀκουμέναις), but they do not ascend to generalized problems (οὐκ εἰς πρόβληματα ἀνίασιν) nor consider (ἐπισκοπεῖν) which numbers are inherently concordant and which are not (τίνες σύμφωνοι ἀριθμοὶ καὶ τίνες οὔ), and why in each case (διὰ τί ἑκάτεροι) (*Rep.* 530e-531c).

If the need to search for concordant numbers does not openly refer to the epistemological process and to its ethical values, the reading of the teleological discourse in the *Timaeus* and the harmony between the subject and the object of knowledge that it underlines, make clear attention to sense phenomenon also in the representation of the universe.

Looking at the description of the world soul in the *Timaeus* and at the application of musical terminology, the image of concord seems to be complementary to the one of harmony, since it represents the execution of the rules that the second displays. As harmony has no sound, since it is the "abstract" rule and lacks sounds to become a real melody, in the same way the harmonic composition of the soul is defined as what is self-moved without speech or sound (37b5-c3 ἐν τῷ κινουμένῳ ὑφ' αὑτοῦ... ἄνευ φθόγγου καὶ ἠχῆς).[19] The kinematical representation of the soul is thus nothing but the explanatory model to whom Plato's reform of astronomy addresses, as its ratios are likewise nothing but the laws of that harmony of the intelligible that harmonics must study.[20]

The first step towards a full understanding of the meaning of σύμφωνοι ἀρυθμοί is, therefore, to draw a close comparison between harmonics and astronomy, especially as it is expressed in the terminology

---

[19] It has to be noticed, however, that this statement appears in a broader context, where the point at issue is the possibility of a true discourse on the soul and the different degrees of truth it achieves whether it refers to the circle of the Same or to the one of the Other.

[20] The same opinion is expressed by Burnyeat (2000: 52) who, speaking of *Rep.* 531, notices that "Soul provides the non-sensible subject-matter for a harmonics of the inaudible." In the following pages, especially in 53 ff and 73, the scholar demonstrates by the comparison between the *Republic* and the *Timaeus* that the structure of the soul presented in the latter meets the need of the former to define a science that inquiries the "real" movements of the universe, just because ἁρμονία indicates the structure of the soul, instead of its movements.

of *Rep.* 531b-c and 529c-e.[21] While the aim of harmonics consists in searching concordant numbers instead of the numbers of concord, speaking about astronomy Socrates urges to consider the movements (φοράς) that real speed and real slowness (τὸ ὂν τάχος καὶ ἡ οὖσα βραδθτής) make according to true number and all true figures (ἐν τῷ ἀληθινῷ καὶ πᾶσι τοῖς ἀληθέσι σχήμασι), and to apprehend them only by reason and thought, but not by sight (λόγῳ μὲν καὶ διανοίᾳ, ὄψει δ' οὔ). Sensible data are thus just παραδείγματα useful for learning what can't be grasped by senses, since it would be "absurd to examine them seriously in the expectation of finding in them the absolute truth (ὡς τὴν ἀληθείαν ἐν αὐτοῖς ληψόμεν) with regard to equals or doubles or any other ratio (ἴσων ἢ διπλασίων ἢ ἄλλης τινὸς συμμετρίας)."

If the turning from sensible data to abstract numbers, as astronomy concerns, is more obviously expressed by the semantic field of truthfulness, in the account on real harmonics it is the very musical terminology that assumes the meaning corresponding to "true".[22] Indeed, while Greek geometry uses drawings and diagrams as "abstract" models to represent every figure reproduction,[23] and so the reference to truth in analyzing them appears natural, the study of musical ratios is different, and the mathematical method of Pythagorean harmonics needs to return to sensible data in order to justify its laws. Nevertheless, the same principles of proportion and ratio, typical of mathematics, as well as the concept of unity, fundamental to define the essence of number and to determine the specific of the δύναμις τῶν μαθημάτων that Plato is looking at, are at the base of music and establish the close relationship between harmonics and mathematics. Unity, proportion, and concord, at the same time, are principles of ethics and politics, as well as of art, so Burnyeat speaks of a

---

[21] The careful analysis of such an analogy is developed by Meriani (2003) and Franco Repellini (2003).

[22] Franco Repellini (2003b: 555 ff) explains that the very close connection between music and numbers makes musical ratios able to play a role in the representation of astronomical truth.

[23] The use of drawing in geometrical demonstration, as well as in defining number, is a specific feature of Greek mathematics. This is variously emphasized by Burnyeat (2000: 27 ff, 38 ff, 61-61); Cattanei (2003: 509 ff); and Franco Repellini (2003: 365 ff), who considers it the main difference between mathematics and dialectic. Moreover, this use of sensible data as a model is the starting point to turn eyes on the intelligible: Franco Repellini (2003: 380-381); Cattanei (2003: 489). A careful analysis of Plato real astronomy and its link with sensible data is developed in relation to geometry and harmonics by Mourelatos (1980).

leitmotif, noticing that such principles appear every time Greeks describe what they admire (Burnyeat, 2000: 54).[24]

It is even more evident in Plato's philosophy, since he finds concord "not only in music, but also in the social order of the ideal city, in the psychic structure of a virtuous individual, and more broadly still, when he is doing physics in the *Timaeus*, throughout the cosmos" (Burnyeat, 2000: 55). The study of this "mathematical harmonics" is, therefore, closely linked to ethics and politics, because it is able to provide the rules that need to be applied in the government of the *polis*, and which, since they are the same rules that govern the parts of the soul, conduct both the individual and the *polis*. So, as Burnyeat concludes, "knowing what numbers are concordant, and why, has a very great deal to do with the task of government, because concord is an important structural value at the lower level of ethics and politics" (Burnyeat, 2000: 56).

Harmonics, as well as astronomy, proceeds from the observation of sound movements perceived by sense. Nevertheless, it must rise to the study of laws that are not simply the expression of observable phenomena, but are relations between pure mathematical entities according to the principles of proportionality and balance on which any possible ratio is based. Searching concordant numbers, rather than the numbers of concord, as well as searching true numbers and figures of the movements of speed and slowness in itself, is nothing more than discerning unity beyond multiplicity, the criterion of proportionality and balance at the base of the idea of good that the philosopher strives for.

A careful reading of the passages of the *Republic* and of the *Timaeus* is therefore needed to demonstrate that concord is a fundamental image at all levels of Platonic universe, and that it is just grounded on the mathematical form given by the philosopher to *cosmos*. The focus on the use of the derivatives from σύμφωνος and on the difference between the image of concord and the one of harmony draws attention to the realisation of the order expressed by the latter, and to the dynamics involved by such

---

[24] The scholar refers to concord with beauty of reason (*Rep.* 401c-d), temperance as concord and attunement between the elements both in the city and in the individual soul (*Rep.* 430e e 432a-b, cf. *Rep.* 431e and 442c-d), musical and gymnastic training as harmonising the elements of the soul (*Rep.* 410a-412b, cf. *Rep.* 441e-442a), justice as an attunement which harmonises the part of the soul (*Rep.* 443d-e), the result of the habituation to attunements and rhythm as good attunement and good rhythm in the soul of the future rulers (*Rep.* 522a).

a realisation, working as a unifying model in the explanation of cosmological, gnoseological and physiological issues.[25]

References

Anton, J.P., ed (1980). *Science and Sciences in Plato*, New York: Eidos.

Barbera, A. (1981). "*Republic* 530C-531C: Another Look at Plato and the Pythagoreans", *The American Journal of Philology* 102 (4): 395-410.

Barker, A. (1978). "ΣΥΜΦΩΝΟΙ ἈΙΘΜΟΙ: A Note on *Republic* 531C1-4", *Classical Philology* 73 (4): 337-342.

—. (1989). *Greek Musical Writings: II Harmonic and Acoustic Theory*, Cambridge: Cambridge University Press.

—. (2007), *The Science of Harmonics in Classical Greece*, Cambridge: Cambridge University Press.

Bontempi, M. (2007). "Vincoli di fuga: misura cosmica e misure visibili per la prospettiva della causa intellligente nel *Timeo*." In Linda M. Napolitano Valditara (ed.), *La Sapienza di Timeo. Riflessioni in margine al «Timeo» di Platone* (pp. 111-146), Milano: Vita e Pensiero.

Brisson, L. (1992). *Platon. Timée-Critias, Traduction inédite, introduction et notes*, Paris: Flammarion.

—. (1997[3]). *Le Même et l'Autre dans la Structure Onthologique du* Timée *de Platon. Un commentaire systématique du* Timée *de Platon*, Sankt Augustin: Academia Verlag.

—. (2003). "How and why do the building blocks of the universe change costantly in Plato's *Timaeus* (52a-61c)?" In Carlo Natali and Stefano Maso (eds.), *Plato Physicus. Cosmologia e antropologia nel Timeo* (pp. 191-205), Amsterdam: Hakkert.

Brisson, L.-F. and W. Meyerstein (1991). *Inventer l'univers. Le problème de la connaisance et les modèles cosmologiques*, Paris: Les Belles Lettres.

Brown, E. (2008). "Contemplative withdrawal in the Hellenistic age", *Philosiphical Studies* 137: 79-89.

[25] It has to be noticed that the very principle of assimilation between sense perception and object perceived is pointed out by Burnyeat with regard to learning abstract entities, since "assimilation means that your soul take on the structure of the abstract realm you study" and "the system we internalise and become assimilated to is the articulation, at the level of mathematical thought (διάνοια), of the world as it is objectively speaking" (Burnyeat, 2000: 70-71). The scholar focuses on the semantic fields of συγγένεια and ὁμοιότης and, referring to *Ti.* 90c-d and *Rep.* 537c, he concludes that just through mathematical sciences is man able to grasp the knowledge of Being (Burnyeat, 2000: 67 ff).

Burkert, M. (1972). *Lore and Science* (english ed. translated with revision from *Weisheit und Wissenshaft: Studien zu Pythagoras, Philolaos und Platon*, Nürnberg 1962), Cambridge, MA: Harvard University Press.

Burnyeat, M.F. (2000). "Plato on Why Mathematics is Good for the Soul." In Timothy Smiley (ed.), *Mathematics and Necessity. Essays in the History of Philosophy*, (*Proceedings of the British Academy, 103)* (pp. 1-81), Oxford: Oxford University Press.

Campese, S. (2007). In Mario Vegetti (ed.), Platone. *La Repubblica*, Traduzione e commento, vol. VII (pp. 399-411), Napoli: Bibliopolis.

Carastro, M. (2006). *La cité des Mages. Penser la magie en Grece ancienne*, Grenoble: Millon.

Casertano, G. (2003). "Cause e concause." In Carlo Natali and Stefano Maso (eds.), *Plato Physicus. Cosmologia e antropologia nel Timeo* (pp. 33-63), Amsterdam: Hakkert.

Cattanei, E. (2003). "Le matematiche al tempo di Platone e la loro riforma." In Mario Vegetti (ed.), Platone. *La Repubblica,* Traduzione e commento, vol. V (pp. 473-539), Napoli: Bibliopolis.

Cozzo, A. (2001). *Tra comunità e violenza. Conoscenza,* logos *e razionalità nella Grecia antica*, Roma: Carocci.

Detienne, M. (1973^2). *Les maîtres de verite dans la Grece archaique*, Paris: Maspero.

Franco Repellini, F. (2003). "La linea e la caverna." In Mario Vegetti (ed.), Platone. *La Repubblica,* Traduzione e commento, vol. V (pp. 355-403), Napoli: Bibliopolis.

—. (2003b). "Astronomia e armonica." In Mario Vegetti (ed.), Platone. *La Repubblica,* Traduzione e commento, vol. V (pp. 541-563), Napoli: Bibliopolis.

—. (2007). "Il fuso e la Necessità." In Mario Vegetti (ed.), Platone. *La Repubblica*, Traduzione e commento, vol. VII (pp.367-397), Napoli: Bibliopolis.

Fronterotta, F. (2007). "Il *Timeo* e la matematica 'embodied'." In Linda M. Napolitano Valditara (ed.), *La Sapienza di Timeo. Riflessioni in margine al «Timeo» di Platone* (pp. 173-206), Milano: Vita e Pensiero.

Huffman, C.A (1993). *Philolaus of Croton: Pythagorean and Presocratic. A commentary on the fragments and testimonia with interpretative essays*, Cambridge: Cambridge University Press.

—. (2005), *Archytas of Tarentum. Pythagorean, philosopher and mathematician king*, Cambridge: Cambridge University Press.

Johansen, T.K. (2006^2). *Plato's Natural Philosophy. A study of the* Timaeus-Critias, Cambridge: Cambridge University Press.

Joubaud, C. (1991). *Le corps humain dans la philosophie platonicienne. Étude à partir du Timée*, Paris: Librairie Philosophique J. Vrin.
Knorr, W.R. (1990). "Plato and Eudoxus on the Planetary Motions", *Journal of History of Astronomy* 21 (4): 313-329.
Lavecchia, S. (2006). *Una via che conduce al divino. La "homoiosis theo" nella filosofia di Platone*, Milano: Vita e Pensiero.
Lloyd, G.E.R. (1978). "Saving the appearances", *Classical Quarterly* 28 (1): 202-222.
—. (1990). "Plato and Archytas in the Seventh Letter", *Phronesis* 35 (2): 159-174.
Mahoney, T. (2005). "Moral Virtue and Assimilation to God in Plato's *Timaeus*", *Oxford Studies in Ancient Philosophy* 28: 77-91.
Maso, S. (2003). "Dal disordine all'ordine." In Carlo Natal and Stefano Maso (eds.), *Plato Physicus. Cosmologia e antropologia nel Timeo* (pp. 243-257), Amsterdam: Hakkert.
Meriani, A. (2003). "Teoria musicale e antiempirismo, Astronomia e armonica." In Mario Vegetti (ed.), Platone. *La Repubblica*, Traduzione e commento, vol. V (pp. 564-602), Napoli: Bibliopolis.
Miller, H.W. (1957). "The Flux of the Body in Plato's Timaeus", *Transactions and Proceedings of the American Philological Association* 88: 103-113.
Mourelatos, A.P.D. (1980). "Plato's 'Real Astronomy': *Republic* 527D-531D." In John P. Anton (ed.), *Science and Sciences in Plato* (pp. 33-73), New York: Eidos.
Natali, C. (2003), "Antropologia, politica e la struttura del *Timeo*." In Carlo Natali and Stefano Maso (eds.), *Plato Physicus. Cosmologia e antropologia nel Timeo* (pp. 225-241), Amsterdam: Hakkert.
Riedweg, C. (2002). *Pythagoras. Leben, Lehre, Nachwirkung*, München: C.H. Beck.
Russel, D.C. (2004). "Virtue as "Likeness to God" in Plato and Seneca", *Journal of the History of Philosophy* 42 (3): 241-260.
Scott, W.C. (1984). *Musical design in Aschylean Theater*, Hannover and London: University Press of New England.
Sedley, D. (1997). "'Becoming like God' in the *Timaeus* and Aristotle." In Tomás Calvo and Luc Brisson (eds.), *Interpreting the* Timaeus-Critias. *Proceedings of the IV Symposium Platonicum. Selected Papers* (pp. 327-339), Sankt Augustin: Academia Verlag.
Thomas, C.J. (2006). "Plato's Prometheanism", *Oxford Studies in Ancient Philosophy* 31: 203-231.
West, M.L. (1983). *The Orphic Poem*, Oxford: Clarendon Press.

ON PTOLEMY'S *HARMONICS* 2.4: DOES THE TEXT REFER TO THE DIAGRAM?

CRISTIAN TOLSA

A first glance at any edition of an ancient Greek mathematical treatise reveals that it is not exclusively made of Greek words forming a continuous text, but also of striking graphical schemes, the so-called diagrams. Greek mathematics breaks the patterns underlying the typical bookroll layout in a very obvious way: inserting diagrams and so dividing the text of the columns in so-called theorems. An ancient book consists otherwise of non-interrupted text running along the columns of a papyrus roll, displayed in very canonical proportions.[1] The modern reader could think that diagrams are not that striking, that mathematics could not be understood without some sort of graphical representation; thus it should be considered natural that it uses diagrams. A general objection to this argument is that other Greek genres whose modern counterpart use graphical representations, like medicine, are written exclusively as text, and following the same layout conventions as any other literary genre. On the other hand, contemporary scholarship in mathematics, for example, rarely uses any sort of graphical representations, but rather *formulae*. My point here is that genres traditionally considered universal like mathematics do not necessarily share similar features in their respective actual practices. In other words, Greek mathematics, Mesopotamian mathematics and contemporary mathematics are not necessarily isomorphic.[2]

---

[1] For a very thorough study of the formal characteristics of the layout of bookrolls see Johnson (2004).

[2] The lettered diagram of Greek mathematics goes at the end of the passage that refers to it. The reference to the diagram in the text is made by means of using the letters of the diagram as labels within a highly fomulaic language: see Netz (1999). On the other hand, the graphical schemes appearing in Ancient Mesopotamian tablets do not share many features with the diagrams appearing in a Greek mathematical treatise. They do not show any similar labels, for example: see Robson (2008).

Claudius Ptolemy, whose activity ranges at least from 125 to 141 CE,[3] uses a mixed style which combines philosophical digressions[4] and mathematically-shaped parts.[5] Famous for being the author of the great astronomical work known as the *Almagest*, he wrote a small treatise on music theory that has survived in three books[6] with the Greek title *Harmonika*. Ptolemy takes up the Pythagorean description of the musical intervals as ratios of numbers, in which the octave, the fifth and the fourth–the so-called first concords (the basic and unanimously accepted harmonic intervals)–correspond respectively to the ratios 2:1, 3:2 and 4:3. These ratios can be visualized as lengths of strings, so that, given a string, another string which measures twice its length sounds an octave lower in relation to the first, another one whose length is one and a half times (3:2) the length of the first sounds a fifth below, and another in the ratio 4:3 sounds a fourth below.

The diagrams appearing in the manuscripts that contain this work show important divergences, as well as the text: some of them change their shape, some appear only in a number of manuscripts. But no critical edition of the diagrams has been published yet. Düring's edition shows no more concern about the diagrams than the criterion of printing only those that are referred to in the text.[7] This is a very reasonable criterion, but it is not sufficient. Clearly, for Düring and for the authors who emended his edition so far, the text is the only important part of the work, while the diagrams play a secondary role: they are not critically revised.[8] There is a genuine interest in restoring the words by Ptolemy, but this interest is not extended to the rest of features which make this work an ancient treatise on music theory. I mean that the work is still considered only as a text to be preserved, and not many efforts are devoted to understanding its appearance in its own cultural context. In other words: classical scholarship still deals in a very scholastic way with ancient scientific texts. But what did this treatise on music theory by Ptolemy look like as an

---

[3] These are the dates of his first and last personally observed astronomical phenomena. See e.g. Pedersen (1974).

[4] See Taub (1993) for the philosophical parts of Ptolemy's work.

[5] Ptolemy introduces a new epistemological–and graphical–tool: the tables.

[6] Compared to the 13 books of the *Almagest*, it is a rather small treatise.

[7] Düring "1930: cii": "Von den Tabellen und Zeichnungen habe ich nur solche mit aufgenommen, auf die Ptolemaios im Texte ausdrücklich hinweist." There is no *apparatus criticus* of the diagrams, so that we don't even know in which manuscripts they appear.

[8] I inspected Alexanderson (1969), Redondo Reyes (2003) and Bobo de la Peña (2006 and 2007), and found no emendation on the diagrams.

object, as a papyrus roll?[9] An answer to this question involves, among other considerations, making a critical edition of the diagrams. We must know whether the ones that appear on the codices are original or not, what their original form looked like and where they appeared breaking the text.[10]

I will here treat the case of one diagram, the one appearing in *Harmonics* 2.4, and I will try to show that even the only criterion of Düring regarding diagrams–that of including only those referred to in the text–is not well applied in this case. He interprets that this diagram is referred to in the text, so he includes it, but a close reading of the text and a simultaneous inspection of the diagram reveal that the reference is in fact to the previous diagram, not to the one appearing in 2.4. So what is this diagram doing there? There is indeed a relation between the diagram and what is being explained in the text: otherwise anyone would have noticed that it is misplaced. But I claim that it is impossible that Ptolemy drew it and put it there. Who and why, then? Here comes into play the commentary that the Neoplatonist philosopher Porphyry (234-ca. 305 CE) wrote on this work by Ptolemy. His commentary is still extant and it was sometimes copied on the same manuscripts as the *Harmonics*. In the passage where he comments chapter 2.4 he draws some diagrams in order to explain what Ptolemy is doing. I claim that the diagram appearing in *Harmonics* 2.4 is by Porphyry.

The narrative of Ptolemy in chapter 2.4 is closely linked to the argument of the preceding chapter.[11] Ptolemy introduces the so-called species or forms (εἴδη) of the first concords. This concept goes back at least to Aristoxenus,[12] and could be formally compared to the modern chord inversions. I will try to explain this in a few lines. In these two chapters the concords are not treated as intervals: they also include the notes in between. The basic harmonic structure of Greek music is the

[9] This point of view is not new, however: it is the focus of what is called history of the book. Another very important issue in situating a text in its own context is the social study of its readership, but this is not the objective of this paper. A good example of this kind study is chapter 7 of Netz (1999).

[10] Regarding the editions of Greek treatises on music theory, we can not fail to recognize the critical notes for the diagrams in the edition of the minor works on music theory by Jan (1895). Regarding geometry, a thorough inspection and critical edition of the diagrams of Archimedes' *On the Sphere and the Cylinder* has been made recently, in Netz (2004).

[11] In fact, the division between 2.3 and 2.4 seems too arbitrary, and it is perhaps one of the reasons of the problem we are discussing, as we will see below.

[12] Barker (1989: 322, n. 26).

tetrachord, and from it the fifth and the octave are constructed. The two external notes of the tetrachord are fixed, and they make a fourth (their ratio is 4:3). The two internal notes are movable, defining what Greek music theorists call *genera*. Then, given a tetrachord in whatever genus, a fifth–whose ratio is 3:2–is obtained by conjoining a tone upon the tetrachord. We get an octave by conjoining another tetrachord under the fifth, so obtaining (4:3) x (3:2) = 2:1, which is the ratio of the octave. Now we come to the species: let us imagine that we put together two tetrachords, say ABCD and DEFG, with D as a common note. Let A be the highest note. Then, ABCD is a fourth, and DEFG is a fourth, as well. In particular, both of them are said to be a first species of the fourth, because their external notes are the fixed notes of a tetrachord. So to speak, the first species of the fourth coincides with a tetrachord. But note that CDEF and BCDE are fourths, too, albeit with a different structure. They do not have the fixed notes of a tetrachord as external notes. The second form of the fourth is obtained from the first by transporting the lowest interval to the uppermost position. In our example, CDEF is a fourth of the second species. The third species is obtained by repeating this process, so that BCDE is the third species. There is no fourth form, since the structure of the resulting fourth would be the same as the structure of the first form. Note that if we apply again this procedure to our example, we get ABCD, which is, as we already know, a first form. The same definition is valid for the fifth: the first form of the fifth will have the disjunctive tone[13] at the highest position, the second form will have it at the second uppermost position, and so on until the fourth form, which will have it at the lowest. The same is true for the octave: the first form of the octave will have the disjunctive tone at the highest position, the second form at the second uppermost position, and so on until the seventh form, which will have it at the lowest.

After explaining what the forms are, Ptolemy shows that the double octave (obtained by putting together two conjunct tetrachords upon an octave) is the smallest scale in which we can find all the species of each one of the first concords. Chapter 2.3 ends up showing in a diagram how we can find all these species in the double octave:[14]

---

[13] It is called "disjunctive tone" because it separates two tetrachords.

[14] The translation appears in Barker (1989: 323). I have only omitted the phrases explaining where the fixed notes lie. I substituted Barker's Latin letters of the diagram for the Greek ones, so we can follow the argument later with Porphyry's commentary. The diagram is adapted from Barker's, on the same page. The fixed notes of the tetrachords are marked with bolder bars than the movable notes. The Greek text in Düring's edition reads: ἐὰν γὰρ ἐκθώμεθα διὰ τεσσάρων τὸ

For if we set out a fourth ΑΒΓΔ, treating A as the highest note, and conjoin with it another similar fourth below, ΔΕΖΗ, then conjoin with this in the same way a tone ΗΘ, conjoin with this, once again, a fourth ΘΚΛΜ, and with that another fourth ΜΝΞΟ (...).The first form of the fourth will be ΜΟ, the second ΛΞ, the third ΚΝ (...). Of the fifth, the first form will be ΗΜ, the second ΖΛ, the third ΕΚ, the fourth ΔΘ (...).Of the octave, the first form will be ΗΟ, the second ΖΞ, the third ΕΝ, the fourth ΔΜ, the fifth ΓΛ, the sixth ΒΚ and the seventh ΑΘ (...).

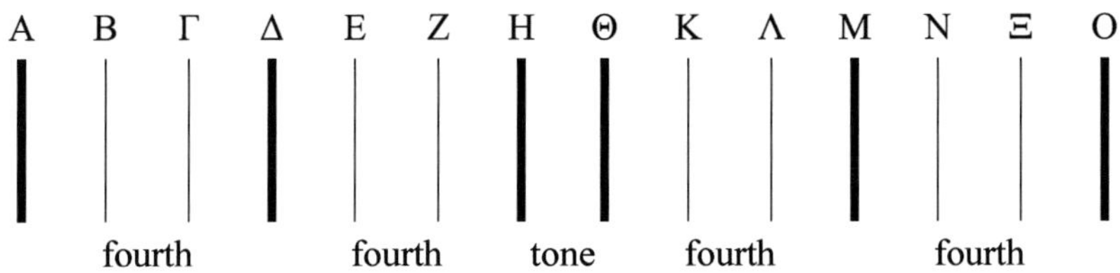

Fig. 1[15]

In chapter 2.4 Ptolemy defines "complete system" as a system (or scale) in which all the species of the first concords can be found. Then the double octave is a complete system. Secondly, he is concerned with underlining that greater scales than the double octave do not contain any other species of the first concords (because all of them appear already in the double octave), while smaller scales contain only some of them. We can say that this chapter makes clear that no other scale than the double octave can be considered a minimal complete system. As a consequence, the scale formed by an octave and a fourth, traditionally considered complete, according to Ptolemy, is not complete at all. Ptolemy goes on specifying which forms of the fifth and the octave will appear in every possible configuration of the scale of octave plus fourth. The configurations vary depending on the order of the octave, the disjunctive

ΑΒΓΔ, τοῦ Α κατὰ τὸν ὀξύτατον φθόγγον νοουμένου, καὶ τούτῳ συνάψωμεν ἕτερον ἐπὶ τὸ βαρὺ διὰ τεσσάρων ὁμοίως ἔχον τὸ ΔΕΖΗ καὶ τούτῳ τόνον ὁμοίως τὸν ΗΘ καὶ πάλιν αὖ τούτῳ μὲν διὰ τεσσάρων τὸ ΘΚΛΜ, τούτῳ δὲ ἕτερον διὰ τεσσάρων τὸ ΜΝΞΟ, (...) τοῦ δὲ διὰ τεσσάρων πρῶτον μὲν εἶδος τὸ ΜΟ, δεύτερον δὲ τὸ ΛΞ, τρίτον δὲ τὸ ΚΝ (...). τοῦ δὲ διὰ πέντε πρῶτον μὲν εἶδος ἔσται τὸ ΗΜ, δεύτερον δὲ τὸ ΖΛ, τρίτον δὲ τὸ ΕΚ, τέταρτον δὲ τὸ ΔΘ (...). καὶ τοῦ διὰ πασῶν πρῶτον μὲν εἶδος ἔσται τὸ ΗΟ, δεύτερον δὲ τὸ ΖΞ, τρίτον δὲ τὸ ΕΝ, τέταρτον δὲ τὸ ΔΜ, πέμπτον δὲ τὸ ΓΛ, ἕκτον δὲ τὸ ΒΚ, ἕβδομον δὲ τὸ ΑΘ (...).

[15] The labels like this are mine.

tone and the fourth.[16] This is the point where the diagram I am concerned with finally appears:[17]

> When it is so positioned that the tone disjoins two conjunct tetrachords from one tetrachord, it will contain the four forms of the fifth, but only four of the seven forms of the octave–those starting from either of the extremes. But when it is so positioned that the tone is at the limit, and the three tetrachords are conjoined, it will contain only one form of the fifth and one of the octave, either the first or the last of both of them. This can be seen from the diagram set out below, if one adds to it one similar tetrachord at one end or the other.

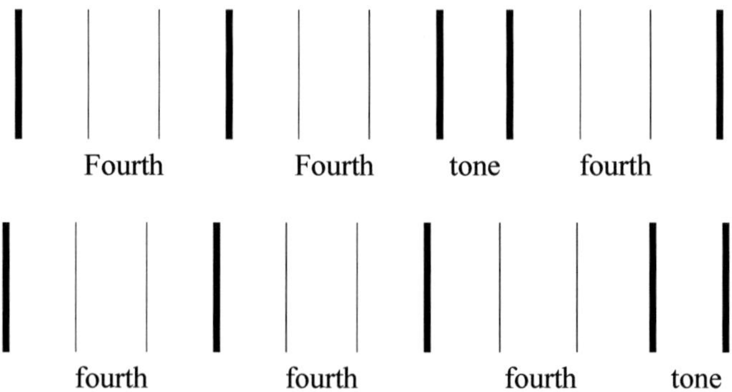

Fig. 2

The two parts of this double diagram–I will call the upper one fig. 2.1 and the lower one fig. 2.2–illustrate the referred cases: when "the tone

[16] Here Ptolemy is probably showing off the technical possibilities that diagrams allow to refute a thesis, since Ptolemy only needs to prove this for the case in which the disjunctive tone is at the bottom (and the three tetrachords are conjunct), for this is the traditional scale of octave plus fourth: the system συνημμένων will be discussed by Ptolemy later in chapter 2.6.

[17] Again in Barker (1989: 324). The diagram is also adapted from there. The Greek text in Düring's edition reads: ἀλλ' ὅταν μὲν οὕτως ἔχῃ θέσεως, ὥστε τὸν τόνον διαζευγνύναι τὰ συνημμένα δύο τετράχορδα τοῦ ἑνός, τὰ μὲν τέσσαρα εἴδη τοῦ διὰ πέντε περιέξει, τῶν δὲ τοῦ διὰ πασῶν ἑπτὰ μόνα τέσσαρα πάλιν, τὰ ἀφ' ὁποτέρου τῶν ἄκρων, ὅταν δ' οὕτως ἔχῃ θέσεως, ὥστε τὸν τόνον ἐπὶ τὸ πέρας εἶναι. καὶ τὰ τρία τετράχορδα συνημμένα ἓν μόνον εἶδος περιέξει καὶ τοῦ διὰ πέντε καὶ τοῦ διὰ πασῶν, ἢ τὸ πρῶτον ἢ τὸ ἔσχατον ἀμφοτέρων, ὡς ἐξέσται σκοπεῖν ἀπὸ τῆς προκειμένης καταγραφῆς ἐπισυνάπτουσιν αὐτῇ καθ' ἑκάτερον τῶν περάτων ἓν ὁμοίως ἔχον τετράχορδον.

disjoins two conjunct tetrachords from one tetrachord" (fig. 2.1) and "when it is so positioned that the tone is at the limit" (fig. 2.2).

But there are some striking facts. First of all, what does the last quoted phrase mean in this context, "if one adds to it [the diagram set out below] one similar tetrachord at one end or the other"? The double diagram (fig. 2) illustrates perfectly both referred situations: there is nothing to be added. Actually, if we add one tetrachord to one end or the other of any of these scales (fig. 2.1 and fig. 2.2), we will not obtain an octave plus fourth, but a double octave. This phrase, then, seems incoherent.

Second striking fact: the Greek text for "the diagram set out below" reads τῆς προκειμένης καταγραφῆς. Then, we should not translate "below" but "above". But the the diagram is printed after the text in Düring's edition, a fact that probably confuses translators, who do not attempt to modify its position in their translation.[18] There is a deeper reason, however, for putting the diagram after the reference in the text: the diagrams are always conventionally put after the text that refers to them in the whole tradition of Greek mathematics.[19] But the text here refers unambiguously to a diagram above. Wallis' edition of the *Harmonics*–the only one before Düring's–shows, interestingly, a somewhat different configuration of this diagram: it appears separated into its two parts (fig. 2.1 and fig. 2.2).[20] Fig. 2.1 comes after the discussion of the first case, while fig. 2.2 comes at the end of the page, and before the reference to it in the text. This is why Wallis can translate, according to the Greek text, "quod videre erit in praecedente descriptione," but at the cost of the surprising position of a diagram (fig. 2.2) before the first reference to it in the text, and unlike the first case (fig. 2.1), in which he puts the diagram after the discussion. Furthermore, in this situation there is a reference to fig. 2.2 but not to fig. 2.1, apart from the problem of interpreting the phrase "if one adds to it one similar tetrachord at one extreme or the other," which remains incoherent.[21]

Let us try to understand what is going on by comparing this passage with similar cases in Ptolemy's work. I found three other instances of τῆς

[18] Redondo Reyes (2003: 194) translates "en el siguiente diagrama"; Solomon (2002: 72) "from the following table"; Düring (1934: 65) chooses the neutral expression "aus der beigefügten Zeichnung", and similarly Raffa (2002: 156) "dal grafico qui presentato."

[19] Netz (2004: 10).

[20] Wallis (1682: 106).

[21] Raffa (2002: 156) follows Wallis in separating fig. 2.1 and fig. 2.2, but he postpones fig. 2.2 after its corresponding text.

προκειμένης καταγραφῆς in the *Almagest*.[22] Ptolemy refers there to figures already used before in the text–diagrams that, say, illustrate a previous theorem–, so it is not the case that he draws previously the figure and he refers to it in the text below. The figure is in those cases reconsidered and altered in order to prove some other result.

I inspected the three oldest and best manuscripts containing the *Harmonics*, representing the two main branches,[23] and fig. 2 appears only attached by a later hand in a margin of one of them,[24] in which space is usually reserved for the figures in the body of the text.

So what is happening here? To answer this question, let us try to read again the quoted text of chapter 2.4, without considering the double diagram (fig. 2) and reading the correct translation "above". We must consider that the preceding diagram is precisely the one in chapter 2.3 (fig. 1). Try to do what Ptolemy says: attach one tetrachord to one end of the diagram. The result is that we get a very similar diagram to fig. 2.2: three conjunct tetrachords with a tone at the extreme (we must only omit the opposite tetrachord). This agrees with the text, in which the phrase seems to refer syntactically only to the second case, in which there are three conjunct tetrachords and a tone at the extreme.[25] In addition, if we accept that the reference is to the diagram in 2.3 (fig. 1), it becomes clear that only in the second case, "when it is so positioned that the tone is at the limit"–i.e. when we have three conjunct tetrachords and a tone–there is a need for modifying the diagram (fig. 1): we have to conjoin another tetrachord to one extreme. On the other hand, we do not have to add nothing to it (fig. 1) to obtain the scale referred to in the first case: two conjunct tetrachords disjoined from another tetrachord by a tone (represented by fig. 2.1). We just need to remove one external tetrachord. Now, we understand why only in the second case Ptolemy comments what we should do with the diagram.

So far, I have tried to prove that the reference in the text is not to the double diagram printed by Düring (fig. 2), but to the diagram in 2.3 above (fig. 1). Now, I will try to explain where Düring's double diagram (fig. 2) comes from and why it stands on the pages of some manuscripts and in the modern editions. The commentary by Porphyry is a very important clue in this regard.

---

[22] Heiberg (1898: 367); Heiberg (1898: 456); Heiberg (1903: 593).

[23] Monacensis gr. 361a, Venetus Marcianus gr. app. cl. VI/X and Vaticanus gr. 191. For the dating and *stemma*, cf. Düring (1930, ix ff).

[24] It appears in Venetus Marcianus gr. app. cl. VI/X f. 27v.

[25] There is a strong pause (ὅταν δ' οὕτως ἔχῃ θέσεως) and so it seems that ὡς ἐξέσται σκοπεῖν refers only to this second case.

The passage concerning the first case–in which there is a tetrachord disjoint from another two by a tone–goes like this:[26]

> If the tone disjoins the two conjunct tetrachords from the one, either the one from the two or the two from the one, however it be, it is clear that the four forms of the fifth and the three of the fourth appear (...) but of the octave there appear only four [species]. And, if the tone disjoins the two higher [tetrachords] from the one lower, like in the first diagram, the last four [octave species] are ΜΔ, ΛΓ, ΚΒ, ΘΑ. And if the tone disjoins the two lower [tetrachords] from the one higher, like in the second diagram, the first four [octave species] are ΟΗ, ΞΖ, ΝΕ and ΜΔ. This makes clear that of "from either limit": because they either begin from A and the lower tetrachord, ΜΝΞΟ, is missing, or from O and the higher, ΑΒΓΔ, is missing.

Porphyry clearly refers to two diagrams, and these can not be those printed in Ptolemy's text in 2.4 (fig. 2), because Porphyry's are lettered and both represent the first case, in which two conjunct tetrachords are separated from another one by a tone. Porphyry analyzes the two possible situations alluded to by Ptolemy. His intention is to clarify this double possibility, and thus he draws two diagrams taken from the diagram in 2.3 (fig. 1), each one resulting in that same diagram but without either external tetrachord: the first one without the lowest tetrachord, ΜΝΞΟ, and the second one without the highest, ΑΒΓΔ. Unfortunately, I have had no occasion to search for these diagrams in the manuscripts containing the commentary,[27] but even if they were not there it is clear that they must have been there originally. They would look like this:

[26] My own translation. The Greek text is Düring (1932: 164 12-23): εἰ μὲν ὁ τόνος διαζεύγνυσι τὰ δύο τετράχορδα ἢ τὸ ἓν ἀπὸ τῶν δύο ἢ τὰ δύο ἀπὸ τοῦ ἑνός, ὁποτέρως ἂν γένηται, τὰ τέσσαρα εἴδη τοῦ διὰ πέντε θεωροῦνται προομολογουμένου ἔτι καὶ τὰ τρία τοῦ διὰ τεσσάρων (...) μόνα δὲ τοῦ διὰ πασῶν τὰ τέσσαρα. πλὴν εἰ μὲν διαζεύγνυσιν ὁ τόνος τὰ δύο ὀξύτερα τοῦ ἑνὸς βαρυτέρου, ὡς ἐπὶ τῆς πρώτης καταγραφῆς, τὰ ὕστερα τέσσαρα τὸ ΜΔ, τὸ ΛΓ, τὸ ΚΒ, τὸ ΘΑ· εἰ δὲ διαζευγνύει ὁ τόνος τὰ δύο βαρύτερα τοῦ ἑνὸς ὀξυτέρου, ὡς ἐπὶ τῆς δευτέρας καταγραφῆς, τὰ πρότερα τέσσαρα τὸ ΟΗ, τὸ ΞΖ, τὸ ΝΕ καὶ τὸ ΜΔ. τοῦτο γὰρ δηλοῖ τὸ ἀφ' ὁποτέρου τῶν ἄκρων· ἢ γὰρ ἀπὸ τοῦ Α ἄρχονται καὶ λείπει τὸ βαρύτερον τετράχορδον τὸ ΜΝΞΟ, ἢ ἀπὸ τοῦ Ο καὶ λείπει τὸ ὀξύτερον τὸ ΑΒΓΔ.

[27] The MS Venetus Marcianus gr. app. cl. VI/X also contains the commentary by Porphyry, but it goes only until chapter 1.5 of the *Harmonics*.

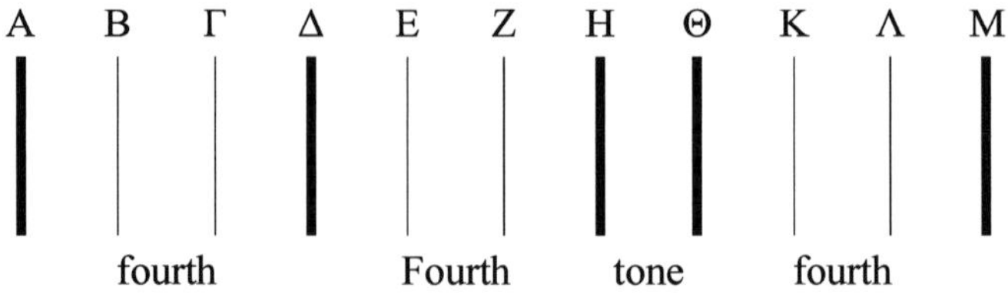

Fig. 3.1

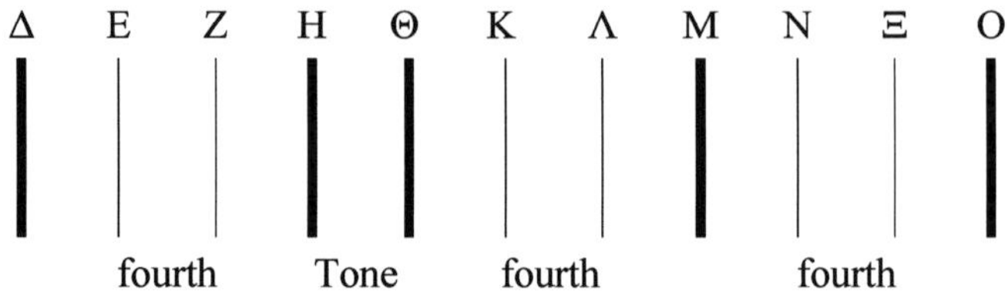

Fig. 3.2

Porphyry performs here a typically scholastic practice:[28] he specifies which are the only four species of the octave that we can find in that scale. Remember that Ptolemy said that we would find "only four of the seven forms of the octave–those starting from either of the extremes." Porphyry specifies which four species: when we remove the lowest tetrachord (in fig. 3.1), they are ΜΔ, ΛΓ, ΚΒ, ΘΑ, while without the highest one (in fig. 3.2) they are ΟΗ, ΞΖ, ΝΕ and ΜΔ.

The next paragraph treats the second case, in which we have three conjunct tetrachords and the tone at the extreme:[29]

---

[28] See Netz (1998) for a study of the commentary as a genre in Late Antiquity and the Middle Ages. Ours is clearly a case of what Netz calls horizontal pedantry, that is, proving easy implied results. In fact, the whole commentary by Porphyry on this chapter represents such a case.

[29] Again my translation. The Greek text is Düring (1932: 164.26-165.4): Ὅταν δ' ἅμα ὦσι τὰ τρία τετράχορδα, ὁ τόνος δ' ἔξωθεν τούτων ἐπὶ τὸ πέρας τὸ ἔσχατον· οὕτω γὰρ συνάγεται τὸ διὰ πασῶν καὶ διὰ τεσσάρων· εἰ μὲν τὸ ὀξύτερον λείπει τετράχορδον τὸ ΑΒΓΔ, τὸ πρῶτον εἶδος θεωρεῖται τοῦ τε διὰ πασῶν καὶ τοῦ διὰ πέντε· τοῦ μὲν διὰ πασῶν τὸ ΗΟ, τοῦ δὲ διὰ πέντε τὸ ΗΜ, ὡς ἐπὶ τῆς πρώτης ἔχει καταγραφῆς. εἰ δὲ κεῖται μὲν ἔσχατος ὁ ΗΘ τόνος, ἄρχονται δὲ τὰ δύο τετράχορδα ἀπὸ τοῦ Α, ἤτοι ΑΒΓΔ καὶ ΔΕΖΗ, τὸ δὲ ΚΛΜ συνάπτεται τῷ Α καὶ τὸ ΜΝΞΟ λείπει· τὸ ἔσχατον θεωρεῖται τοῦ τε διὰ πέντε τὸ ΔΘ καὶ τοῦ διὰ πασῶν τὸ ἔσχατον τὸ ΘΑ, ὡς ἐπὶ τῆς δευτέρας ἔχει καταγραφῆς.

> When there are the three tetrachords at the same time, and the tone outside from them at the farthest limit: in this way he [Ptolemy] puts together the octave and the fourth. If the higher tetrachord, ΑΒΓΔ, is missing, we observe the first species of the octave and of the fifth: of the octave, ΗΟ; of the fifth, ΗΜ, like in the first diagram. And if the tone ΗΘ is at the farthest, and the two [conjunct] tetrachords begin from Α, namely ΑΒΓΔ and ΔΕΖΗ, and ΚΛΜ is joined at Α, and ΜΝΞΟ is missing, then we observe the last [species] of the fifth, ΔΘ, and the last of the octave, ΘΑ, like in the second diagram.

This is the case where Ptolemy refers to the "above diagram", from which we should obtain the result by "conjoining a similar tetrachord to either extreme of it." Porphyry shows again the two possibilities–"at one end or the other"–, and illustrates them again with a diagram for each one: "if the higher tetrachord is missing" and "if the tone ΗΘ is at the farthest, and the two [conjunct] tetrachords begin from Α."

The wording for the first diagram is somewhat ambiguous: Porphyry does not make clear that he joins the tetrachord ΔΕΖΗ to the other extreme of the diagram, but he seems to be actually doing this: he observes the first species of the octave and the fifth (the ones with the disjunctive tone at the uppermost position) in ΗΟ and ΗΜ, which agrees with that procedure. In fact, this is the only logical possibility. The diagram can only look like this:[30]

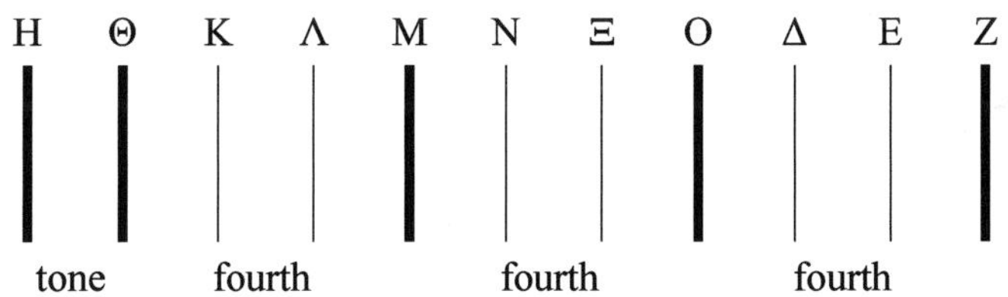

Fig. 4.1

The wording explaining the second diagram is much clearer. Here, as Ptolemy says, only the last species of the fifth and the octave (the ones

[30] Technically, if we "cut and paste" from diagram 2.3, we should join the tetrachord ΔΕΖΗ to Ο, obtaining Δ=Ο, but this would result in the repetition of the letter Η, which appears at the other extreme of the scale. This is avoided by Porphyry by renaming the tetrachord to ΟΔΕΖ.

with the disjunctive tone at the lowest position) are to be found. Porphyry specifies that they are respectively ΔΘ and ΘA:[31]

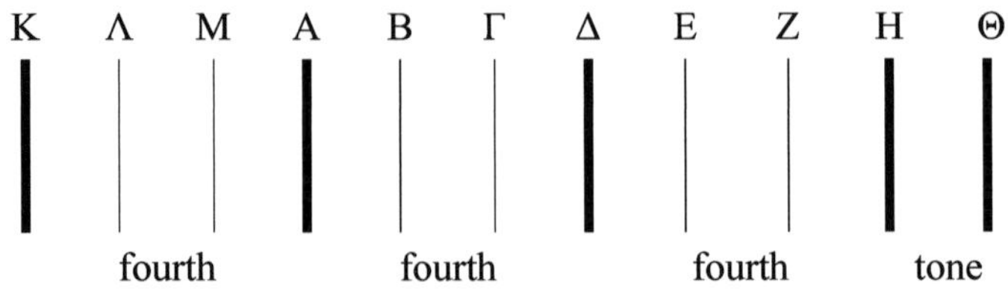

Fig. 4.2

Now, we recognize among these last four diagrams (figs. 3.1, 3.2, 4.1 and 4.2) the two printed by Düring in the chapter 2.4 of Ptolemy's *Harmonics* (figs. 2.1 and 2.2): fig. 2.1 corresponds to fig. 3.1 and fig. 2.2 corresponds to 4.2. The double diagram in Düring's edition is non-lettered, but this is fairly reasonable, because the text does not deal with it.

But why did fig. 2 pass into the edition of the *Harmonics*? There are many reasons for this, but it seems highly probable to me that the division in chapters[32] plays a big role here. As I said above, many diagrams were copied to the manuscripts containing the *Harmonics*–which in some cases contained also the commentary of Porphyry–, but they did not necessarily pass into the body of the text: that's why the editors have to make a selection.[33] The fact that chapter 2.4 is separated from chapter 2.3, and thus has no diagram, while there is an explicit reference to a diagram in

[31] It occurs the same as in the previous case: we should join the tetrachord ΘKΛM to A, with M=A, but that would result in the repetition of the letter Θ, which is avoided by Porphyry by renaming the tetrachord to KΛMA.

[32] It is most likely that both the division in chapters and their headings are not original. Cf. Netz (1998: 273): "it should be realized that in all probability Ptolemy's text totally lacked chapter divisions (not an Ancient practice), let alone chapter headings. Whereas the original Ptolemy was a discursive text that proceeded smoothly and without breaks through the entire astronomical corpus, the Late Ancient tradition created a Ptolemy progressing from one discrete, well-signalled point of argument to the next."

[33] As I said above, it is even a doubtful question if the diagram that Wallis and Düring print in chapter 2.4 appears in the body of the text in any manuscript. Another example of that could be the case of the table in chapter 1.16, which is absent from the three abovementioned manuscripts but printed in Wallis and Düring.

the text, made it possible for a diagram in the margin to pass into the body of the text, as though it were naturally "the diagram" of this chapter.

And what is the reason why this double diagram (fig. 2) remained in the text? The answer to this question is rather obvious: the figure in itself is not striking; it shows the resulting diagrams of the operation alluded to by Ptolemy in the text. It is the equivalent of a marginal note entering the body of the text. As a matter of fact, Porphyry drew it as a note to understand Ptolemy. If we do not read carefully the explicit reference "from the diagram above, if one adds to it one similar tetrachord at one end or the other," we encounter no difficulty. But once we recognize that the text is not referring to it, we are confronted with a diagram without any explicit reference in the text which, at the same time, can be traced back to the exegetic study by Porphyry.[34] I think that these are sufficient reasons to consider that the double diagram printed in Düring's edition of Ptolemy's *Harmonics* 2.4 (fig. 2) is not original.

References

Alexanderson, B. (1969). *Textual Remarks on Ptolemy's* Harmonica *and Porphyry's* Commentary, Göteborg.

Barker, A. (1989) *Greek Musical Writings, Volume II: Harmonic and Acoustic Theory*, Cambridge.

Bobo de la Peña, M. (2006). "Algunas consideraciones críticas y exegéticas en torno al texto de la *Harmonía* de C. Tolomeo (I)", *Emerita* 74/2, 341-58.

—. (2007). "Algunas consideraciones críticas y exegéticas en torno al texto de la *Harmonía* de C. Tolomeo (II)", *Emerita* 75/1, 137-52.

Düring, I. (1930). *Die Harmonielehre des Klaudius Ptolemaios*, Göteborg.

—. (1932). *Porphyrios. Kommentar der Harmonielehre des Ptolemaios*, Göteborg.

Düring, I. (1934). *Ptolemaios und Porphyrios über die Musik*, Göteborg.

Heiberg, J. L. (1898). *Claudii Ptolemaei opera quae extant omnia:* Syntaxis Mathematica I, Leipzig.

—. (1903). *Claudii Ptolemaei opera quae extant omnia:* Syntaxis Mathematica II, Leipzig.

Jan, K. v. (1895). *Musici scriptores graeci*, Stuttgart-Leipzig.

Johnson, W. A. (2004). *Bookrolls and Scribes in Oxyrhynchus*, Toronto.

[34] I do not take into consideration the possibility that the four diagrams I presented as Porphyry's are by Ptolemy, since undoubtedly in that case they would have been referred to in Ptolemy's text.

Netz, R. (1998). "Deuteronomic Texts. Late Antiquity and the History of Mathematics", *Revue d'histoire des mathématiques* 4, 261–288.

—. (1999). *The Shaping of Deduction in Greek Mathematics. A Study in Cognitive History*, Cambridge.

—. (2004). *The Works of Archimedes, Vol. 1. The Two Books On the Sphere and the Cylinder*, Cambridge.

Pedersen, O. (1974). *A Survey of the Almagest*, Odense.

Raffa, M. (2002). *La scienza armonica di Claudio Tolomeo*, Messina.

Redondo Reyes, P. (2003). *La 'Harmónica' de Claudio Ptolomeo: edición crítica con introducción, traducción y comentario*, Murcia.

Robson, E. (2008). *Mathematics in Ancient Iraq. A Social History*, Princeton, N.J.

Solomon, J. (2002). *Ptolemy Harmonics: Translation and Commentary*, Leiden.

Taub, L. C. (1993). *Ptolemy's universe : the natural philosophical and ethical foundations of Ptolemy's astronomy*, La Salle, Ill.

Wallis, J. (1682). *Claudii Ptolemaei Harmonicorum libri tres*, Oxford

EUCLID'S SWAN SONG: EUCLID'S *ELEMENTS* IN EARLY MODERN EUROPE

ANTONI MALET

Introduction

Francesco Maurolico's famous 1540 letter to the influential man of letters, Cardinal Pietro Bembo, outlined his plans for the mathematical renaissance. Published in 1543 as a preface to his *Cosmographia*, it featured a comprehensive, almost universal list of classical mathematical texts the edition and translation of which Maurolico advocated. For many of them Latin versions already existed, but Maurolico criticized these for being incompetently done, as regards their mathematics or their Latin or both. Divided into four sections, Maurolico's list vividly conveys the enormous task facing those like him who wanted to digest in its entirety the classical mathematical heritage. The four-section structure reveals the meta-mathematical categories that came to organize mathematical knowledge, replacing the centuries old *quadrivium*. The first section, devoted to geometry, includes first Euclid's *Elements* (Euclid is "our prince of mathematicians"), and then Apollonius's *Conics*, Euclid's *Data*, and Archimedes's work on the measure of the circle, the sphere and cylinder. Works by Theodosius, Serenus, Menelaus, Dionysodorus, and Maurolico himself (on regular bodies) are also included.

The second section includes works on optics (by Euclid, Ptolemy, Pecham, Roger Bacon and Maurolico), music (Boethius, Guido d'Arezzo), theoretical mechanics (Maurolico, Archimedes; for treatises on machines, see below), arithmetic (Boethius, Jordanus, Maurolico), and algebra (no author mentioned). The third section covers astronomy, cosmography and geography. It is a long section, running the gamut from Ptolemy to Peuerbach and Regiomontanus–but not Copernicus. The fourth and final section is devoted to instruments and machines. It includes pseudo-Aristotle's mechanical problems, Vitruvius's and Heron's works (supplemented with recent literature on hydraulic machines), and treatises on the construction of quadrants, astrolabes, sundials, and clockworks.[1]

Notice the structure of the sections. The first one is devoted to foundational, general, theoretical and geometrical works. The second is devoted to mixed or applied mathematics–except for astronomy, which gets its own section along with cosmography and geography, all of them disciplines characteristic of that century of geographical discoveries. Notice the logic of this structure–from general geometrical works through the traditional disciplines that apply mathematics to physical subjects down to mathematical descriptions of machines. Notice finally the contrast with the *quadrivium*, in which two abstract, general sciences, independent of each other–geometry and arithmetic–had each their own subordinated sciences. In Maurolico's division of the mathematical sciences, arithmetic and algebra belong in the second, "applied" section, thereby suggesting that arithmetic is no longer thought of as being independent of geometry. In the light of subsequent developments, this may be taken as an early sign of the process that will eventually erase the opposition between the classical category of number or discrete quantity, the object of arithmetic, and that of geometrical magnitude or continuous quantity, the proper stuff of geometry. This topic not only looms large in 17th-century discussions of the nature of the mathematical sciences, but also, as we shall see here, it decisively shaped the reception of Euclid's *Elements* in early modern Europe.

Taking into account that all the important titles in Maurolico's list of geometrical books were competently translated in the decades following the publication of Maurolico's letter, one might say that by the mid seventeenth century the reception of the mathematical classical heritage (as outlined by Maurolico) had been successfully completed. On the other hand, however, substantial evidence suggests that most seventeenth-century commentators of Euclid's *Elements* misunderstood and modified basic, fundamental notions of classical mathematics. In what follows, we will focus on early modern interpretations of the Euclidean notions of number and continuous magnitude, ratio and proportionality. They offer a case study of conceptual change in mathematical notions.

---

[1] F. Maurolico, *Cosmographia* (Venice, 1543), pp. 3rd to 7th of the unpaginated preface titled "Reverendissimo Domino D. Petro Bembo, ..."; compare with Rose (1975: 163-164).

The status of Euclid in the 16th and 17th centuries

As is well known, between Campanus's edition of Euclid's *Elements* (mid 13th century, first printed in 1482) and Commandino's (1572) most, not to say all translators and commentators spoke of its author as Euclid of Megara, the philosopher contemporary of Plato who lived *ca.* 400 BCE. Commandino's clarification of the mathematician's identity notwithstanding, Euclid's historical person–including his authorship of the *Elements,* as they were actually known in the 16th century–remained elusive.[2] In his mathematically cogent, innovative, and influential Italian translation of the *Elements* (1543), Tartaglia maintains that no faithful version of Euclid's original *Elements* is available. In particular, he claims, the *Elements* originally presented the "highest and most admirable order imaginable," yet the "moderns" (*sic*) "corrupted [and] annihilated" it. He thinks the two versions of the *Elements* available by the early 16th century (Campanus's and Zambert's) have "disordered" and "debased" the original work. He concludes that the genuine arithmetic and geometry of Euclid "are almost lost". Tartaglia does not elaborate on the reasons for such decadence, other than to stress the language barrier that separates the moderns from the ancients.[3] It is therefore implied that Tartaglia's rendering of the *Elements* cannot be true to the original, since no faithful version exists, but that he will make the best possible sense of the inherited textual tradition. By assuming the text of the *Elements* to be at best an approximation to the mathematical knowledge of the ancients, Tartaglia felt entitled to interpret the *Elements* available to him. He did modify its contents with happy results, for he was the first to correct the faulty reading of Euclid's definitions of ratio and proportionality to be found in all Medieval Latin versions of the *Elements*.[4]

Clavius's edition of Euclid's *Elements* (1574) stresses too the unreliability of the versions then available.[5] Maurolico made similar comments in his editions of the *Elements* (only partially printed in 1575) and of Apollonius's *Conics*.[6] By the turn of the 17th century the topic of

---

[2] See I. Bulmer-Thomas's and J. Murdoch's contributions to Gillispie ([ed.], 1970-1980); Euclid, edited by T.L. Heath (1956, vol. I, p. 3-4).

[3] N. Tartaglia, *Euclide megarense philosopho: solo introduttore delle scientie mathematice: diligentemente reassettato, et alla integrita ridotto... Secondo le due Tradottioni* (Venice, 1543), f. 3r.

[4] See Murdoch (1963; 1968) and Drake (1973).

[5] C. Clavius, *Commentaria in Euclidis Elementa Geometrica* (Mainz, 1611, repr. Hildesheim: Olms-Weidmann, 1999), p. 10.

[6] Rose (1975: 163, 165-6).

the lost mathematical knowledge of the ancients reappears forcefully and with an original twist in Simon Stevin. In his innovative work on arithmetic and algebra, *L'Arithmetique* (1585), Stevin first confronted head on Euclid's notion of number as *arithmoi*, a finite collection of discrete units (more about this below).[7] The background for Stevin's innovative (one is tempted to say revolutionary) conceptualization of numbers as continuous magnitude is found in his evocation of a golden age lost in the beginning of times, *De Wysentyt*.[8] First published in Dutch in 1608 and translated by Stevin's student, Albert Girard, into French as *Le siècle sage*, Stevin's *age of the sages* introduces a historical time of exceptional learning the reality of which is suggested "with certainty" by a number of signs. These include the Arabic mathematical learning that cannot be traced back to Greek sources along with branches of knowledge that only shortly before Stevin's time had resurfaced, such as algebra, alchemy, and magic.[9] However, these signs did not inform Stevin about where or when or among whom that extraordinary learning was found in ancient times. In particular, Stevin construes Arabic algebraic knowledge as evidence that men in the past mastered arithmetic in ways unknown to the Hebrews, the Greeks or the Romans, all of whom he disparages as being "no arithmeticians worth the name".[10] Since the arithmetic of the learned age fell into decline and Euclid passed it down corrupted and full of imperfections, Stevin urges the reader to turn to sources other than Euclid, particularly to Arabic sources, to improve arithmetical knowledge.[11] Stevin's *Arithmetique*, with its innovative notion of number as measure and its criticism of Euclidean *arithmoi*, is Stevin's own contribution to the task of restoring the *prisca mathematica* of the sages.

Stevin was not idiosyncratic in assuming that Hermes Trismegistus and other men of extraordinary learning taught humankind philosophy and mathematics that became afterwards corrupted or lost. Tartaglia's long list

---

[7] On the conceptual innovations in Stevin's *Arithmetique*, see Malet (2006; 2008).

[8] *De Wysentyt* was originally published as an introductory text to the geographical section of Stevin's cosmography, in part I of the 2-volume *Wisconstige gedachtenissen* (1605-1608). Under the title *The Age of the Sages*, it has been partially reproduced with English translation in *The Principal Works of Simon Stevin*, D.J. Struik et al. (eds.), 5 vol. (Amsterdam, 1955–1966), III, pp. 591-623. A. Girard's translation, titled *Le siècle sage*, is found in *Les Oeuvres Mathematiques de Simon Stevin*, A. Girard (ed.) (Leyde, 1634), part II, pp.104-128.

[9] Stevin, *Principal Works*, III, 593, 599-607.

[10] Stevin, *Principal Works*, III, 599.

[11] Stevin, *Principal Works*, III, 601.

of past mathematicians opens with Hermes Trismegistus, "philosopho, sacerdote, & Re d'Egitto". [12] Stevin discussed his views about the age of the sages with the learned jurist Hugo Grotius–Grotius had in his youth helped translate into Latin some of Stevin's works–and got from him a list of classical authors holding similar beliefs.[13] The influential reformer and likely source of inspiration for Stevin, Pierre de la Ramée, awarded different roles to Euclid in his different histories of mathematics, but he always assumed–as was quite generally assumed at the time–that Euclid was not the author of the demonstrations of the propositions in the *Elements*. In any case, he believed the original *Elements* lost, while the ones available in the Middle Ages and the 16th century were but a second-hand, unreliable corrupted version of the "good", "natural" mathematics of the golden age.[14]

The sixteenth century produced better editions and translations of Euclid's *Elements* than the ones available at any time since antiquity. First class mathematicians and humanists, including Commandino, Tartaglia, and Clavius, prepared remarkable editions with commentaries that contributed to fix a text free of gross mathematical errors and inconsistencies. As shown above, however, the *Elements* was not above suspicion or criticism, and in fact most editions modified definitions, corrected proofs, reinterpreted substantial results–in sum its contents were questioned and occasionally amended. As we shall see now, the seventeenth century radicalized this trend while, on the contrary, no one seemed interested in pursuing the exegetical efforts of Commandino or Clavius.

We have listed in the Appendix a large sample of editions and translations of the *Elements* originally published between 1600 and 1700 by some fifty odd different authors; some sixteenth-century editions were still popular in the seventeenth century, like Commandino's and Clavius's, but these have been deliberately omitted. A few seventeenth-century editions were due to creative, well-known mathematicians, such as Isaac Barrow, Nicholas Mercator or André Tacquet. However, the Appendix

---

[12] Hermes, who opens the list, is followed by (in this order) Pythagoras, Plato, Plotinus, Aristotle, Averroes, Hypocrates, Euclid, Ptolemy, Archimedes, Apollonius, Jordanus, and Vitruvius; see Tartaglia, *Euclide*, f. [4]r (sic, the page is actually numbered "iii").

[13] Stevin, *Principal Works*, III, 607

[14] Ramus changed his views on the history of mathematics and on the nature of the mathematics of the ancients during his intellectual development, but once he got familiarized with Euclid's *Elements* he did not have a high opinion of it; see Goulding (2006), Popper (2006), Cifoletti (1996).

shows most versions of the *Elements* originally published between 1600 and 1700 to be due to obscure and otherwise unknown teachers and mathematical practitioners, including military engineers. Not surprisingly, we find quite a number of Jesuits among the authors listed–George Fournier, Claude Richard, Claude Dechales, and Tacquet, among them. Some of the editions turned out to be extremely popular and went through multiple reprints or new editions. Thus, Denis Henrion's was first published in 1615 and knew no less than 9 reprints before 1700. George Fournier's, had 5 reprints between 1643 and 1665. Isaac Barrow's, first published in Latin in 1655 then published in English in 1660, was perhaps the most popular Euclid in England; it went through no less than 11 reprints (counting both Latin and English versions) up to 1687. The Oratorian Bernard Lamy's version knew 7 editions between 1685 and 1758. Highly popular all over Europe were Dechales's and Tacquet's editions, both originally published in Latin in 1660 and 1654, respectively, and then reprinted no less than 3 times and 12 times up to 1690 and 1745; they were translated into English in 1685 and 1714, in which language they knew no less than 10 and 8 editions, the last ones apparently in 1748 and 1791. Dechales's Euclid was also translated into Italian at least once (1749) and French–12 times between 1672 and 1778.

These figures give some of Euclid's editions and translations the status of mathematical best sellers, which is of course consistent with the suggestion that seventeenth-century editions were done with the main purpose of turning the *Elements* into a teaching aid. Many titles openly declare this purpose, thus Dechales's *Euclidis Elementorum libri octo : Ad faciliorem captum accommodati* (1660); Martini Gestrini's *In Geometriam Euclidis demonstrationum libri sex in quibus Geometria planorum traditur et brevibus notis perspicue explicatur* (1642); or Johannes Lanz's *Euclidis Elementorum geometricorum libri sex priores. Nova interpretatione in usum studiosæ iuventutis* (1617). Around three quarters of these editions shorten the *Elements* by leaving out the books that were deemed too difficult and/or without interest from a practical perspective–the arithmetical books (VII to IX), the book on irrational quadratics (X), and sometimes even one or both of the books on three-dimensional bodies (XI and XII). Consistently, many titles and prefaces prominently trumpeted that in order to improve the *Elements* the authors had variously abbreviated, rewritten, rearranged, and reformed the work–thus Thomas Rudd's *Euclides Elements of geometry: the first VI books: in a compendious form contracted and demonstrated* (1651), G. Alfonso Borelli's *Euclides restitutus; sive, Prisca geometriae elementa, brevius & facilius contexta, in quibus precipue proportionum theoriae nova, firmiorique methodo proponuntur* (1658), or William Alingham's *Geometry epitomiz'd : being a compendious collection of the most useful propositions in the first, third, fourth, fifth and sixth books of Euclid: together with their uses, in several practical parts of the mathematicks: also, Euclid's second book and doctrine of proportion, algebraically demonstrated* (1695).

Alterations in the order and the demonstrations of the *Elements* were introduced on the understanding that Euclid's original demonstrations were lost or poorly preserved. In the last decades of the century, the Oratorian Bernard Lamy (1640-1715) put it in language somehow reminiscent of Stevin's:

> From Euclid, we just have the propositions of the *Elements*: the demonstrations are by Proclus. Therefore, in order to offer Euclid's works, it is enough to convey [the enunciations of] his propositions. Experience will show that simply the order in which I offer them facilitates their demonstrations; this is why I am confident that Euclid will be learned here in a much easier way than in any of his Commentators.[15]

Alterations and rearrangements in the *Elements* were introduced alike by competent and learned mathematicians and by obscure mathematical practitioners, with one remarkable exception that we will address below. Authors rewrote substantial parts of the *Elements* because they thought the work could be made easier and clearer. In the words of the Jesuit Claude M. Dechales (1621-1678):

> ... most of those who learn Euclid's *Elements* very often feel dismayed as they don't understand the usefulness of the propositions ... so difficult: so I have considered ... convenient ... to render them as easy as possible,... This has compelled me to modify some demonstrations...[16]

---

[15] "Nous n'avons de lui [Euclide] que les propositions qui sont dans ses *Elemens*: les démonstrations sont de Proclus. Ainsi pour donner un Euclide, il n'est question que de rapporter ses propositons. L'experience fera voir que le seule ordre que je leur donne, en facilite la démonstration; c'est pourquoi j'espere qu'on aprendra ici Euclide avec beaucoup plus de facilité qu'en aucun de ses Commentateurs", B. Lamy, *Les Elemens de Geometrie, ou de la mesure de l'étenduë; qui comprennent les Elemens d'Euclide* (4e ed., Paris, 1710), Preface, unnumbered page marked âvj verso.

[16] "la pluspart de ceux qui apprennent les *Elemens* d'Euclide, s'en dégoûtent fort souvent, pour ne pas sçavoir, à quoy servent des Propositions ... si difficiles: j'ai crü ... tres-à-propos ... de les rendre les plus faciles qu'on pourroit; ... C'est ce qui m'a obligé de changer quelques demonstrations ...", C.F. Milliet Dechales, *Les Elemens d'Euclide expliquez d'une maniere nouvelle et tres-facile* (Paris, 1677), Avant-propos.

Euclid's text was criticized at least on three counts. First, it lacked 'method', as Euclid's sole preoccupation was to place propositions in the sequence required to prove one after the other (no further elaboration). Lamy–echoing the Port Royal theologian Pierre Nicole's complain in the Preface to Arnauld's *Elemens of Geometrie* of 1667–stressed that for geometry to be of pedagogical value it had to be presented with the right method, but the *Elements* lacked it. The consequence was 'confusion':

> it is necessary that it [geometry] be dealt with method, something Euclid does not. He just took pains to place propositions in such a way that some act as proofs for the others, and he did it right. Truth can be found in the *Elements*; but also so much confusion... [that the *Elements* make spirits habituated] to disorder and confusion, just as Monsieur Nicole complains.[17]

Secondly, authors criticized many of Euclid's demonstrations for being unnecessarily convoluted and so shortened and simplified them. Finally, they blamed Euclid for not properly defining some fundamental notions of his work, and these included at the very least the cluster of notions related to ratio and proportionality. Thus Dechales:

> proportion is a likeness [that is, equality] of ratios. But it is difficult to understand what this likeness between ratios consists of. ... Euclid did not offer a proper definition, that is, one that explained the nature [of the likeness]; he felt it was enough to give us a signal by which we may know when the quatities have the same ratios. And it is precisely the obscure quality of this definition that makes this book [the 5th] so difficult. I will try to amend this fault.[18]

In sum, by the late seventeenth century the *Elements* was not an authoritative source offering insight in controversial matters, or guidance

---

[17] "il faut qu'elle [la géométrie] soit traitée avec methode, ce que ne fait pas Euclide. Il n'a pensé qu'à ranger ses Propositions de maniere qu'elles se servissent des preuves les unes aux autres; en quoi il a reussi. La verité se trouve dans ses *Elemens*; mais au reste il y a tant de confusion ... [que les *Elemens* accoûtument l'esprit] au desordre et à la confusion, comme s'en plaint Monsieur Nicole", Lamy, *Elemens de Geometrie*, Preface, âvj recto.

[18] "la proportion est une similitude de raisons. Mais on a de la peine à entendre en quoy consiste cette similitude de raisons. ... Car Euclide n'en a pas donné une definition juste, et qui en expliquât la nature; s'étant contenté de nous donner une marque par laquelle nous puissons connoistre, si les quantitez avoient une mesme raison. Et c'est l'obscurité de cette definition qui a rendu ce Livre difficile. Ie tâcheray de supléer à ce defaut", Dechales, *Elemens d'Euclide*, p. 225

about the right understanding of contested notions. Most mathematical teachers and authors apparently considered the *Elements* as a text they could and should improve. This opened the door to debates about some fundamental notions of mathematics, to which we turn.

Conceptual change: ratios as a case study

As is well known, Euclid's *Elements* contains no equivalent to the notion of a segment length. More generally, the geometrical objects in the *Elements* have no measures, and in consequence are operationally different from numbers.[19] While the rectangle "built" at right angles on two straight-line segments is indirectly related to the modern notion of product, Euclidean segments cannot be multiplied or divided. Similarly, powers and roots do not exist as such for Euclidean magnitudes, although they are related to the notion of continued proportion between magnitudes. Euclidean numbers, on the other hand, are always (positive) integer numbers, defined as collections of units. The unit, in turn, is not mathematically defined but is a logical a priori concept. In the *Elements*, operations with numbers (for instance, the determination of proportional means) are not always allowed, since it is necessary for the answers to be meaningful–that is to say, integer numbers. As far as we know, the neat, consistent separation between Euclidean numbers and continuous magnitudes was preserved in Latin medieval translations of the *Elements*. By the late 17th century, however, the dividing line between numbers and continuous magnitudes was largely gone.

The most important antecedent for what we may call the arithmetization or quantification of continuous magnitudes is found in Arabic medieval mathematics, where the classical notions of number and magnitude come very close together. As a matter of course, magnitudes are there handled under the assumption that they have numerical measures. Arabic medieval algebra takes for granted that arithmetical notions and operations can be applied to geometrical magnitudes. To do so, some specific resources are needed, such as computational rules for numerical roots (or "surd" numbers), but Arabic algebra does provide the necessary algorithms.

The quantification of continuous magnitude in 16th- and 17th-century Europe was neither theoretically elaborated nor properly grounded. We find no trace of it in Medieval Latin editions of the *Elements*. Apparently,

---

[19] For a detailed discussion of the differences between Euclidean numbers and continuous magnitudes, see Bos (2001: 120-127).

it first came to the Latin West mostly by way of the applied mathematics taught in *abacus* schools and set forth in *abacus* manuscripts, the obvious heirs of Arabic algebras. In ways that we do not know well yet, sixteenth-century mathematicians came to embrace a new understanding of number and magnitude very much influenced by the abacus tradition. We have dealt elsewhere with the modifications three important sixteenth-century editions of the *Elements* introduced in the notions of number, magnitude, and their operations.[20] They did so by way of commentary and interpolation while essentially preserving the letter of the *Elements*, which is perhaps the reason why the notions of ratio and proportionality were left largely untouched in 16th-century editions of the *Elements*. We shall focus just on the alterations introduced during the 17th century in the notions of ratio and proportionality and some of the debates they prompted.

As is well known, "ratio" is defined in Euclid's *Elements*, Book V 3, as "a sort of relation in respect of size between two magnitudes of the same kind". Already in ancient times commentators pointed out the ambiguity of this definition.[21] Ratios, however, were safely handled in mathematical demonstrations by way of Euclid's equimultiple definition of equality of ratios (Book V, definition 5):

> Magnitudes are said to be in the same ratio, the first to the second and the third to the fourth, when, if any equimultiples whatever be taken of the first and the third, and any equimultiples whatever of the second and fourth, the former equimultiples alike exceed, are alike equal to, or alike fall short of, the latter equimultiples respectively taken in corresponding order.[22]

Grasping the mathematical sense of this intricately worded definition requires the gifts of a trained mathematician. However, it was not only for the mathematical apprentice to complain. Starting in the 1640s, if not before, influential mathematicians raised doubts about the adequacy of Euclid's definitions. According to Galileo, writing probably in 1641, Euclid's definition of equality of ratios was unsuitable because of its difficulty and unintelligibility.[23] He would rather have the equality of

---

[20] Malet (2006).

[21] Heath ([ed.], 1956: II, p. 114, 116-119).

[22] Heath ([ed.], 1956: II, p. 118).

[23] In fact, he was dictating to Torricelli what is now known as the *Principio di giornata aggiunta*, or the 'Fifth Day of the Discorsi'. It first appeared in print in V. Viviani, *Qvinto libro degli Elementi d'Evclide: ovvero Scienza vniversale delle proporzioni spiegata colla dottrina del Galileo* (Florence, 1674). We have used E. Giusti's edition of Galileo's text in Giusti (1993); references to the unintelligibility of Euclid's definition on p. 281-2.

ratios defined, first, between commensurable magnitudes through the condition that the antecedent of each ratio is the same "multiple" of the consequent–and here "multiple" is understood to refer to fractional numbers too, as in Galileo's example: "if the first [magnitude] contains three and a half times the second one, and the third contains three and half times the fourth one, then it is to be understood that these magnitudes are proportional".[24] However, when Galileo turns to ratios in general, he falls into a circular definition. He explicitly assumes the axiom that for any three magnitudes (say A, B, C, he does not bother stating conditions of existence for the ratios, such as the homogeneity of magnitudes), there is one fourth magnitude D such that (A:B) = (C:D). Now, given four magnitudes (say P, Q, R, T), Galileo defines the two ratios (P:Q) and (R:T) equal if T coincides exactly with the fourth proportional to P, Q, and R.[25]

Somehow echoing Galileo's criticism of the difficulty and indirectness of Euclid's definition, many seventeenth-century authors suggested alternative definitions of ratio and equality of ratios–the list includes but is not restricted to Mersenne, Tacquet, Saint-Vincent, Galileo, Torricelli, Borelli, Wallis, Hobbes, Dechales, Lamy, Arnauld, and Prestet. The Jesuit, highly competent mathematician André Tacquet (1612-1660) details the difficulties he finds in Euclid's definitions before advancing a new definition of equality of ratios (first published in 1654) that was to appear in other authors.[26] According to Tacquet, who brings in the authority of his teacher, the Jesuit Gregory of Saint Vincent, Euclid's definition of equality of ratios does not explain the "nature" (sic) of equal ratios, but only provides a property the two ratios have in common–that is, that the equimultiples of their antecedents and consequents behave as the definition requires.[27] Tacquet suggests that he and his readers already

[24] "se la prima [magnitude] contenga tre volte e mezzo la seconda, et anco la terza contenga tre volte e mezzo la quarta, allora s'intenderà che queste grandezze siano proportionali".

[25] Giusti (1993: 282-284); Galileo's definition and quotation for ratios between commensurable magnitudes, on p. 283; his definition for the general case, on p. 284; on the circularity of Galileo's definition, see Giusti's comments on *ibid*, pp. 76-78.

[26] On Tacquet's understanding of proportionality, compare with Palladino (1991).

[27] "Hence ... it is true that this definition does not explain the nature of equal ratios, but just a particular property of them" ("Nam ... certum est ea definitione non naturam aequalium rationum, sed affectionem solummodo aliquam explicari"), see A. Tacquet, *Elementa Euclidea Geometriae Planae, ac Solidae*, W. Whiston, ed. (Venice, 1781), p. 83; I have checked Tacquet's text in this 1781 edition with

know what makes two ratios equal. Accordingly, he thinks it necessary to demonstrate that Euclid's definition fully characterizes equality (as tacitly understood by Tacquet and his readers). That is, it is necessary to prove that the equimultiple property is found in equal ratios (in Tacquet's tacit sense) and only in them–something that neither Euclid nor any of his followers had previously done. Alternatively, says Tacquet, we can take Euclid's definition as a new way of construing equality between ratios (different to the tacit understanding); in this case, from now on "two ratios are equal" only means that the equimultiples of their antecedents and consequents behave in the manner required by Euclid. In this case, we shall be secure of the truth of theorems received, and yet the power of the demonstrations cannot properly show us the equality of the ratios.[28] He provides the example of proposition VI 1, which proves the ratio of two triangles of equal height to be equal to the ratio of their bases. Tacquet gives credit to Euclid for proving that the ratios are "equal"–but "equal" here only means the property of the equimultiples. He highlights that the proof does not allow us to think the two ratios truly equal, since it has not been demonstrated that the property of the equimultiples necessarily entails the true equality of ratios (in Tacquet's tacit sense).[29] In essence, Tacquet's point is a metaphysical one. He blames Euclid's definition of equality of ratios for being "external" to the objects concerned, that is, for telling us when we are allowed to write $\rho_1 = \rho_2$ (for any two ratios, ρ_1 and ρ_2), while providing no information about the mathematical objects ρ_1 and ρ_2. According to Tacquet, it is not enough–as Euclid does–to choose a characteristic property or feature two ratios have in common and turn it into a mark of equality. Equality can only be predicated of two objects when we know each of them, and can conclude out of this knowledge that they are equal.

To mend this state of affairs, Tacquet rewrites Euclid's Book V and provides new definitions. Ratio he defines *à la Euclid*, and yet he introduces a subtle but crucial modification: "Ratio or proportion is a sort

---

that of an earlier one of 1665 and they are identical. W. Whiston's comments to the 1781 edition are clearly marked as additions to Tacquet's text.

[28] Tacquet, *Elementa Euclidea Geometriae* (1781), p. 83-84.

[29] "we cannot infer, however, from the ratios of these triangles, that the ratios of their bases are certainly and absolutely equal, since it is not demonstrated that the property of the equimultiples is necessarily linked to true equality of ratios" ("non colligemus tamen rationes illas triangulorum, basium rationibus vere et absolute aequales esse, cum demonstratum non sit affectionem illam multiplicium cum absoluta et vera rationum aequalitate necessario esse conexam"), in Tacquet, *Elementa Euclidea Geometriae* (1781), p. 84

of relation according to quantity between two magnitudes of the same kind" ("Ratio sive proportio est duarum ejusdem generis magnitudinum mutua quaedam secundum quantitatem habitudine"). Tacquet has silently substituted the Euclidean original notion of size by the notion of quantity–which tacitly opens the door to a numerical consideration of continuous magnitudes. Galileo had done the same thing.[30] Tacquet's most substantial modification is to be found in his definition of equality of ratios, which rests on the quantification of the way one of the terms of any given ratio contains the other. Tacquet knows that for ratios between magnitudes in general he cannot talk of the *number* of times one magnitude contains the other–this is acceptable only for ratios between commensurable magnitudes. Tacquet's (obviously weak) option is to rely on the notion of magnitudes "contained in the same way":

> Two ratios (A to B, and C to F) are alike, equal; when the antecedent of the first (A) contains its consequent (B) equally or in the same way (that is, neither more nor less) as the antecedent of the second (C) contains its consequent (F).[31]

Tacquet adds that for rational ratios, numbers can explain the way in which antecedents contain or are contained by consequents. If ratios are irrational, however, that notion–that is, the way in which one term contains the other–neither can nor should be further explained or exposed. Given an irrational ratio (A : B), he claims there is a definite, unique way for one of the magnitudes (say, A) to contain the other, since A contains B differently than any other magnitude C contains B. Since numbers cannot express this characteristic way, one should not further ask how to express it.[32]

As quoted above, Dechales echoed Tacquet in blaming Euclid for not explaining the "nature" (sic) of the equality of ratios and for giving instead a "mark" by means of which to recognize whether or not the ratios are

---

[30] Tacquet, *Elementa Euclidea Geometriae* (1781), p. 87. Galileo assumes what he calls *Euclid's* definition: "the proportion between two magnitudes [is] a particular rapport or relationship between them regarding their quantity" ("la proportione fra due grandezze [is] un tale rispetto, o relatione fra di loro per quanto appartiene alla quantità"), in Giusti (1993: 282).

[31] "Duae rationes (A ad B, et C ad F) sunt similes, aequales, eadem; cum unius antecedens (A) aeque seu eodem modo (hoc est nec magis, nec minus) continet suum consequens (B), quo alterius antecedens (C) continet suum consequens (F)", Tacquet, *Elementa Euclidea Geometriae* (1781), p. 87. Tacquet adds to the definition the case in which the consequents are bigger than and equally contain the antecedents.

[32] Tacquet, *Elementa Euclidea Geometriae* (1781), p. 88.

equal. Writing in 1660, Dechales adds that Euclid's definition is so intricate that it cannot pass for a principle–he is here assuming the Aristotelian requirement that principles of knowledge must be self-explanatory and universally received. Echoing again Tacquet, Dechales suggests that a theorem is needed that demonstrates that equal ratios verify the equimultiple property of Euclid's definition.[33] Playing with Tacquet's idea for the equality of ratios, Dechales defines the proportionality of four magnitudes by requiring that the first be a similar part or a similar whole in relation to the second as the third is in relation to the fourth. He refines this definition by adding an informal limit process: the ratio between A and B is the same as the ratio between C and D, if A contains one hundredth, one thousandth, one ten-thousandth, etc, of B as many times as C contains one hundredth, one thousandth, one ten-thousandth, etc of D, and so on for all imaginable aliquot parts of B and D. Notice the big problem in this definition, as well as in Tacquet's, that there is no general, abstract way of comparing parts of continuous magnitudes in relation to the whole. Whether a continuous magnitude A is the "same part" of B as C is of D, in Euclid's *Elements* can be ascertained only by having the four magnitudes embedded in a geometrical construction. The introduction of decimal fractions in the procedure did not circumvent the problem (more about this below).

Both Tacquet and Dechales were Jesuits, but Antoine Arnauld (1612-1694), the Port Royal theologian, mathematician, and Jesuits's enemy in theology, philosophy and morals, held mathematical views similar to theirs. If anything, Arnauld in his *Nouveaux Elemens de Geometrie* (1667) is more radical and more explicit than they were in assuming that continuous magnitudes are "quantities" and therefore amenable to arithmetical operations. He is also more radical in questioning the organization of the *Elements*, although his idea of method seems to be psychological in nature rather than logical. According to Arnauld, the axiomatic-deductive method is not the "natural" (sic) order for the sciences. The natural order he compares to a light that makes everything in it perspicuous because every idea illuminates the rest. In particular, he advocates assuming everything that is known by "natural light". One of Arnauld's assumptions is that "multiplication and division apply to all magnitudes and not only to numbers".[34] He defines the ratio of two magnitudes as "the manner [*manière*] in which one magnitude is contained

[33] Dechales, *Elemens d'Euclide*, p. 226.

[34] A. Arnauld, *Nouveaux Elemens de Geometrie* (Paris, 1667), see the Preface for the claim that all magnitudes are quantities; Preface and p. 1, on the "natural" order and the axiomatico-deductive method; p. 3 for the "Quatrieme supposition".

in the other", and defines equal ratios as those in which the antecedents contain (or are contained by) the consequents "in equal manner": "Two ratios are called equal when the antecedents contain in equal manner the consequents, or are contained in equal manner in the consequents".[35]

Interestingly, Arnauld stops short of saying that the quotient of antecedent by consequent is a measure of the manner in which the former contains the latter, and consequently he does not use the quotients to define the equality of ratios. In fact, the definition of the quotient between continuous magnitudes poses a mathematical problem that requires the notion of ratio (more about this below). He acknowledges, however, that it is often difficult to judge whether antecedents equally contain consequents (or vice versa). He attempts a practical improvement of his definition by asking that *all* aliquot parts of the antecedents be equally contained in the consequents–and apparently he does not recognize that he is applying again the troublesome notion of "being contained" to general continuous magnitudes.[36]

Many more 17th-century works drop Euclid's definitions of ratio and proportionality, criticize them for reasons of both form and content, and offer instead new definitions similar to the ones presented here. In 1666 Isaac Barrow devoted his Cambridge lectures to ratio and proportionality, which we will briefly consider next. Barrow paid particular attention to the views of Tacquet and Wallis, but his lectures contain a detailed and thoughtful analysis of the criticism addressed to Euclid's definitions by the likes of Tacquet, Dechales, and Wallis, as well as of the alternative definitions they set forth.[37] Barrow's comments are evidence that there

---

[35] "Deux raisons sont appellées égales quand les antecedens contiennent également les consequens, ou sont également contenus dans les consequens", Arnauld, *Nouveaux Elemens de Geometrie*, p. 23 (for the defintion of ratio), p. 26-27 (for the definition of equal ratios).

[36] Arnauld, *Nouveaux Elemens de Geometrie*, p. 29: "Two ratios are called equal when *all* the aliquot parts of the antecedents are equally contained in the consequents" ("Deux raisons sont appellées égales quand toutes les aliquotes pareilles des antecedens sont chacunes également contenües dans chaque consequent"). For any antecedent A, the suite of all its aliquot parts is A/10, A/100, A/1000, etc. Arnauld stresses that his second definition is grounded on the first, it being nothing else than a technical trick to facilitate computations (see p. 31).

[37] I. Barrow, *Isaaci Barrow, mathematicae professoris lucasiani Lectiones habitae in scholis publicis Academiae Cantabrigiensis*, Annis Domini 1664, 1665, 1666 (London, 1683); new editions with slightly different titles appeared in 1684 and 1685. It was anglicized as *The usefulness of mathematical learning explained and demonstrated*, John Kirkby, trans. (London, 1734).

were no explicit mathematical grounds to abandon Euclid's definitions of ratio and proportionality.

Barrow on ratio and proportionality

The new definitions of ratio and proportionality all somehow assumed that for every ratio it is possible to consider "how many times" or "in what manner" the ratio's terms *contain* each other. This assumption requires, first, that the relation of "being contained" holds generally between homogeneous magnitudes. Secondly, it requires a procedure to exactly determine how many times one continuous general magnitude contains another. Notice that these are significantly stronger requirements than Euclid's Definition V 4, that the terms of any ratio be capable when multiplied of *exceeding* one another. It is not obvious how two weights, speeds, illuminations, or forces may *contain* one another, nor how the number of times one contains the other may be determined, unless *it be assumed that general magnitudes are measured or numerically expressed.* If this assumption is not made–as the letter of the *Elements* requires–there is no general way to determine "how many times" a continuous magnitude A is contained in or contains a homogeneous magnitude B. In this case, the definition of the ratio (A : B) in terms of "how many times" one of the terms contains the other amounts to something like "the ratio of A to B is what comes from the comparison of the size of A to the size of B". That was Tacquet's and Saint Vincent's position.[38] In Barrow's view this approach was inconsistent in that it purported to assign a quantity to every ratio but failed to determine it properly.[39] He stressed, furthermore, that it can never be conceived what it means for the antecedent of an irrational ratio to contain or be contained by the consequent.[40]

On the other hand, if it is assumed that all magnitudes are measured or numerically expressed, then in any ratio (A : B) the antecedent A may be said to contain the consequent B exactly the number of times determined by the quotient A/B. That was Wallis's position, already in 1657. He defined the equality of ratios by means of the equality of their quotients.[41] Answering to one of his critics, Wallis defended the mathematical

---

[38] A. Tacquet, *Elementa geometriae planae ac solidae*, 2nd ed. (Antwerp, 1665), pp. 131-154, and in particular, p. 131 (definition of ratio), p. 132 (definition of equality of ratios).

[39] Barrow, *Usefulness of Mathematical Learning*, pp. 411-412, 428-9.

[40] Barrow, *Usefulness of Mathematical Learning*, p. 390.

[41] J. Wallis, *Mathesis universalis* (Oxford, 1657), in J. Wallis, *Opera Mathematica*, I, (Oxford, 1695), pp. 183-196.

legitimacy of the quotient A/B–whether A and B be commensurable or not–by comparing the quotient $3/\sqrt{2}$ to the difference $3-\sqrt{2}$, whose mathematical meaning Wallis is confident nobody questions. Hence his conclusion that $3/\sqrt{2}$ is as well defined a quotient as $3-\sqrt{2}$ is a well-defined difference.[42] Wallis overlooks here that the difference of segments has a clear, well-defined counterpart in the *Elements*, but the quotient has not. Barrow points out that for general magnitudes the quotient A/B is only defined as the magnitude C such that (A:B) = (C:1). Therefore, we are in full circle, since the definition of quotient requires the definition of equality of ratios.[43]

The assumption that all magnitudes are measured or numerically expressed seems to circumvent this difficulty. Barrow never took this idea seriously. He stressed that magnitudes can only be compared one to another geometrically, for what they are.[44] He also stressed that the measure of magnitudes is defined in terms of the *ratio* between any given magnitude and the chosen unit–therefore it requires the notion of equal ratios. If the notion of measured magnitudes is used to define the equality of ratios, the argument is circular. He also dismissed the idea that ratios could be identified with quantities of some kind, because circularity is also involved.[45] He pointed to the impossibility of determining which quantities correspond to irrational ratios.[46] In the 17th-century mathematical context (crucial) quantities are often defined by ratios–roots in general, the ratio between circumference and diameter, for instance. Ratios therefore cannot be determined by assuming their terms to be quantities because quantities are often determined by ratios.

Barrow's position has a peculiar anti-philosophical slant in that he links criticism of Euclid's definition by the likes of Tacquet and Dechales to the philosopher's infatuation with the perfect definition of general abstract notions.[47] The disagreements among mathematicians about the definition of ratio, Barrow presents as a consequence of putting too much emphasis on the definition of words. He points out that Euclid's definition of ratio, which he calls "metaphysical", is never used throughout the *Elements*.[48] Barrow contrasts this metaphysical definition with the

[42] J. Wallis, *Due Correction for Mr Hobbes* (Oxford, 1656), pp. 70-71.
[43] Barrow, *Usefulness of Mathematical Learning*, p. 434.
[44] Barrow, *Usefulness of Mathematical Learning*, pp. 336, 342.
[45] Barrow, *Usefulness of Mathematical Learning*, p. 370.
[46] Barrow, *Usefulness of Mathematical Learning*, p. 370.
[47] Barrow, *Usefulness of Mathematical Learning*, pp. 319-320.
[48] Heath ([ed.], 1956: II, 114, 116); Barrow, *Usefulness of Mathematical Learning*, p. 320.

mathematically clever equimultiple definition of equality of ratios, which has eliminated all ambiguity in handling ratios.[49] Barrow also attacks the view that Euclid failed to reveal the true "nature" of the equality of ratios.[50] He refuses the tacit arithmetization of ratios that lead many authors to endow them with a quantified "value". Barrow claims that ratios have no essential nature other than what can be deduced from Euclid's mathematical definition of equality. Facing a choice between sound mathematics and tacit knowledge–embodied in the alleged intelligibility of "customary names" (*sic*)–Barrow prefers Euclid's difficult but mathematically sound definition over definitions apparently more intelligible but mathematically untenable.[51]

After Barrow

While Barrow's criticism is mathematically compelling to the modern reader, it went unheeded.[52] If anything, the numerical understanding of proportionality grounded on adopting quotients of antecedents by consequents gained favour and arguments in support of it became bolder and more elaborate.

With the Oratorian mathematician and theologian Bernard Lamy's *Les Elemens de Geometrie, ou de la mesure de l'étenduë; qui comprennent les Elemens d'Euclide* (at least 8 editions between 1685 and 1758) the process of rewriting Euclid's definitions of ratio and proportionality reaches a sort of breaking point. Lamy, who defines the ratio of two magnitudes as the "manner" (*manière*) in which the first one contains or is contained by the second, assumes the division of two magnitudes always to exist, whether or not this division can be expressed in numbers.[53] Hence, for any two magnitudes, A and B, their quotient, A/B, is the "expression" of their ratio and is called its "exponent" (*exposant*):

---

[49] Barrow, *Usefulness of Mathematical Learning*, p. 350.

[50] Barrow, *Usefulness of Mathematical Learning*, pp. 403 (for Ramus), 405, 409-410 (for Tacquet), 418 (for Wallis).

[51] Barrow, *Usefulness of Mathematical Learning*, pp. 406-411.

[52] T. L. Heath extolled Barrow's insightful analysis of 17th-century critiques of Euclid's definitions of ratio and equality of ratios; see Heath ([ed.], 1956: II, 121).

[53] "Raison d'une ligne à une ligne, d'un plan à un plan, d'un solide à un solide; c'est la maniere dont une ligne contient ou est contenuë dans celle avec qui on la compare; [etc.]", see B. Lamy, *Les Elemens de Geometrie, ou de la mesure de l'étenduë* (Paris, 1710), p. 152.

> It is by division that the manner in which one magnitude is contained in another magnitude is to be found out. Thus, in order to express the relation or the ratio of one line to another, as of line A to line B, one divides B by A, and writes B/A … This expression exposes or expresses the ratio of A to B, that is how many times, and in which manner, A is in B, or which part B is of A, which is called the exponent of the ratio of A to B.[54]

The equality of ratios is then defined by the equality of their exponents. Lamy's definitions are ultimately grounded on the notion of "general magnitude" that he had presented in a previous, also highly popular work, *Traité de la Grandeur en general, Qui comprend l'Arithmetique, l'Algebre, l'Analyse, et les Principes de toutes les Sciences qui ont la Grandeur pour objet* (at least 9 editions between 1680 and 1765). As its title suggests, the *Traité de la Grandeur* deals with what is common to all magnitudes. Lamy defines magnitude (*grandeur*) as everything that can be increased and diminished and has parts, and lists some of its particular kinds–among them, length, volume, time, weight, speed, motion, sound.[55] He makes two important points about general magnitude, one concerning its epistemological superiority over extended or geometrical magnitude, the other concerning its arithmetical nature.

Lamy claims that mathematical reasoning is most easy when it is most general, which happens when its object is general magnitude. Hence, the study of general magnitude is like a "universal mathematics" (*mathématique universelle*) and must be the very beginning of the study of mathematics.[56] Euclid's *Elements* are not appropriate as an introduction to mathematics in general because it deals with particular kinds of magnitudes–those we find in bodies (*corps*) and are therefore extended– and the properties of body magnitudes are more difficult than those of magnitude in general. A good treatise of general magnitude is therefore to be considered the true "elements" of mathematics–indeed, Lamy's *Traité de la Grandeur* was titled anew *Elémens des Mathématiques* from its second edition of 1689 on. The most general properties of general magnitude include their addition, subtraction (of a lesser magnitude from a greater one), multiplication and division, and the opening "algebraic"

---

[54] "C'est par la division qu'on connoît la maniere qu'une grandeur est contenuë dans une autre grandeur. Ainsi pour exprimer le rapport, ou la raison d'une ligne à une autre, comme de la ligne A à la ligne B on divise B para A, écrivant … B/A … Cette expression expose ou exprime la raison de A à B, c'est à dire combien des fois, et de quelle maniere A est dans B, ou quelle partie B, est de A; ce qui se nomme l'Exposant de la raison de A, à B", Lamy, *Elemens de Geometrie*, p. 153.

[55] Lamy, *Elémens des Mathématiques* (Paris, 1704), pp. 2-3.

[56] Lamy, *Elémens des Mathématiques*, p. 3.

propositions of Book II of Euclid's *Elements*.[57] Lamy taught how to operate with general magnitudes by representing them by letters.

Lamy's claim about the arithmetical nature of general magnitude is grounded, first, on an understanding of numbers as names for parts of continuous magnitude, rather than as Euclidean collections of units: "Numbers are properly nothing but names that express the parts that are considered in ... magnitude".[58] This includes fractions and in particular decimal fractions. Secondly, Lamy assumed that continuous magnitude maps onto an arithmetical continuum, that is to say, that any part of a continuous magnitude can be numerically identified. In this way, all that can be predicated about numbers also applies to continuous magnitude. This is not demonstrated or argued, it is simply stated:

> Discrete quantities or numbers can be said to include continuous quantities, and so what is taught about discrete quantities or numbers can be equally applied to them.[59]

As mentioned above, Lamy defines the equality of ratios by the equality of the quotients of antecedents by consequents.[60] He acknowledges that some critics point out that in ratios that cannot be expressed as ratios of finite (whole) numbers, the "manner" in which the antecedent contains or is contained by the antecedent is unknown. This criticism is irrelevant, according to Lamy, because drawing consequences from or about equal ratios does not require the exact knowledge of how the terms of each ratio contain each other. In order effectively to prove that antecedents are "in the same manner" contained in consequents in different ratios it is not necessary for us to be able to express that "manner" accurately. Instead, Lamy loosely suggests a procedure reminiscent of anthyphairesis in which the equality of ratios ($b : c$) and ($d : f$) follows when the remainder in the division b/c, divided in as many equal parts as wished (say 1000), is shown to be comparatively contained in c as much as the remainder of d/f, divided in the same number of parts, is in f, "and so on infinitely" (*ainsi à l'infini*):

---

[57] B. Lamy, *Traité de la Grandeur en general* (Paris, 1680), pp. 48-68, 69-131.

[58] "Les nombres ne sont proprement que des noms qui expressent les parties que l'on conçoit dans ... la grandeur", Lamy, *Elémens des Mathématiques*, p. 4.

[59] "on peut dire que la quantité discrete ou les nombres comprennent la quantité continuë, et que ce que l'on enseigne de la quantité discrete ou des nombres s'y peut appliquer", Lamy, *Elémens des Mathématiques*, p. 5.

[60] Lamy, *Elémens des Mathématiques*, pp. 171-172.

> [Even] Ignoring, I say, the manner in which *b* is contained in *c*, I could show how *d* is to *f* as *b* to [*c*], if I had the means to demonstrate that *d* is in fact contained in *f* as *b* is in *c*. It would not be necessary for me to be able to express exactly the manner,... It would be enough for me to show that, if I assume *c* and *f* divided into a thousand parts, if *b* is a thousandth part of *c*, *d* would also be a thousandth part of *f*; and if dividing *c* by *b* I obtain a remainder, dividing *f* by *d* I would also obtain a remainder; and if I consider this remainder from *c* divided as well in a thousand parts, and the remainder from *f* also divided in a thousand parts, *b* would be to each thousandth part of the remainder from *c* as *d* to the thousandth part of the remainder from *f*; and thus to infinity. When this can be shown, it is evident that the manner in which *d* is contained in *f* is the same as that in which *b* is contained in *c*: therefore this definition of ratio ... works for every ratio, for those that can be expressed by numbers as well as for those that cannot[61]

Again, the problem with such a definition is circularity, since it rests on the *comparison* of remainders. It reveals, on the other hand, Lamy's (and other's) tacit assumption that magnitudes will be compared by means of their arithmetical expression involving decimal approximations. Lamy did not further elaborate on the algorithm implicit in his definition, which we recognize to be a poorly worded, sketchy reference to an antyphairetic procedure.

Lamy's ideas were refined and freshly articulated in the works of his fellow Oratorians, Jean Prestet (1648-1691) and Charles-René Reyneau (1656-1728). Prestet's first edition of his *Elemens des Mathematiques* (1675) anticipated Lamy in claiming both the epistemological superiority of arithmetical magnitude upon extended or geometrical magnitude, and in identifying magnitudes in general with their numerical expression–in his

---

[61] "[Même] Ignorant, dis je, la maniere dont *b* est contenu dans *c*, je pourrois faire voir que *d* est à *f* comme *b* à [*c*,] si j'avois un moyen de demontrer que effectivement *d* est contenu dans *f* comme *b* l'est dans *c*. Il ne seroit pas necessaire que je pûsse dire precisément cette maniere, ... Il suffiroit que je fisse voir que si on supposoit *c* et *f* divisez en mille parties, si *b* étoit une milliéme partie de *c*, *d* seroit aussi une milliéme partie de *f*; et que si divisant *c* par *b* il restoit un reste, divisant *f* par *d* il y auroit aussi un reste; et que si on concevoit ce reste de *c* divisé en mille parties, et le reste de *f* divisé aussi en mille parties, *b* seroit a chaque milliéme partie du reste de *c* comme *d* a la milliéme partie du reste de *f*; ainsi à l'infini. Lors, dis je, que cela se peut démontrer, il est évident que la maniere dont *d* est contenu dans *f* est la même que celle dont *b* est contenu dans *c*; ainsi cette définition des Raisons ... convient à toutes les Raisons, tant à celles qu'on peut exprimer par nombre, qu'à celles qui ne le peuvent", Lamy, *Elémens des Mathématiques*, pp. 162-163.

book, magnitude (*grandeur*) and number (*nombre*) became synonymous.[62] Prestet's treatment of ratio and proportionality was poorly argued in his 1675 *Elemens*. In his greatly enlarged, new version of 1689, titled anew *Nouveaux Elemens des Mathematiques*, he provided a detailed definition of equality of ratios by anthyphairesis. The ratios (A:B) and (a:b) are equal if A contains B as many times as a contains b; and B contains as many times the remainder C of A as b does the remainder c of a; and so on until the remainders are zero, or to infinity if there are always non-zero remainders.[63]

We found in Lamy and Prestet a cluster of ideas that will prove resilient and influential–general (as opposed to extended) magnitude as the object of mathematics, the arithmetical nature of general magnitude, the methodological superiority of algebra, the characterization of ratios by the quotients of their terms. These ideas are taken up by C.-R. Reyneau–author of the highly influential calculus textbook, *Analyse démontrée* (1708)–in his textbook on arithmetic and algebra, *La Science du calcul des Grandeurs en general*.[64] We can stop here, since this is evidence enough that by 1700 a new understanding of ratio and equality of ratios had successfully replaced Euclid's.

Conclusion

In the second half of the 18th century, the use of quotients for handling ratios was fully legitimated and reference to anthyphairesis was no longer needed. The 18th-century comments added to Tacquet's version of the *Elements* well illustrate the final stage in the development of the notions of ratio and proportionality that we saw in its early stage in the 1650s and 1660s. William Whiston's notes to Tacquet's edition of Book V of the *Elements* presents the notions of ratio and proportionality as a matter of common understanding, almost innate ideas, which Euclid muddled up because he handled them in a roundabout, obscure way (*per*

---

[62] Prestet, *Elemens des Mathematiques*, Preface, p. 3-6. We cannot dwell on Prestet's notions here, but see Asselah (2007), Robinet (1960).

[63] J. Prestet, *Nouveaux Elemens des Mathematiques ou Principes generaux de toutes les sciences qui ont les grandeurs pour objet* (Paris, 1689), pp. 167, 169.

[64] C.-R. Reyneau, *La Science du calcul des Grandeurs en general, ou les Elemens des Mathematiques*, 2 vol. (Paris, 1714-1736). Reyneau does not provide a formal definition of general magnitude, but points out that it can be expressed in lines, numbers and letters (Preface, i-ii, xii-xv); for the theme of the superiority of numbers and letters, see p. xviii-xxi; p. 18 (definition of ratio, *rapport geometrique* or *raison geometrique*), 23-24, 26-27 (definition of proportionality).

ambages quamplurimas).[65] While Tacquet did not dare to introduce quotients of general magnitudes (see above), his 18th-century commentator states as a matter of course that "the ratio of two quantities is rightly expressed by the quotient coming from the division of the antecedent by the consequent". The claim is grounded on the explicit assumption that any ratio can be expressed by numbers–certain and determined numbers for ratios of commensurable magnitudes, or by numbers that approximate the true ratio to infinity:

> There is no ratio of commensurable quantities that cannot be exactly expressed by fixed numbers, nor is there any ratio of incommensurable magnitudes that cannot be exactly expressed by numbers that approximate to infinity the true ratio.[66]

The new 18th-century consensus about ratios is perhaps best exposed in Euler's *Algebra*, first published in 1770. He introduces the basic notions of mathematics in its opening chapter, "Of mathematics in general".[67] Mathematics deals with magnitudes, or quantities, that are all expressed in numbers. Quantities are determined or measured by choosing a unit and then finding the ratio between the given magnitude and the unit. While the notion of magnitude is defined ("whatever is capable of increase or diminution") and given pride of place, the notion of ratio or proportion between two magnitudes is taken for granted and used without explanation in introducing the measure of quantities–as usual, quantities are measured or determined by the ratio they keep with the chosen unit. No elaboration is offered for the claim that the ratio or proportion to unit "is always expressed by numbers", and that "all magnitudes" can be expressed by numbers. It is concluded that numbers are nothing but the ratios of magnitudes to an (arbitrarily chosen) unit:

---

[65] "I, certainly, consider this notion [of proportionality] as a matter of common understanding, almost innate to the human mind" ("Verum hanc doctrinam animis humanis cum communi ratione quasi congenitam existimo"), in A. Tacquet, *Elementa Euclidea Geometriae ... illustrata a Guilielmo Whiston*, 2 vol. (Venice, 1781), I, 84.

[66] "Neque ulla est quantitatum commensurabilium ratio quae non posit per numeros certos; neque sane incommensurabilium ulla quae non possit per numeros ad veram rationem in infinitum approximantes certissime exprimi", Tacquet, *Elementa Euclidea Geometriae ... illustrata a Guilielmo Whiston*, I, 84.

[67] L. Euler, *Vollständige Anleitung zur Algebra*, 2 vol. (St. Petersburg, 1770); all quotations come from John Hewlett's translation, *Elements of algebra* (New York : Springer-Verlag, [1984]).

> This proportion is always expressed by numbers; so that a number is nothing but the proportion of one magnitude to another arbitrarily assumed as the unit.

Since every magnitude is expressed by numbers, the foundation of all the mathematical sciences must be laid in a "complete treatise on the science of numbers".[68] A whole section of Euler's *Algebra* deals with the applications of arithmetic and geometric proportionality to the addition of progressions, interest computation, etc. Here, in opposition to arithmetical ratio (which is the answer to the question, "By how much is A greater than B?"), Euler defines geometrical ratio as the answer to the question, "How many times is A greater than B?" The division of A by B provides the answer and the resulting quotient expresses the ratio. Euler defines the equality of ratios by the equality of their quotients.[69] It appears, therefore, that Euler's notions of ratio and equality of ratios are very similar to Lamy's. Notice that they incorporate all the logical and mathematical difficulties that Barrow took pains to detail. The difficulties, however, did not deter Euler. Neither did they get in the way of the quantification of continuous magnitude, which was tacitly taken for granted.

To conclude, in the 17th century, on ground prepared by 16th-century editors and commentators of Euclid, the claim to "reform" the *Elements* became mainstream, in the sense that the list of 17th-century authors introducing changes in it, particularly in Book V, is impressive, while the list of authors defending the consistency of the *Elements* is very short. The goal of achieving a philologically faithful reconstruction of Euclid's text was all but forgotten.

From the end of the 17th century and throughout the next, the *Elements* were mostly used as an introductory textbook for beginners–its contents were shortened and some of its basic notions modified to fit in with the work's new role. Outstanding among the changes introduced in the *Elements* were the new notions of ratio and equality of ratios. Simultaneously, other works appeared that found new foundations for the mathematical sciences in the methods of algebra and in a new arithmetical rather than geometrical notion of magnitude. Interestingly, Barrow's

---

[68] Euler, *Elements of algebra*, Chapter I, *passim*, but particularly §§3, 4 and 5; quotations in §§4 and 5, respectively, pp. 1-2.

[69] Euler, *Elements of algebra*, Section III, which includes 13 chapters; ratios (arithmetical and geometrical) are first introduced in §378, pp. 126-7; for the definition of geometrical ratio, see §440, p. 146; for the quotient of antecedent by consequent as expression of the ratio, see §§441 to 445, pp. 146-147; for the definition of equality of ratios, see §§461, 462, p. 152.

critique of the major inconsistencies he perceived in the new notions of ratio and equality of ratio was never answered. As we saw, the new arithmetical understanding of general magnitude leads to conceptual circularity, since the notion of measurable magnitude needs to assume that ratios are known through quotients–but quotients require that magnitudes are measured and known numerically. These notions, however, became the new consensus about ratio and proportionality in the 18th century. The life of Euclid's *Elements* came to an end before a fully consistent mathematical alternative was available.

Appendix: Some seventeenth-century editions of Euclid's *Elements*

The following list of seventeenth-century editions and translations of Euclid's *Elements* comes from the online catalogues of the Bibliothèque national de France:

http://catalogue.bnf.fr/servlet/ListeNotices?host=catalogue

and COPAC:

http://copac.ac.uk/search?rn=742&au=euclid&date=1600-1700&sort-order=date

consulted on 30 November 2011, limiting the dates of publication between 1600 and 1700 (both included).

This list is by no means meant to be comprehensive. It is not grounded on a systematic search through catalogues of world-class libraries. It excludes all seventeenth-century reprints or new editions of works originally published in the sixteenth-century–for instance, the many 17th-century reprints of the popular editions by Clavius, Peletier, Commandino, Candalle, and others, first published before 1600. It also excludes works that focus on one particular book of the *Elements*–such as Florimond du Puy's translation and commentary on Book X (printed 1612); it also excludes editions, which just provide a list of definitions and enunciations of propositions–such as Mersene's or the anonymous *Compendium* published in Amsterdam (1673). This list is just meant to illustrate the nature and character of the seventeenth-century editions and translations of the *Elements* through a representative sample of them. Translations into different languages of the same edition have been listed separately.

1603 Dybvad, Christoffer 1572-1622. *In geometriam Euclidis prioribus sex Elementorum libris comprehensam demonstratio linealis*. Gelderland. [also 1605]

1605 Errard, Jean. *Les nevf premiers livres des Elemens d'Evclide*. Paris. [also 1629]

1606 Dou, Jan Pieterszoon, 1572-1635. *De ses eerste boecken Euclides : van de beginselen ende fondamenten der geometrie waer by-gevoeght zijn eenige nuttigheden, uyt de selve boecken getrocken; mitsgaders de specien in geometrijsche figueren, als 't maken veranderen, t' samenvoeghen, aftrecken, vermenigvuldigen ende deelen*. Leiden. [also, 1681]

1609 Rhodius, Ambrosius, 1577-1633. *Euclidis Elementorum libri XIII : Succinctis & perspicuis demonstrationibus comprehensi*. Wittenberg. [also 1634, 1661]

1610 Marius (Mayr), Simon, 1573-1624. *Die ersten sechs Bücher Elementorum ... in welchen die Anfäng und Gründe der Geometria ordenlich gelehret, und gründtlich erwiesen werden*. Ansbach.

1610 Dounot, Didier. *Les elemens de la geometrie d'Evclides Megarien, traduicts & restitues a leur ancienne breuete, selon l'ordre de Theon*. Paris. [also 1613]

1612 Gracilis, Stephanus. *Euclidis elementorum libri XV*. Köln.

1615 Henrion, Denis. *Les quinze livres des Elements d'Euclide. Traduicts de Latin en François: avec un sommaire et abbregé de l'algèbre, qui sert à faciliter l'intelligence du dixiesme liure*. Paris. [also, 1621, 1623, 1630, 1632, 1649, 1676, 1683, 1685]

1617 Lanz, Johann. *Euclidis Elementorum geometricorum libri sex priores. Nova interpretatione in usum studiosæ iuventutis in lucem dati a Ioanne Lanz*. Ingolstadt.

1620 Cataldi, Pietro A. *I primi sei libri de gl'elementi d'Euclide ridotti alla prattica*. Bologna.

1620 Malapertius (Malapert), Carolus, 1581-1630. *Euclidis elementorum libri sex priores quorum demonstrationes tum alibi sparsim tum medias, libro quinto ad faciliorem captum accomodavit Carolus Malapertius*. Ducai. [also 1625, 1633]

1622 Le Mardelé, P. *Les quinze livres des elements geometriques d'Euclide ... traduicts de Grec en Francois, et augmentez de plusieurs figures et demonstrations*. Paris. [also 1632, 1645]

1625 Brunn, Lucas. *Euclidis Elementa practica, oder Auszug aller Problematum und Handarbeiten auss den 15 Büchern dargegeben*. Nuremberg.

1628 Duinus, Catharinus. *Euclidis Elementorum libri VI*. Ferrara.

1629 Grienberger, Christoph, 1561-1636. *Euclidis sex primi Elementorum geometricorum libri cum parte undecimi ex majoribus Clavii commentariis... contracti*. Roma.

1629 Montalbani, Ovidio, 1601-1671. *Speculum euclidianum... totam Euclidis planimetriam... repraesentans*. Bologna.

1634 Curtius, Sebastian, 1576-1659. *Die sechs ersten Bücher Euclidis, des hochgelährten weitberümbten griechischen Philosophi vnd Mathematici.* Amsterdam.

1634 Herigone, Pierre. *Six premiers livres des elements d'Euclide. In Cursus mathematicus nova, brevi, et clara methodo demonstratus, per notas reales & universales, citra usum cuiuscunque idiomatis, intellectu faciles.* Paris. [also 1639, 1644]

1637 Carduchi, Luis. *Elementos geometricos de Euclides.* Alcalá.

1642 Gestrini, Martini E. *In Geometriam Euclidis demonstrationum libri sex in quibus Geometria planorum traditur et brevibus notis perspicue explicatur.* Upsala.

1643 Fournier, George, 1595-1652. *Euclidis sex primi elementorum geometricorum libri, in commodiorem formam contracti et demonstrati.* Paris. [also 1644, 1654, 1655, 1665]

1644 Cyriaque de Mangin, Clément. *Les six premiers livres des Elements d'Evclide: demonstrez par notes, d'vne methode tres-brieve & intelligible. Auec les principalles parties des mathematiques, expliquées succintement sans notes.* Paris.

1645 Richard, Claude. *Evclidis Elementorvm geometricorvm libros tredecim ... tam secundum antiquos, quam secundum recentiores geometras, nouis vbique ferè demonstrationibus illustrauit, & multis definitionibus, axiomatibus, propositionibus, corollariis, & animaduersionibus, ad geometriam rectè intelligendam necessariis locupletauit Clavdivs Richardvs.* Anvers.

1651 Ricci, F. Giovanni. *De gli Elementi di Euclide li primi sei libri.* Bologna.

1651 Rudd, Thomas, 1584?-1651. *Euclides Elements of geometry: the first VI books: in a compendious form contracted and demonstrated.* London.

1654 Meurs, Jacob van. Elementa geometriae planae ac solidae. Anvers.

1654 Tacquet, André. *Elementa geometriae planae ac solidae (libri 1-6, 11-12).* Anvers. [also 1665, 1672, 1683, 1694, 1700]

1655 Barrow, Isaac, 1630-1677. *Euclidis Elementorum libri XV. breviter demonstrati.* Cambridge. [also 1657, 1659, 1675, 1676, 1678, 1685, 1687]

1658 Borelli, G. Alfonso. *EUCLIDES restitutus; sive, Prisca geometriae elementa, brevius & facilius contexta, in quibus precipue proportionum theoriae nova, firmiorique methodo proponuntur.* Pisa. [also 1659, 1679, 1695]

1660 Barrow, Isaac, 1630-1677. *Euclide's Elements: the whole fifteen books compendiously demonstrated.* London. [also 1675, 1678, 1685, 1686]

1660 Dechales, Claude-François Milliet, 1621-1678. *Euclidis Elementorum libri octo: Ad faciliorem captum accommodati.* Lyon. [also 1675, 1677, 1683, 1685, 1690, 1696, 1700]

1661 Leeke, John; Serle, George. *Euclid's Elements of geometry: in XV books ; with a supplement of divers propositions and corollaries.* London.

1663 Borelli, G. Alfonso. *Evclide rinnovato, overo, gl'antichi Elementi della Geometria rodotti à maggior breuità, e facilità.* D. Magni, trad. Bologna.

1666 Mercator, Nicolaus, 1620-1687. *Euclidis elementa geometrica [books 1-6, and 11-12] novo ordine ac methodo fere demonstrata*. London. [also 1678]

1671 Guarino, Guarino. 1624-1683. *Euclides adauctus et methodicus mathematicaq[ue] vniuersalis: ... , quæ ne dum propositionum dependentiam, sed & rerum ordinem obseruat. Et complectitur ea omnia, quæ de quantitate tum discreta, tum continua abstracta speculari queunt. Resectis superfluis demonstrationibus, & requisitis omnibus profusè coadunatis. Singulis quoque tractatus nouis propositionibus adaucti sunt, & aliqui etiam ex integro adornati*. Turin

1672 Dechales, Claude-François Milliet, 1621-1678. *Huict livres des Elements Deuclide rendus plus faciles*. Lyon. [Subsequent editions titled *Les Elemens d'Euclide: expliquez d'une maniere nouvelle & tres-facile*, 1677, 1682, 1683, 1690, 1700]

1672 Mohr, Georg. *EUCLIDES Danicus, bestaende in twee deelen... Handeldt van de meetkonstige werckstucken, begrepen on de ses eerste boecken EUCLIDIS... Geest aenley ding om verscheyde werckstucken te maecken*. Amsterdam.

1673 Melder, Christianus. *Euclidis elementorum sex priores libri*. Lyon.

1673 Zaragozà (Saragossà), José. *Euclides novo-antiquus singulari methodo illustratus*. València.

1680 Vitale, Giordano, 1633-1711. *Euclide restituto overo gli antichi Elementi geometrici / ristaurati, e facilitati ... ; Libri XV. Ne i quali principalemente si dimostra la compositione delle proportioni secondo la definitione datane dal suo antico autore*. Roma. [also 1686]

1680-81 Moore, Jonas, et al. *A new systeme of the mathematicks: containing I. Arithmetick, as well natural and decimal, as in Species or the principles of algebra; II. Practical geometry, together with the first six books of Euclid's Elements, as also the eleventh and twelfth ... III. Trigonometry ... IV. Cosmography ... V. Navigation ... VI. The doctrine of the sphere ... VII. Astronomical tables ... VIII. A new geography ... for the use of the mathematical school in Christ-Hospital*. London.

1682 Rohault, Jacques, 1618-1672. *Oeuvres posthumes* [contains Euclid's Elements, Bks I to VI]. Paris.

1684 Polo, Johann Francisco. *Sex priora Euclidis geometrica elementa*. Bologna.

1685 Dechales, Claude-François Milliet, 1621-1678. *The elements of Euclid: explained and demonstrated in a new and most easie method*. London. [also 1696, 1700]

1685 Lamy, Bernard, 1640-1715. *Les elemens de geometrie, ou De la mesure du corps : qui comprennent les elemens d'Euclide & l'Analise*. Paris [also 1695]

1686 [No editor given] *De gli elementi di Euclide li primi sei libri*. Bologna.

1688 Fernández de Medrano, Sebastián, 1646-1705. *Los seis primeros libros, onze y doze de los Elementos geometricos del famoso philosopho Euclides, amplificado por ...* Brussels.

1689 Kresa, Jacob, 1645-1715. *Elementos geometricos de Euclides: los seis primeros libros de los planos; y los onzeno, y dozeno de los solidos: ... traducidos, y explicados.* Brussels.

1690 [No editor given] *Elementi piani, e solidi d'Euclide : agl'illustrissimi sig. dell'accademia de Nobili.* Florence.

1691 Astorini, Elia. *Elementa Euclidis ad usum Novae Academiae nobilium Senensium nova methodo, & succincte demonstrata.* Siena.

1691 Coets, Henrik, 1670-1730. *Euclidis Elementorum sex libri priores / magnam partem novis demonstrationibus adornati opera & studio.* Leiden. [also 1692]

1693 Rondelli, Geminiano, 1652-1739. *Planorum, ac solidorum Euclidis elementa facilioribus auctorum demonstrationibus.* Bologna.

1694 Pirckenstein, Anton Ernst Burckhard von. *Teutsch redender Euclides, oder acht Bücher Messkunst : auss eine neue und ganzz leichte Art zu nutzen allen Generalen, Ingeniern, Natur- und Warheit-Kundigern, Bau-Meistern, Kunstlern und Handiwercken in Teutscher sprach eingerichtet und bewiesen.* Wienn.

1695 Alingham, William, fl. 1694-1710. *Geometry epitomiz'd : being a compendious collection of the most useful propositions in the first, third, fourth, fifth and sixth books of Euclid : together with their uses, in several practical parts of the mathematicks : also, Euclid's second book and doctrine of proportion, algebraically demonstrated, with some of the most useful problems required in practise.* London. [also 1700]

1697 Reyher, Samuel, 1635-1714. *In Teutscher sprache vorgestellter Euclides: dessen VI erste Bücher auf sonderbahre Art.* Hamburg.

1698 Larrando de Mauleón, Francisco. *Libro primero (- sexto) de los Elementos Geometricos de Euclides.* [Barcelona].

References

Asselah, K. (2007). *Arithmétique et algèbre dans la seconde moitié du XVIIe siècle français: les "Élémens & nouveaux élémens de mathématiques" de Jean Prestet*, Ph.D. Diss., Université Paris 7.

Bos, H. (2001). *Redefining Geometrical Exactness*, New York: Springer.

Cifoletti, G. (1996). "The creation of the history of algebra in the sixteenth century". In C. Goldstein, J. Gray and J. Ritter (eds.), *L'Europe mathématique. Histoires, Mythes, Identités* (pp. 121.142), Paris: Editions de la Maison des Sciences de l'Homme.

Drake, S. (1973). "Velocity and Eudoxian proportion theory", *Physis* 15: 49-64.

Gillispie, C.C., ed. (1970-1980). *Dictionary of Scientific Biography*, 16 v., New York : Charles Scribner's Sons.

Giusti, E. (1993). *Euclides reformatus. La teoria delle proporzioni nella scuola galileiana*, Turin: Bollati Boringhieri.

Goulding, R. (2006). "Method and Mathematics. Peter Ramus's Histories of the Sciences", *Journal of the History of Ideas* 67: 63-85.

Heath, T.L., ed. (1956). Euclid, *The Thirteen Books of the Elements*, 3 v., New York: Dover.

Malet, A. (2006). "Renaissance notions of number and magnitude", *Historia Mathematica* 33: 63-81.

—. (2008). "Just before Viète: Numbers, polynomials, demonstrations, and variables in Stevin's *L'Arithmetique* (1585)". In P. Radelet-de Grave *et al.* (eds.), *Liber amicorum Jean Dhombres* (pp. 211-229), Louvain: Université de Louvain and Brepols.

Murdoch, J.E. (1968). "The Medieval Euclid: Salient aspects of the translations of the *Elements* by Adelard of Bath and Campanus of Novara", *Revue de Synthèse*, ser. 3, nos. 49-52: 67-94

—. (1963). "The Medieval Language of Proportions". In A.C. Crombie (ed.) *Scientific change*, pp. 237-271, London.

Palladino, F. (1991). "Sulla teoria delle proporzioni nel Seicento: due macchinazioni notevoli: le Sezioni dei razionali del galileiano G.A. Borelli, le Classi di misure del gesuita A. Tacquet", *Nuncius* 6: 33-81.

Popper, N. (2006). "Abraham, Planter of Mathematics: Histories of Mathematics and Astrology in Early Modern Europe", *Journal of the History of Ideas* 67: 87-106.

Robinet, A. (1960). "Jean Prestet ou la bonne foi cartésienne (1648-1691)", *Revue d'histoire des sciences* 13: 95-104.

Rose, P. (1975). *The Italian Renaissance of Mathematics*, Geneva: Droz.

Struik, D.J. *et al.*, eds. (1955-1966). *The Principal Works of Simon Stevin*, 5 vol, Amsterdam.

PART III

ARTS AND SCIENCES OF LIFE AND HEALTH

PHILOSOPHY AND TELEOLOGY: THE CREATION OF THE MARROW AND THE HEAD IN PLATO'S *TIMAEUS*[*]

JORGE CANO

Having placed the three parts of soul in their bodily chambers, Plato begins with the statement that the marrow is the "starting point" for all the different tissues and elements of the body (τούτοις σύμπασιν ἀρχὴ μὲν ἡ τοῦ μυελοῦ γένεσις, *Tim.* 73b2). A. E. Taylor (1928: 518) interpreted this ἀρχὴ as "the central thing in the body not only because, as T. supposes, it supplies matter for the other 'tissues', but because the 'cables' by which the *psyche* is moored to the body are attached to it"[1] Taylor's interpretation does not refer, though, to the relationship established between the soul and the bodily substance marrow, one of the most important points in the body-soul relationship on the physiological plane. This is what we will attempt to analyze here, how god creates the most physically perfect element found in human beings and implants in it the three kinds of soul, what the characteristics of this element are, and how in marrow Plato forges a synthesis of the physiological theories regarding what should be considered the principal organic element within the body. Within the general scheme of the *eikos logos* disclosed by Timaeus, marrow thus appears as the most perfect proof of divine intervention in the physiological configuration of the human body.

J.-F. Pradeau (2008: 68) interprets ἀρχὴ as origin, the chronological starting point in the making of tissue and the basic material from which the rest is created. He adds, "Mais, nous y avons assez insisté, il existe une autre *arche* qui, dans ces mêmes pages du *Timée*, vient précisément s'attacher au corps, c'est l'âme." Additionally (*ibid.*, p. 57), these bonds that join the soul to the body represent a physiological approach to the eschatology of the soul presented in *Gorgias* (493a-b), *Phaedrus* (246d-

---

[*] English translation by Nedra Rivera Huntington.

[1] L. Brisson (2001: 188) translates it as "point de départ" and adds, "Dans le texte grec, on lit *arkhe*. Cela ne signifie cependant pas que la moelle est le principe de tout le corps."

249d) and *Phaedo* (81e2-4). The reincarnation of the soul in a body means being re-chained to it and from this condition there arises an anatomical and physiological description and specific vocabulary (*endesmos*, *endeo*). Marrow is the perfect mixture created by the demiurge from the basic triangles, those which already demonstrate the *chora* yielding to reason to the greatest extent possible (56c4-7), the triangles which produce fire, water, air and earth most exactly (73b5-7).[2] It is for this reason as well, as the result of being the most perfect mixture of the four basic elements, that marrow is also ἀρχή of the body, as in it the very same elements that will be used to create the rest of the body are found in an earlier state. The resulting mixture is in proportion (73c1: μειγνὺς δὲ ἀλλήλοις σύμμετρα),[3] so that marrow acquires a similar excellence in the *krasis* as blood had in Empedocles: both are excellent organic materials within their mixture, and therefore those which are used to transmit intelligence and/or sensation. The marrow is in turn divided into two varieties: one spherical, "(the ploughland, as it were) that was to contain the divine seed and called it the brain."[4] In contrast, the part that contains the mortal soul is described as elongated and circular in shape: "called by the general name

---

[2] On this matter, J.-F- Pradeau (2008: 68) points out a clear difference between the triangles before there was order and those which god created in ordering the universe, through form and number, as indicated in 53b: "to a state of the greatest possible beauty and goodness, in which they were not before" (trans. D. Lee, 2008: 45)

[3] It is relevant to mention 69b2-5, a passage in which συμμετρία was the result of god's activity in ordering the universe: πρὸς ἄλληλα συμμετρίας ἐνεποίησεν; ἀνάλογα καὶ σύμμετρα εἶναι). For *symmetros*, cf. L. Brisson (2001: 266, n. 591). Σύμμετρα, on the other hand, is a word that appears in *Anonymus Londinensis* referring to the theory of health of Petron of Aegina: ὅταν, ἃ σύμμετρα, ἡ κοιλία μὴ λ[αβοῦσ]α, πλεί̤ω̣ δέ, μὴ κατεργάσ̣ηται αὐτά, συμβαίνει νόσους γί(νεσθαι). For Petron, illnesses emerged from food residue: when the stomach does not receive a proportional amount, a suitable measure, it cannot digest the intake and residue is produced, which degenerates into pathologies. According to the papyrus, for Petron, the other cause of illness was a disproportion between the opposite basic elements (heat-dryness versus cold-moisture): ἢ ἀπὸ τ(ῶν) στοιχείων τ(ῶν) προειρημέν(ων), ὅταν ἀνώμαλα ἦι, νόσους ἀπεργάζεται. Neither theory is at all alien to the *Timaeus*, neither the idea of συμμετρία nor the need to remedy an imbalance (Tracy, 1969: 28). Regarding the importance of a *metrios* configuration of body and soul, we should not overlook the famous passage from *Theaetetus* (191c-d) about the soul as a wax tablet: some are better than others because of the state of the mixture (191d1-2: ἔστι δ' οἷς μετρίως ἔχοντος).

[4] 73c8-d1: ἐπωνόμασεν τοῦ μυελοῦ ταύτην τὴν μοῖραν ἐγκέφαλον, trans. D. Lee (2008: 70).

'marrow'."[5] The difference between the figure which is completely spherical (73c7-8: ἐν αὐτῇ περιφερῆ πανταχῇ) and that which is circular and elongated (73d4: στρογγύλα καὶ προμήκη) is not in the least a trivial one.[6] Here, excellence emerges from the similarity to the most perfect geometric figure, the sphere.[7] The circular nature of the marrow is linked to the circular motion of the soul, not only in the brain, which is spherical (73c8), but also the vertebrae, in which the marrow is described as circular and elongated in shape (73d3-4).[8] The elongated shape of the marrow would correspond to the motions which are not perfectly circular that it makes once placed inside the human body (43c-44b).

Here it is worth noting two interesting areas of overlap: One, the precedence given to the brain over the other parts by identifying it with the divine (73c7: τὸ θεῖον σπέρμα), an idea which is clearly encephalocentric in nature, but which signifies a teleological shift within the general scheme of the *Timaeus*. Two, we seem to able to conclude from the text that marrow is continuous–there is no separation in the marrow between the types of soul–and also common to mortals of all kinds: animals and humans. A physiological, although not gnoseological, analogy may therefore be inferred for all beings of *physis*, an idea which, Empedocles defended in his identification between sensation and knowledge. For Manuli and Vegetti (2009: 103-104), there is an area where encephalocentric and Empedoclean models overlap: the precedence of marrow in creating the body is a continuation of the basic thesis of the encephalocentric theory that, as we have seen, supported the embryological precedence of the brain. However, the fact that marrow is produced by mixing the four elements, balanced mathematically in

---

5 This task of name-giving to the world and its beings by the demiurge is interesting. It recalls the figure of God in *Genesis*. Similarly, at 78e2-3, the processes of exhalation and inhalation receive these names from the demiurge: τούτῳ δὲ δὴ τῷ γένει τὸν τὰς ἐπωνυμίας θέμενον ἀναπνοὴν καὶ ἐκπνοὴν λέγομεν θέσθαι τοὔνομα.

6 For *promēkē*, cf. also *Theaetetus*, 148a.

7 Cf. 33b and 55c, where the different shapes that can be put together from the basic triangles are described: tetrahedron, cube, octahedron and icosahedron, with the sphere being the only one to encompass all of them within it. C. Joubaud (1991: 56) points out: "Il faut noter la perfection de la forme, perfection liée à la fonction dont s'acquitte l'organe en question et aussi perfection de l'adéquation de l'un à l'autre."

8 J.-F. Pradeau (2008: 70): "Le vocabulaire géométrique ici employé qualifie la moelle des vertèbres (les portions de moelle plutôt, puisqu'elle est divisée en plusieurs figures identiques) comme des disques qui composent un même prisme ellipsoïde."

proportion (73c1) and geometrically in the sphere (73c7-8), makes this basic, organic substance similar to the biological model of Empedocles, primarily that of the blood as an excellent model of *krasis* in agreement with a numerical ratio. In contrast to fr. 13 of Philolaus, in which the brain was the organ which differentiated human beings from the other animals, here it appears as a part of the very thing which unites them, which is also connected to the theory of reincarnation and transmigration of the soul expounded at the end of the *Timaeus* (89e-92c). Both the body and the soul have an element of boundary, of ordering, given that in the deep make-up of the bodies which form them there is an underlying mathematical *logos* intended by the demiurge, and the result is harmony, a state of both psychic and physical health. This is important to defend two very significant points for the present paper: 1) the criterion of truth for the *eikos logos* and 2) that the restorative and healing possibilities of the *homoiosis theo*–through imitating stability by means of sight (astronomy) and hearing (music)–are linked on the physiological plane with the fact that there is no absolute physical separation, no break, between the brain and the rest of the marrow, although obviously, a clear hierarchy is established between what contains the θεῖον σπέρμα and that around which the body is constructed (73d6-7: περὶ τοῦτο σύμπαν ἤδη τὸ σῶμα ἡμῶν ἀπηργάζετο). A human being is in fact one unit in three psychic-physiological parts, and as such a unit, he has a single purpose: to imitate that which is better than him and which serves as a model in order to achieve the best condition, both internal and external (42d1-2: λόγῳ κρατήσας εἰς τὸ τῆς πρώτης καὶ ἀρίστης ἀφίκοιτο εἶδος ἕξεως). This imitative imperative is directed at both the immortal soul, which must copy the perfectly stable revolutions of the world/order, and the body, which must carry out its own imitation of the movement/order that is the *chora* (88c-89a).

Bonds are extended from the portion containing the mortal part[9] to the entire soul and the body is constructed around it (73d6: καὶ καθάπερ ἐξ ἀγκυρῶν βαλλόμενος ἐκ τούτων πάσης ψυχῆς δεσμοὺς περὶ τοῦτο σύμπαν ἤδη τὸ σῶμα ἡμῶν ἀπηργάζετο).[10] It should not be forgotten that just a few lines earlier, he stated that it is the bonds of life which give

[9] Cf. F. Lisi (1992: 235, n. 90: "*peri touto* is not, as A. E. Taylor (*Commentary*, p. 523) states, the soul, nor is it, as F. M. Cornford (*Cosmology*, p. 294, n. 2) holds, the brain and the marrow, but rather simply the marrow. It is clear that the body is not constructed around the head."

[10] It is relevant to mention here the information given by Diogenes Laertius (VIII, 31) about Pythagoras: δεσμά τ' εἶναι τῆς ψυχῆς τὰς φλέβας καὶ τὰς ἀρτηρίας καὶ τὰ νεῦρα.

humankind its strong roots (73b3-5: οἱ γὰρ τοῦ βίου δεσμοί, τῆς ψυχῆς τῷ σώματι συνδουμένης, ἐν τούτῳ διαδούμενοι κατερρίζουν τὸ θνητὸν γένος). Around the marrow a bony covering is constructed for protection. The compound that serves as its protector is the result of mixing earth with marrow, which then undergoes a cooking and cooling process until it hardens.[11] As C. Joubaud points out (1991: 57), the way the mixture is processed is linked to the manufacture of metals using heat for melting and moulding and water for cooling. This same material is used to shape the sphere that contains the brain (73e6-8). According to A. E. Taylor (1928: 526-527),[12] Plato may be echoing an idea of Empedocles, as Aetius (A 78 A 78; Inwood 98: DK 98) states that, for him, flesh comes from a mixture of equal parts of the four elements; nerves, from the mixture of earth and fire with double the amount of water in proportion; animal claws, from the solidifying effect of cold air on the nerves; and bones, from the mixture of two parts water and earth and four of fire inside the earth. Another fragment of Empedocles (Inwood 62/96)[13] stresses this earthly origin of the bones based on water (Nestis) and heat (Hephaestus). Likewise, in a fragment that has been expanded thanks to the Strasbourg papyrus (Martin Primavesi, b+fr. 76 D), Empedocles states that animals' hard protective features–horns and shells–are concentrations of earth on the skin.

---

[11] In the medical treatise *On Fleshes,* the process of creating hard tissues involves consuming the liquid and fat formed in the earth using heat (3, 5: Καὶ πολλῷ χρόνῳ θερμαινόμενον, ὁκόσον μὲν ἐτύγχανεν ἐκ τῆς γαίης σηπεδόνος λιπαρόν τε ἐὸν καὶ ὀλίγιστον τοῦ ὑγροῦ ἔχον, τάχιστα ἐξεκαύθη καὶ ἐγένετο ὀστέα). As in the *Timaeus,* it is similar to the process of baking clay in an oven: when he talks about the formation of the various hollow organs (c. 3, 4), such as the throat, the esophagus and the intestines to the anus, for the solidification and drying process, he uses the verb *eksoptēthen* (ὑπὸ τοῦ θερμοῦ ἐξοπτηθὲν, μήνιγξ ἐγένετο), which appears in Herodotus (IV, 164) for the baking of pieces of clay in an oven (ἐν τῇ καμίνῳ τοὺς ἀμφορέας ἐξοπτῆσαι). *On Fleshes* also shares the belief that the spinal cord comes from the brain (4, 7.8: καὶ ὁ μυελὸς ὁ καλεόμενος νωτιαῖος καθήκει ἀπὸ τοῦ ἐγκεφάλου).

[12] "We see from fr. 34, *alphiton hydati kollesas* (quoted by Aristotle, *Meteor*. A, 382a1), that Empedocles also had used the illustration of the baker in some unspecified context. It looks as though the words may have belonged to the passage about the making of bones. T. may well be following Empedocles in introducing his thought at this point."

[13] ἡ δὲ χθὼν ἐπίηρος ἐν εὐστέρνοις χοάνοισι / τὼ δύο τῶν ὀκτὼ μερέων λάχε Νήστιδος αἴγλης, / τέσσαρα δ' Ἡφαίστοιο· τὰ δ' ὀστέα λευκὰ γένοντο / Ἁρμονίης κόλλῃσιν ἀρηρότα θεσπεσίηθεν.

Construction of the body around the marrow primarily involves bone tissue, sinews and flesh (74b): the bones–the spinal column–as an enclosure for the marrow, the sinews for flexibility and movement, and lastly, the flesh as protection against cold, heat and falls, as well as a temperature regulator in summer and winter (74c). Flesh is made up of water, fire and earth and "adding a ferment of acid and salt to the mixture; the sinews he made from a mixture of bone and unfermented flesh, producing a substance intermediate between the two and adding a yellow colour."[14] Unlike Empedocles (A 78; Inwood 98: DK 98), Plato does not add air to the flesh mixture. A. E. Taylor (1928: 531) emphasizes that the marrow is formed from the four root elements and that in relation to illnesses he states that there is air in the flesh, although a lesser amount. In *Places in Man* there is a similar explanation (c.4): the nerves are dry and compact, their nature lies somewhere between bone and flesh, but they are moister and fleshier than bone, and drier and bonier than flesh.[15] God wraps the sinews around bones and covers them all with flesh. The distribution of these tissues established by Plato is clearly related to the parts of the soul and their areas of responsibility: the soul appears here as an element linked to movement. The parts of the body which contain the greatest amount of soul are covered with very little flesh, while those least endowed with soul have the thickest and densest flesh (74e). The joints of the bones also lack flesh to prevent them from hindering bending and movement. The flesh made the body "insensitive because of the thickness," thus "hampering memory and sharpness of mind."[16] A mass of flesh around the bone interferes with sensitivity and makes the mind dull, as it impedes movement. Timaeus explains the absence of flesh on the head and spinal column with the first reason, and the lack of flesh on the joints with the second. Thus, the bones of the arms, legs and hips are those which have the most flesh, as they contain the least amount of soul in their marrow. For A. E. Taylor (1928: 532), unlike Archer Hind (1888: 271-272), there is no difference between marrow and bone marrow, only that the bones contain a proportionately small amount of the *myelos*–and therefore *psyche*–with which the brain and spinal column are made.[17]

---

[14] trans. D. Lee (2008: 71).

[15] In his edition of the treatise (1998: 115), E. M. Craik points out that: "Both *phlebes* and *neura* are essentially 'links' as in the Pythagorean term 'bond of life' applied to both, along with arteries (DK 58 B 1: DL 8, 31: FGH 3 fr. 140). Here they are regarded as intermediate between flesh and bone in color and texture."

[16] trans. D. Lee (2008: 72).

[17] In contrast, the author of *On Fleshes* does establish a distinction, in the idea that the marrow which comes from the brain, the spinal marrow, lacks fat, like the

Everything that contains intelligence is made up of very little flesh, except the tongue. "For out of necessity the nature with which we are born and which grows with us in no way allows that thickness of bone and abundance of flesh be combined with quickness of perception."[18] Plato believes that the proximity of the less noble elements constitutes a risk of polluting the main element, following the same logic that leads him to explain the separation of the head from the trunk by means of the neck and the division of the thorax in two in order to separate the three kinds of soul (69e-70a). The proportion of flesh and bone in each part of the body therefore follows a divine plan linked to the *dynameis* of the soul. Centuries later, Galen resurrects this idea: although the body is divided into parts, "The function of all of them is to serve the soul, as the body is an instrument of the soul" (*On the Use of the Parts* I, 2). Plato concludes (75b2-c3) that the composition of the head is formed in this way with a view to improving the performance of the soul, rather than a longer and more comfortable life for human beings.[19] A connection should be made between this and the theory of transmigration of the soul and the imperative which determines it: the imitation of the movements of the soul of the world and the *homoiosis theo*, both of which involve an excellence

---

brain: "For this reason, the name 'marrow' is not appropriate, as it is not the same as the other marrow, that which is found inside the other bones. Indeed, it is the only marrow that has membranes, while the rest of the marrow does not. Clear evidence of this can be seen if on one hand, one roasts the nervous and viscous parts, and on the other, the remaining parts. All the remaining parts will roast quickly, but the nervous and viscous parts cannot be roasted, as they have very little fat, while the more oily and fatty parts roast quickly." Cf. also *On Diseases* IV, 36 and *On Regimen in Acute Diseases*, 3.

[18] 75a1-b2: τὸ περὶ τὴν τῶν ἰσχίων φύσιν τά τε περὶ τὰ τῶν βραχιόνων ὀστᾶ καὶ τὰ τῶν πήχεων, καὶ ὅσα ἄλλα ἡμῶν ἄναρθρα, ὅσα τε ἐντὸς ὀστᾶ δι' ὀλιγότητα ψυχῆς ἐν μυελῷ κενά ἐστιν φρονήσεως, ταῦτα πάντα συμπεπλήρωται σαρξίν· ὅσα δὲ ἔμφρονα, ἧττον–εἰ μή πού τινα αὐτὴν καθ' αὑτὴν αἰσθήσεων ἕνεκα σάρκα οὕτω συνέστησεν, οἷον τὸ τῆς γλώττης εἶδος– τὰ δὲ πλεῖστα ἐκείνως· ἡ γὰρ ἐξ ἀνάγκης γιγνομένη καὶ συντρεφομένη φύσις οὐδαμῇ προσδέχεται πυκνὸν ὀστοῦν καὶ σάρκα πολλὴν ἅμα τε αὐτοῖς ὀξυήκοον αἴσθησιν.

[19] μάλιστα γὰρ ἂν αὐτὰ πάντων ἔσχεν ἡ περὶ τὴν κεφαλὴν σύστασις, εἴπερ ἅμα συμπίπτειν ἠθελησάτην, καὶ τὸ τῶν ἀνθρώπων γένος σαρκώδη ἔχον ἐφ' ἑαυτῷ καὶ νευρώδη κρατεράν τε κεφαλὴν βίον ἂν διπλοῦν καὶ πολλαπλοῦν καὶ ὑγιεινότερον καὶ ἀλυπότερον τοῦ νῦν κατεκτήσατο. νῦν δὲ τοῖς περὶ τὴν ἡμετέραν γένεσιν δημιουργοῖς, ἀναλογιζομένοις πότερον πολυχρονιώτερον χεῖρον ἢ βραχυχρονιώτερον βέλτιον ἀπεργάσαιντο γένος, συνέδοξεν τοῦ πλείονος βίου, φαυλοτέρου δέ, τὸν ἐλάττονα ἀμείνονα ὄντα παντὶ πάντως αἱρετέον.

in intelligence and a naturally philosophical soul (*Theaetetus* 176a-b, 191c-e). This composition (σύστασις) is therefore a physiological requirement which becomes a divine sign of what type of life is preferable for all beings at all times (75c3: παντὶ πάντως αἱρετέον): "a shorter and better life was for everyone in all respects preferable to a longer but poorer one."[20] Life, therefore, is not determined by biological parameters, but rather metaphysical ones. In this point lies the profound difference established between Plato and the rest of the medical physiology which has come down to us. Plato's intent transcends the spheres of *tekmerion* and *to phaneron*, which were expounded in relation to the medical treatises, to interpret corporeality as a sign of the supremacy of intelligence over *tyche*, as he establishes in the section devoted to the *chora* and its stabilization through mathematics and harmony. However, it should be pointed out that what is interesting about the exposition of the *Timaeus* is that it is precisely this anatomical and organic description which makes such a conclusion possible (Steel, 2001: 111-112, about the tongue and sight).

To sum up: marrow is the ἀρχὴ and from it the brain is separated as the most divine part which contains τὸ θεῖον σπέρμα (73c). The immortal soul and the brain–while the most excellent part of the marrow–are however different, as they have been crafted using two different methods–the immortal part of soul: 41d; the marrow: 73d–and from different elements. Around this is formed the body, which is basically made up of bones, flesh and sinews. The marrow is in turn protected by the spinal column, which runs along the trunk from top to bottom. The flesh is concentrated in the places least endowed with the soul–as either the intellectual element or the sensory element (74e), as well as in places with greater movement, the joints. The head therefore has the least concentration of flesh, as it is precisely this part of the body where the immortal part of the soul is contained. The lack of movement of the head, which, as Plato points out, has no joints (75c4-5: ἅτε οὐδὲ καμπὰς ἔχουσαν), makes the presence of flesh and sinews unnecessary, above all the latter. Thus the head is more perceptive and intelligent, but also much weaker (75c6: εὐαισθητοτέρα μὲν καὶ φρονιμωτέρα, πολὺ δὲ ἀσθενεστέρα). As in all combinations of intelligence and necessity, it is possible to explain this morphology of the head as a situation in which there is always an underlying fragility, a weakness in the structure, for in the physical world, god always operates in such a way as to "lead most of the things that come about to the best result" (48a2-3: νοῦ δὲ ἀνάγκης

[20] trans. D. Lee (2008: 72).

ἄρχοντος τῷ πείθειν αὐτὴν τῶν γιγνομένων τὰ πλεῖστα ἐπὶ τὸ βέλτιστον ἄγειν).[21] As Cornford indicates (1937: 297-298):

> In order to shield the marrow and the life it contains, Reason uses, as indispensable means, the solidity of bone and the softer covering of flesh. But the necessary constitution of these integuments, as such, tends to defeat another purpose: that sensations shall be easily and quickly transmitted to the marrow of the brain. The necessary means to that is thinness both of bone and flesh. So two sets of necessary means "refuse to coincide". Necessity cannot be wholly persuaded by Reason to serve both its purposes, and Reason has to sacrifice the less important.[22]

The mouth becomes another example of the combination of necessity and virtue (75d5-e5),[23] as it allows entry to the necessary, food, and exit to the stream of speech which, when subordinated to *phronesis* (75e4: ὑπηρετοῦν φρονήσει) is the greatest virtue of these organs.[24]

---

[21] trans. D. Lee (2008: 38).

[22] Within this specification of marrow-intelligence versus flesh-sensation, the position of the tongue is noteworthy (75a5-6). Passage 66b7-c7 establishes a theory regarding taste which contradicts Theophrastus' assertion (*On Sensation* 5, 1-3) that, of the individual sensations, Plato only concerned himself with hearing and sight (Πλάτων δὲ ἐπὶ πλέον μὲν ἧπται τῶν κατὰ μέρος, οὐ μὴν εἴρηκέ γε περὶ ἁπασῶν, ἀλλὰ μόνον περὶ ἀκοῆς καὶ ὄψεως). After taste, smell is also analyzed (66d-67a), although it is true that they are not dealt with as extensively as sight or hearing. In the *Timaeus* there are four different classes of taste, each linked to the predominant presence of one of the four elements. The passage has certain textual problems, cf. the interpretation by G. A. Grube (1930: 72-75, 1930), which disputes Taylor's (1928: 464-470). For Manuli and Vegetti (2009: 112), the connection between the tongue and the heart (65c6-d4) is a clear example of the Platonic synthesis of the encephalocentric (hearing) and cardiocentric (taste) models.

[23] τὴν δὲ δὴ τοῦ στόματος ἡμῶν δύναμιν ὀδοῦσιν καὶ γλώττῃ καὶ χείλεσιν ἕνεκα τῶν ἀναγκαίων καὶ τῶν ἀρίστων διεκόσμησαν οἱ διακοσμοῦντες ᾗ νῦν διατέτακται, τὴν μὲν εἴσοδον τῶν ἀναγκαίων μηχανώμενοι χάριν, τὴν δ' ἔξοδον τῶν ἀρίστων: ἀναγκαῖον μὲν γὰρ πᾶν ὅσον εἰσέρχεται τροφὴν διδὸν τῷ σώματι, τὸ δὲ λόγων νᾶμα ἔξω ῥέον καὶ ὑπηρετοῦν φρονήσει κάλλιστον καὶ ἄριστον πάντων ναμάτων.

[24] Aristotle (*On the Parts of Animals* II 659b27) echoes this very assertion and points out that the *raison d'être* of the lips and the tongue is to make us capable of *logos*, of intelligent discourse. As C. Steel points out (2001: 111-112): "If we wish to know the purpose of the mouth, our answer cannot be that we require an opening to take in food. Indeed, this is a necessity: the gods had to foresee it when creating mortal beings. However, they transformed this biological system for

Lastly, the presence of hair on the head follows the same reasoning as the make-up of the rest of the head: to avoid a mass around the skull and the brain which would hinder the natural movements of the soul. Hair (76b-d) emerges from perforations made in the skin by the divine.[25] The natural moisture of the brain (76a) makes its way out into the air through them. The slow movement of the exudation causes them to solidify when they come into contact with the air, coil up and take root beneath the skin. However, although it may appear so, the process is not random, but rather controlled by the divine maker[26] (76c5-d3: τούτῳ δὴ λασίαν ἡμῶν ἀπηργάσατο τὴν κεφαλὴν ὁ ποιῶν, χρώμενος μὲν αἰτίοις τοῖς εἰρημένοις, διανοούμενος δὲ ἀντὶ σαρκὸς αὐτὸ δεῖν εἶναι στέγασμα τῆς ερὶ τὸν ἐγκέφαλον ἕνεκα ἀσφαλείας κοῦφον καὶ θέρους χειμῶνός τε ἱκανὸν σκιὰν καὶ σκέπην παρέχειν, εὐαισθησίας δὲ οὐδὲν διακώλυμα ἐμποδὼν γενησόμενον). As noted in the case of flesh, the need for something light to shield the head ensures protection against the weather and a perception which is not dulled as it would be by flesh. As Taylor (1928: 539) indicates, the hair appears to be sort of felt: "*pilesis* itself suggests this, as the word properly means the conversion of wool into felt by packing layers of it close together." Curiously, Plato does not explain the rest of the covering hair as "hindering quickness of perception,"[27] which would help to solve the problem of whether τὸ θεῖον refers to the demiurge or the fire of the brain. Finally, it is understood that it is in this passage about the morphology of the head where the creation of the nails appears, as this is a process analogous to the emergence of hair, as a result of the hardening and drying of the mass of skin, sinews and bones found on the tips of the fingers and toes. Looking ahead, nails are created in

consuming food in small pieces into an amazing instrument at the service of intelligent life. Without the mouth we would be without logos."

[25] Here I am following L. Brisson (2001: 270), in contrast to F. M. Cornford (1937: 300), who understands 76b1-2 (τοῦτο δὴ πᾶν τὸ δέρμα κύκλῳ κατεκέντει πυρὶ τὸ θεῖον) as "Now this skin was pricked all round with fire by the divine part" and adds the note: "The fire in the brain, forcing its way upwards to seek its like. Cf. *to theion*, 69d, and *to theion sperma*, 73c." Although it would be more interesting to link this very small aspect of the microcosm to the physiological perspective of the *Timaeus* and identify fire as the most divine element in the cosmos due to its relationship with light and heat and its function as a determining factor in seeing (40a, 45b-46b), it is not well understood why the divine part of the soul acts in this way on the surface of that which contains it. In addition, for L. Brisson, τὸ θεῖον would correspond to 76c5-6: τούτῳ δὴ λασίαν ἡμῶν ἀπηργάσατο τὴν κεφαλὴν ὁ ποιῶν.

[26] ὁ ποιῶν also in 31b2.

[27] Trans. D. Lee (2008: 73).

anticipation of later reincarnations: "For those who framed us knew that later on women and the other animals would be produced from men, and that many creatures would need claws and hoofs for different purposes; so they provided the rudiments of them in men at their first creation."[28]

In conclusion regarding the assumption of an encephalocentric bias on the part of Plato in the *Timaeus*, it is worth pointing out that its central position in the organic system makes it the organic and physic centre, as it is the site of the immortal soul; subordinate to it are the heart and liver, home to the two mortal parts of the soul. It is given precedence on various levels: marrow is the first part of humans shaped by god (73b8); it is also fundamentally superior, as it is made from the most perfect mixture of the four elements. In addition, in the important section on reproduction, Plato puts forth an encephalo-myeloid theory of semen, which emerges from the lower part of man's marrow (91a-b).[29] Although the brain is the centre of rational activity, Plato echoes the physiological theories expounded in the medical corpus and limits the activity of the brain, which ceases to be the primary node of the blood system and the centre of sensory activity. Plato therefore adopts an encephalocentric solution which was very much in keeping with the anatomical theories circulating at the time, primarily those which gave the heart a determining role in the vascular system. As Manuli and Vegetti (2009: 108) point out: "Da un punto di vista più generale, però, è evidente che Platone rifiuta dell'encefalocentrismo

[28] 76d-e: ὡς γάρ ποτε ἐξ ἀνδρῶν γυναῖκες καὶ τἆλλα ντο οἱ συνιστάντες ἡμᾶς, καὶ δὴ καὶ τῆς τῶν ὀνύχων χρείας ὅτι πολλὰ τῶν θρεμμάτων καὶ ἐπὶ πολλὰ δεήσοιτο ᾔδεσαν, ὅθεν ἐν ἀνθρώποις εὐθὺς γιγνομένοις ὑπετυπώσαντο τὴν τῶν ὀνύχων γένεσιν.

[29] Timaeus maintains a certain ambiguity with regard to whether the woman contributes to reproduction with semen or not. Taylor's comment (1928: 637) is still very much to the point in isolating the problem: "The *gone* of animals is thus, according to the theory, an actual portion of the spinal *myelos* which gets out at this newly formed exit. The question about the part played by male and female in procreation keenly interested the fifth century biologists, who were particularly concerned to determine whether both *proientai sperma* or not, and whether the *sperma* is secreted from every part of the body or not." The treatise *On Generation* supports a pangenetic doctrine of semen, that it belongs to both sexes, not just the male, just as *On the Diseases of Women* I, 8 upholds. Pangenesis also appears in *On Airs, Waters and Places* 14, 5 and *On the Sacred Disease* 5, and was likewise postulated by Democritus (DK B 32 y 124), Anaxagoras (DK 59 A 111) and Empedocles, according to Aristotle's critique (*On the Generation of Animals* I 722b8). Plato adopts an encephalo-myeloid doctrine of semen. Among the defenders of blood as the origin of semen, primarily in the area of Magna Græcia, there were various theories (Manuli-Vegetti, 2009: 107).

l'aspetto rígidamente monocentrico, che non consentiva di rappresentare il rapporto cervello-corpo con una serie di mediazioni, in cui l'attività sensitiva, emotiva e volitiva fosse delegata ad organi subordinati." It is possible to state that the views of Philolaus of Croton serve to mediate between Hippocratic encephalocentric theories and the "hemo-centric" ideas of Empedocles, allowing him to isolate the head as the acropolis of the body and connect it directly with the divine according to the general teleological scheme of the *Timaeus*. As happens with arithmetic and geometry, other *mathemata*, medical physiology serves as proof and evidence of the mark of the demiurge underlying the entire universe. As C. Steel (2001: 109) points out, Plato's intent is not merely that of a naturalist, but rather his physiological teleology ultimately prevails as an ethical imperative.

References

Archer-Hind, R. D. (1888). *The* Timaeus *of Plato*, Londres: MacMillan.

Ashbaugh, A. F. (1988). *Plato's Theory of Explanation: A Study of the Cosmological Account in the Timaeus*, Albany: SUNY Press.

Brisson, L. (1997). "Perception sensible et raison dans le *Timée*". In L. Brisson and T. Calvo, (eds.), *Interpreting the* Timaeus - Critias*: proceedings of the IV Symposium Platonicum*, (pp. 307-316), Sankt Augustin: Academia Verlag.

—. (1998). *Le Même et l'Autre dans la structure ontologique du* Timée *de Platon*, Sankt Augustin: Academia Verlag.

—. (2001). Platon, *Timée, Critias. Présentation et traduction*, Paris: Flammarion.

Burnyeat, M. F. (2005). "*Eikos logos*", *Rhizai: journal for ancient philosophy and science* 2: 143-165.

Cornford, F. M. (1937, 1966). *Plato's cosmology. The Timaeus of Plato translated with a running commentary*, London: Routledge and Kegan Paul ltd.

—. (2001), Platon, *Timée, Critias. Présentation et traduction*, Paris: Flammarion.

Craik, E. M. (1998), Hippocrates, *Places in man, Greek text and translation with introduction and commentary*, Oxford: Oxford University Press.

Esposito, A. (1991). "Il sangue nel *Timeo* di Platone". In F. Vattioni, (ed.), *Atti della VI settimana Sangue e Antropologia nella teologia*, (pp. 443-450), Roma: Pia Unione Preziosissimo Sangue.

Grube, G. M. A. (1930). "The sense of taste in *Timaeus* 65b-66c", *Classical Philology* 25/1: 72-75.

Huffman, C. (1993, 2006). *Philolaus of Croton, Pythagorean and Presocratic, a commentary on the fragments and testimonia with interpretative essays*, Cambridge: Cambridge University Press.

Inwood, B. (2001). *The poem of Empedocles. A text and translation with an introduction*, revised ed., Toronto-Buffalo-London: University of Toronto Press.

Joubaud, C. (1991). *Le corps humain dans la philosophie platonicienne: étude à partir du* Timée, Paris: Ed. J. Vrin.

Lisi, F. L. (1992). Platón *Timeo*, introd., trad. y notas, *Diálogos VI*, (pp. 127-261), Madrid: Biblioteca Clásica Gredos.

—. (2005). "A alma do mundo e a alma humana no *Timeu*. Apontamentos para uma reinterpretação da psicologia platônica", *Hypnos* 14: 57-69.

—. (2005). "El alma humana en el *Timeo*", *Études Platoniciennes* II: 155-174.

Lloyd, G.E.R. (1968). "Plato as a natural scientist", *The Journal of Hellenic Studies* 88: 78-92.

—. (1991). *Methods and problems in Greek science*, Cambridge: Cambridge Univ. Press.

Manuli, P. and M. Vegetti (2009). *Cuore, sangue e cervello. Biologia e antropologia nel pensiero antico*, Milan: Editrice Petite plaisance.

Martin, A. and O. Primavesi (1999). *L'Empédocle de Strasbourg*, Berlin-New Cork: Walter de Gruyter.

Pradeau, J.-F. (2008). *La communauté des affections: études sur la pensée éthique et politique de Platon*, Paris: J. Vrin.

Robinson, T. M. (1995). *Plato's psychology*, Toronto-Buffalo-London: University of Toronto Press.

Steel, C. (2001). "The moral purpose of the human body. A reading of *Timaeus* 69-72", *Phronesis* XLVI/2: 105-128.

Taylor, A.E. (1928, 1962). *A commentary on Plato's Timaeus*, Oxford: Oxford University Press.

Tracy, Th.J. (1969). *Physiological theory and the doctrine of the mean in Plato and Aristotle*, The Hague-Paris: Muton.

PHARMAKA: MEDICINE, MAGIC AND FOLK MEDICINE IN THE WORK OF EURIPIDES*

SARA MACÍAS

Introduction

The word *pharmaka* was used to refer to preparations of herbal, mineral or animal origin, occasionally linked to the recitation of charms and spells.[1] *Pharmaka* were used as remedies for ailments of both a physical and spiritual nature.[2] It is, therefore, a term that falls under the scope of the field of medicine, either scientific or folk, as well as magic.

With regard to the effects produced by the *pharmaka* and the intent with which they are used, they can refer to antidotes and medical remedies which are beneficial for the health, as well as magical remedies for various types of problems. However, on the other hand, this is also the name given to poisons which can end a person's life and spells for the purpose of subjecting someone to another's will.[3]

On many occasions, it is difficult to distinguish folk medicine from magic or folk healing. Leaving aside the remedies which are clearly linked to magic (for example, as we will see, *pharmaka* that produce invisibility or charms that cause objects to move on their own), we take the intent with which the remedies are used as an indicator: if the intent is to heal a sick

---

* English translation by Nedra Rivera Huntington.

[1] Cf. for example, Artelt (1937), Derrida (1981), Lloyd (2003), Carastro (2006). Plato *Chrm.* 155e speaks of the existence of a *pharmakon* composed of a herb and reciting an incantation. The remedy is only made effective by the combined effect of the two elements, as separately, neither the herb nor the incantation would be of any use.

[2] These are remedies for what in Greek is referred to by the broad term *nosos*, which encompasses what we today call diseases and also feelings such as love, understood as a disease (Euripides *Hp.* 767), or attitudes such as being disrespectful to the gods (Euripides *Or.* 10). A *nosos* could even affect a city and not just an individual (Euripides *Tro.* 26-27).

[3] This is the case of love potions and spells, which seek to force the will of the beloved so that they will return the feeling. Cf. Betz (1986) and Winkler (1990).

person, we are looking at an example of medicine, whether scientific or folk, but if it involves doing harm or subjecting someone to another's will, this would be magic.

In the works of Euripides, the term *pharmaka* appears on quite a few occasions, used with different connotations. Below, we will consider some of the most significant passages.

Magic

Poisons

Euripides uses the word *pharmaka* to describe the poisons utilized by Medea, in the tragedy of the same name, to end the life of the future wife of Jason.[4] According to mythological tradition, Medea was viewed as a powerful sorceress or wizardess, on a par with Circe, her aunt.[5] Due to both her status as sorceress and the intent with which she uses her knowledge, it may therefore be seen as more closely related to magic than medical science.

Earlier authors such as Pindar used terms formed from *pharmaka* to refer to Medea, names which continued down to Apollonius Rhodius. However, in addition to also being applied to Circe, they are used to refer to doctors as well. We can clearly see the two aspects of the word, which was considered valid for designating both magical and medical remedies.[6]

As an example of the effects produced by Medea's *pharmaka* in the Euripidean tragedy, verses 1197-1202 are highly significant:

> Her eyes no longer kept their wonted form nor did her shapely face. From the top of her head blood dripped, mingled with fire, and her flesh dropped from her bones like resin from a pine torch, torn by the unseen jaws of the poison (*pharmakōn*), a dreadful sight.[7]

---

[4] On five occasions, *pharmaka* is used to designate the poisons of Medea: Euripides *Med.* 385, 789, 806, 1126 y 1201.

[5] On Circe and *pharmaka,* cf. Carastro (2006: 141-159).

[6] Pindar (*Pi.* 4.221) uses the verb *pharmakoō* to indicate how Medea prepares her remedies and poisons. Pindar (*Pi.* 223) also applies the adjective *pampharmakos* to Medea and Apollonius Rhodius (3.27) calls her *polupharmakos*, the same adjective used in the *Odyssey* (10.276) to describe the wizardess Circe, but which in the *Iliad* (16.28) is used for doctors who treat wounds on the battlefield.

[7] With some little changes, all the translations of Euripides' pieces that we present in this work are from Kovacs (the tragedies) and from Collard-Cropp (the fragments) at *Loeb Classical Library*.

This is the account given by the messenger who tells how Creusa has died of the poison with which Medea had impregnated the gifts she offered her. However, according to Euripides, Medea was enough of an expert in the use of *pharmaka* to do evil, as we have just seen, and also to resolve health problems. This is the case of Aegeus, whom Medea offered a solution to his infertility problem in exchange for his protection, using some *pharmaka* that she was familiar with:

> But receive me into your land and your house as suppliant. As you grant my request, so may your longing for children be brought to fulfilment by the gods, and may you yourself die happy! You do not know what a lucky find you have made in me. I will put an end to your childlessness and cause you to beget children, for I know the remedies (*pharmaka*) to do it (Euripides *Med.* 713-718).

In this case, Medea's remedies seem closer to folk healing, or even folk medicine, than magic. However, they may be understood to belong to this final category given that Aegeus' sterility had been classified as incurable by the doctors of the time.

The term *pharmaka* also appears in *Ion* on five occasions[8] to designate the poison with which Creusa will attempt to take the life of the young man she thought was her stepson, but was in fact her son. This poison, as Creusa states in verse 1015, is a drop of Gorgon blood containing the venom of its serpents, which came into her hands through her ancestor Erichthonius, son of Athena, possessor of the head of Medusa. Ion, spurred by a bad omen, pours the poisoned wine from his cup. It is then consumed by a dove which immediately suffers its terrible effects:

> Then he took a special cup (…) and gave it to him (*sc.* to Ion) filled with wine, slipping into it the deadly poison (*pharmakon*) they say my mistress gave him, meaning to kill this newfound son (…) one of the servants uttered a word of evil omen (…) he (*s.c.* Ion) took it as a sign and gave the order to fill another mixing bowl. The first libations to the god he poured out upon the ground (…) one dove settle where the newfound son had poured out his drink, and no sooner had she tasted the wine than her feathered body was shaken and convulsed like a Bacchant, and she uttered a cry of distress (…) she gasped away her life and died, her red legs and feet all limp (Euripides *Io* 1185-1208).

In a passage from *Cretan Women,* Euripides includes the rather widespread idea that many women even ended up poisoning their own

[8] Euripides *Io.* 616, 845, 1185, 1221, 1286.

husbands. Once again in this text, the word used to refer to the poisons is *pharmaka*: "Well, go on and get married, get married, and then die either through poisons (*pharmaka*) or plots from your wife!" (Euripides Fr. 464 Kannicht).

In addition, the Euripidean corpus also contains mentions of the use of *pharmaka* which, while they do not cause death, have harmful effects such as infertility and miscarriage.[9] This is the case of Andromache, in the tragedy of the same name. In the prologue, she describes her situation thus:

> But ever since my master married Hermione, spurning my servile bed, I have been hounded by cruel abuse from her. She says that with secret poisons (*pharmakois*) I make her childless and hated by her husband, and that I wish to take her place in the house, forcibly casting her out as wife (Euripides *Andr*. 29-35).

In Euripides' time there was legislation against those who used *pharmaka* to harmful effect; it was treated as a sentence for practising black magic, sometimes death. Thus, our tragedian has his heroine make a reference to "submitting to justice":

> For of my own accord, willingly and taking no refuge at an altar, I shall stand trial (*dikēn*) to determine whether I am administering poisons (*pharmakeuomen*) to your daughter and making her womb infertile, as she claims. My judge shall be your son-in-law, for in his eyes no less than yours I deserve punishment if I afflict him with childlessness (Euripides *Andr*. 355-360).

This type of accusation can also be found in other authors from the 5th and 4th centuries BCE, reflecting the legal reality of the period, when there seemed to exist a charge of *pharmakeia* for those who caused harm using poison.[10] We even have evidence in the form of a fifth-century BCE decree from Teos which states that anyone who goes against the wellbeing of the city by means of the use of *pharmaka* is punished and cursed, and a second-century CE oracle of Apollo in which it was interpreted that the plague at Ephesus had been caused by *wizards* and their *pharmaka*.[11] Thus

[9] Euripides *Androm*. 32, 157, 205, 355.

[10] Hdt. 2.181, Pl. *Lg*. 933d, Arist. *MM.* 1188b 31-37, D. 25, 79ss.

[11] Cf. Carastro (2006: 153-156). The decree is no. 104 in Van Heffenterre-Ruzé (1994). The oracle is *SEG* 41.981.

Plato in his utopia of the ideal city prescribes the existence of laws against assaults committed using *pharmakeia*.[12]

In the work of Euripides, *pharmaka* like poisons are used on every occasion by women, normally foreigners. The combination of feminine nature, considered innately malicious, together with the status of foreigner, odd because of her different customs and culture, was considered a potential threat by Greeks of the period. Medea meets both conditions and so both Jason and Creon fear her dangerous reaction. Andromache was also a stranger in Greek lands, but she differs from Medea in the fact that, while suspicion of Medea was well founded, there were no grounds to suspect her: Andromache is in no way to blame for Hermione's infertility. The fact that Euripides mentions the use of justice in the case of an accusation of causing infertility by means of *pharmaka* could be proof of the firm belief in the effectiveness of such magical practices in the 5th century BCE.

Love potions

Also within the sphere of magic, *pharmaka* refers to the famous love potions. Like many other authors, Euripides echoes the existence of magical remedies for lovesickness. We find this in *Hippolytus*, where the nurse, who does not know the reason Phaedra is so affected, initially makes a clear distinction between two types of complaint: those that are perhaps indecorous, which men must not know about (as is the case, an unspeakable love) and which can be remedied by women through magic, and those others that belong to the sphere of medicine and must be treated by a doctor:

> If you suffer (*noseis*) some unmentionable malady, here are women to help set your disease (*noson*) to right. If your misfortune (*sumphora*) may be spoken of to men, speak so that the thing may be revealed to doctors (Euripides *Hipp*. 293-296).

In these verses we find a play on words which should perhaps be highlighted: The unspeakable illness which must be set right by women is called *nosos*, which, although it may refer to suffering in general, is habitually used to refer to disease and therefore the physical plane. However, the illness for which a solution if provided by the doctors is here called *sumphora* ("misfortune," "affliction"), alluding rather to a state of mind or the sentimental plane. It would seem more logical to use the

[12] Pl. *Lg*. 932e-933e.

terminology the other way round. Perhaps this play on words by the nurse is intended to illustrate that both areas of knowledge, the magical and the medical, are complementary, as the problems which one cannot resolve fall under the sphere of the other.

Later in the work, when the nurse knows that the cause of Phaedra's misery is that she has fallen in love with her stepson, she suggests a solution: here the nurse once again uses the word *nosos*, repeatedly, to refer to love[13] and proposes spells and charms as remedies, which she refers to as *pharmaka*:

> Bear up under your love: it was a god that willed it. And if you are ill (*nosousa*) with it, use some good measures to subdue your illness (*noson*). There are incantations and words that charm: some remedy for your illness (*pharmakon nosou*) will be found. Men will be slow to invent such contrivances if we women do not find them (Euripides *Hipp.* 476-481).

A little later, the nurse offers Phaedra an infallible solution to Phaedra's "disease": some potions that will subjugate the will of Hippolytus and make him return his stepmother's feelings:

> Nurse: I have bewitching potions (*philtra*) of love within the house–I just thought of it this very moment–that will free you from this disease (*nosou*) without disgrace to you or harm to your mind, if only you do not flinch. We must get some token from the man you love, a lock of hair or a piece of clothing, then compound from the twain a single blessing.
> Phae.: This remedy (*pharmakon*), is it an ointment or a potion? (Euripides *Hipp.* 509-516).

Phaedra uses the word *pharmakon* to refer to love potions. Based on these passages, we realize that in Euripides' view, magic was to be used as a remedy for things that were beyond the province of medicine, but the generic word used to refer to the solutions which the two arts offered was common to both. Additionally, we have already seen that those who practise magic are usually women (Medea, Andromache); in this case it is the nurse who puts it into practice. In addition, she mentions the field of love-related magic as something almost exclusive to women, while medicine is said to belong to the doctors, who are men.

---

[13] Regarding the use of the word *nosos* in connection with the forbidden love felt by Phaedra in the tragedy *Hippolytus*, cf. Lloyd (2003: 95-97).

Magical remedies for problems of any type

Magical remedies were used to resolve any kind of problem, not just medical or love-related. And so in *Orestes* we have a Phrygian slave offering an example of magic being used to become invisible through *pharmaka*, which in this context seems to refer to some sort of potion, perhaps even accompanied by charms: “But she <*s.c.* Helen> was nowhere to be seen throughout the house [...] either because of potions (*pharmakoisin*) or magicians’ contrivance or stolen away by the gods” (Euripides *Or.* 1494-1498).

There is an example in *Cyclops* in which a magical incantation is mentioned as a solution to almost anything: the satyrs, who are afraid of the monster, suggest using an incantation of Orpheus to “remotely” help Odysseus drive a firebrand into Polyphemus’ eye: “But I know an incantation of Orpheus so wonderful that the fire brand all on its own will march up to his skull and set the one-eyed son of Earth on fire” (Euripides *Cyc.* 646-648).

The text clearly makes fun of certain spells circulating under Orpheus’ name in order to get a laugh out of the audience, as is appropriate in a satyr drama. These incantations, which were almost certainly marketed and sold, could be used for any purpose. And so we have Euripides using exaggeration for comic effect: the satyr states that there is even an incantation to make a firebrand move on its own and blind the Cyclops by itself. This indicates that such spells must have been very famous in the 5th century BCE, for if the audience had not been familiar with them, the author would not have succeeded in getting a laugh out of these in attendance.

But evidence of the existence of these spells attributed to Orpheus is not only found in Euripides’ time; this is a type of literature which must have gained great importance and survived over a long period, as we find various references to it from after the 5th century BCE.[14]

[14] Diogenes of Sinope (*TrGF* I 88 F 7. 10ss Snell), Apul. *Apol.* 30. 11, Ath. Al. cod. Reg. 1993 f. 317 (*Patrol. Gr.* 26 p.1320)

Folk healing, folk medicine and scientific medicine

Medical remedies and antidotes

The term *pharmaka* also refers to medical remedies, related less to magic than folk healing or the scientific medicine of the period. In *The Bacchae*, Teiresias tries to convince Pentheus, who will not listen to reason, to accept Dionysus and not put himself at risk by confronting a god. Pentheus persists in his obstinacy and Teiresias 'diagnoses'[15] him as suffering from incurable madness: "You are mad and most painfully so: even drugs (*pharmakois*) could bring no cure, nor are they needed to fall ill (*noseis*)" (Euripides *Ba.* 326-327).

Teiresias speaks of both aspects of *pharmaka*: those capable of curing madness, and contrarily, those which can cause it. The first may be understood as antidotes to madness, perhaps more closely related to medicine than sorcery, and the second as poisons or potions that cause madness, and therefore belong to the sphere of magic. In this passage, Euripides plays with this duality, and it is not possible to categorically differentiate between the two areas.[16]

In other passages, we find metaphorical references to antidotes or drugs which make it possible to improve the health. Thus in *Andromache*, the protagonist, indignant at the slanderous allegations made against her by Hermione, speaks of a hypothetical antidote to the venom of bad women, which would be comparable to that of serpents:

> It is strange that some god has given man remedies against snakes of the wild, yet where something worse than snake or fire is concerned, no one has yet found an antidote (*pharmaka*) against wicked woman (Euripides *Andr*. 269-273).

Also metaphorically, there is mention of "remedies" or "medicines" which can cure an "ill" speech: "But unjust argument, being diseased (*nosōn*) in itself, requires clever medicines (*pharmakōn*)" (Euripides *Phoen*. 471-472).

---

[15] Cf. Roux (1972: 359).

[16] This complicated passage has given rise to various interpretations and even suggestions correcting the text passed down in the manuscripts. Cf. the comments of Dodds (1944: 112-113), Winnington-Ingram (1969: 53), Roux (1972: 359), Thomsom (1979), Seaford (2001: 178-179), Lloyd (2003: 93-94).

The tablets of Orpheus

One of the most significant passages is found in verses 965-970 of *Alcestis*:

> I have found nothing stronger than Necessity, nor is there any remedy (*pharmakon*) for it in the Thracian tablets set by the voice of Orpheus, nor in all the remedies (*pharmaka*) which Phoebus harvested in aid of trouble-ridden mortals and gave to the sons of Asclepius (Euripides *Alc.* 965-970).

Euripides mentions certain "Thracian tablets" attributed to Orpheus and that they contained remedies. Although it is not clear what type of remedies these may have been, they seem to be medical in this case, as these tablets are paired with the remedies which Phoebus gave to the heirs of Asclepius (that is, to the doctors). In both cases, the tragedian uses the noun *pharmakon.* It is therefore possible to think[17] that the tablets of Orpheus must contain a type of medicine related to magic and folk healing, a folk medicine,[18] while the remedies of the heirs of Asclepius would be the scientific medicine of the period.

Using a medical metaphor, Euripides expresses the inexorable nature of what has been stipulated by Necessity, as if it were some incurable disease which was impossible to cut short with medicine of any kind. With great pathos, the chorus reflects on the inability to bring Alcestis back to life, as natural law, represented here by Necessity, will not yield to any human remedy, either medical or magical.

Additionally, the fact that in this context both Orpheus and Asclepius are mentioned may give rise to a secondary association of ideas: both mythical characters, according to their respective legends, tried to bring somebody back to life and were punished by the gods for it.[19] This would

---

[17] Cf. Martín Hernández (2003: 57-58).

[18] Later (1st century CE), Orpheus' relationship to medicine grew broader: he was considered the author of medical books, but always very much folk medicine mixed with magic (these books dealt with the preparation of both ointments and curative drugs and poisons, as well as talismans and love potions), as we are informed by Pliny in certain passages of his *Natural History* (*NH.* 20. 32; 28. 34; 28. 43) and by Galen (*De antid.* 14. 144 Kühn).

[19] Orpheus tried to bring his wife back to life and the gods punished him for his audacity in two ways: in the end his efforts were not successful and he died a tragic death at the hands of the Thracian women. The connection between the death of the bard and his attempt to break the law of death was first demonstrated in Pl. *Smp.* 179d. On this subject, Cf. Macías (2008: 42-65). As regards Asclepius, as Pindar tells us (*Pi.* 3. 54-60; cf. et. Philostr. *Her.* 33), moved by greed, he dared

intensify even more the lack of hope felt by the chorus when faced with the impossibility of Alcestis coming back to life.

Two scholia refer to these "tablets set down by the voice of Orpheus", but they are very brief comments and do not specify what the content of these texts was. Also, it is necessary to add that there is some uncertainty as to whether the scholiasts were providing information that does not correspond to the reality of the 5th century BCE, but rather later periods.

The scholium to these verses explains that Euripides described the tablets as "Thracian" because he was referring to some tablets of Orpheus that, according to Heraclides, were found on Mount Haemus in Thrace:

> "Set down by (the word of) Orpheus": Orpheus was a poet and fortune teller. Philochorus, in the first book of *On fortune telling*, explains that there existed poems of his that read thus: 'in truth, I am not unlucky in choosing oracles, but rather the thoughts in my breast are the truth'. And the physicist Heraclides says that there exist some tablets of Orpheus, stating thus: 'In the temple of Dionysus founded in the place called Haemus in Thrace, where they say that there are some written tablets (of Orpheus)' (Sch. E. *Alc*. 968 [III 239, 3 Schwartz]).

The scholiast adds that Orpheus was the author of oracles, basing this on a statement by Philochorus. From this information, it may be deduced that there must have existed in Thrace some tablets attributed to Orpheus which contained, above all, oracles. However, it is impossible to determine whether Euripides is referring to these, as we do not even know if they would have existed in the tragedian's time.

Another scholium, this one to the tragedy *Hecabe*, talks about the existence of an oracle of Dionysus on Mount Pangaion, where some tablets with inscriptions of Orpheus were kept. According to the scholiast, these are the tablets to which Euripides refers in the passage of *Alcestis* under discussion: "There is an oracle of Dionysus, some say at Pangaion, others at Haemus, where the inscriptions on the tablets of Orpheus spoken of in *Alcestis* are found" (Sch. E. *Hec*. 1267 [I 89, 12 Schwartz]).

Therefore, we have seen how Euripides' text links the tablets of Orpheus with medicine by placing them on the same level as the remedies of the heirs of Asclepius, while the scholia focus their comments on the idea that these tablets contained oracles and texts related to fortune telling. It is true that there are oracular temples consecrated to Asclepius where healing practices were also carried out. Perhaps, given that Dionysus was

---

bring a man who was already dead back to life and as punishment for his transgression, he was struck down by a Zeus' thunder bolt.

on occasions considered a healer and oracular divinity, in Thrace there may have been a temple similar to those of Asclepius or the oracle of Delphi, where some supposed oracles of Orpheus were kept, containing, among other things, medical remedies.[20] However, we lack any further information to support this hypothesis: although the scholiasts could have been giving real news of the existence of some tablets of Orpheus, either on Mount Haemus or Pangaion, we do not know if they are referring to the situation during their period or during the period of Euripides.

On the other hand, it is clear that the scholiasts demonstrate their erudition in the comments and we should not forget that, if Euripides has his chorus mention tablets of Orpheus, it is because they must have been familiar to the audience. It seems unlikely that the majority of the Athenian spectators would be aware of the existence of some tablets on a Thracian mountain.

Although the scholiasts do not clarify anything with regard to the content of the tablets and seem rather to relate them to oracular poetry, in my opinion, based on the tragic text, the simplest and most credible answer is that they were incantations, spells, charms and prescriptions used to cure any type of problem, even medical. As they circulated in association with the prestigious name of Orpheus, whose homeland was Thrace, by extension, the tablets would also be called Thracian.

Hippocratic medicine

An obscure fragment of *Bellerophon* contains a defence of Hippocratic medicine,[21] in which examination and diagnosis are the predominant features. It states that the task of the doctor is to tend to the observation of the patient before administering any medicine:[22]

> As to illness (*noson*), a doctor too must cure it after examining it, not by giving remedies (*pharmaka*) by rote, in case they do not suit the illness (*nosōi*). Human illnesses (*nosoi*) are some of them self-inflicted, others come from the gods, but we treat them using traditional methods (*tōi nomōi*). This is what I want to say to you, however: if gods do anything shameful, they are not gods (Euripides *Fr.* 286b Kannicht).

---

[20] Cf. Martín Hernández (2006: 246)

[21] On Hippocratic medicine in the tragedy, cf. Guardasole (2000), Kosak (2004).

[22] This statement can be found in the Hippocratic Corpus, for example, Hipp. *aër.* 11= 218, 10-13 Jouanna= II 50-52 L.

The translation of *tōi nomōi* in this context is highly contentious: we accept the interpretation of Nestle,[23] who believes that Euripides uses this word to refer to folk medicine, that is, remedies involving incantations and purifications which were usually prescribed for illnesses of divine origin. This would be a contrast between scientific medicine, which recommends observation in order to later carry out an effective treatment, and folk medicine, which aims to treat divine illnesses through ritual, religious and magical procedures. Although neither the position of the tragedian nor the meaning of this fragment, which has been inundated with interpretations,[24] is clear, we may attempt to venture our own: following Nestle, we might believe that this passage shows a rationalist Euripides, a defender of Hippocratic science who proposes as a solution for illnesses which are believed to be caused by the gods, a treatment similar to that which would be used for other illnesses, that is, diagnosis and observation for subsequent medication. It would seem that Euripides believes that the intervention of religion or magic would not be considered viable in these cases and ends by denying, although in a veiled way, that these illnesses were of divine origin, as if the gods caused them, they would not be gods.

We also have another fragment, 917 Kannicht, in which the influence of the theories of Hippocratic medicine is easy to make out, as Clement of Alexandria noted on stating:

> The doctor Hippocrates writing that "in truth it is advisable to observe the season, the region, the age and the illness", in a *rhesis* in hexameters,[25] Euripides states that "the man who knows how to practise medicine correctly must look to the lifestyles (*diaitas*) of a city's inhabitants and to their land (*gēn*) when he examines their illnesses (*nosous*)."

Clement relates the Euripidean text we are discussing to one from Hippocrates in *Aphorisms* (I 2= IV 458, 5-10 L.).[26] Both passages defend

[23] Nestle (1938: 27-28), as well as Guardasole (2000: 84). In opposition, there are those who believe it refers to the "rules of the practice" which govern medical activity. Cf. Jouan and Van Looy (2000: 27, n. 57), who make a brief but highly clarifying presentation of the problems presented by this fragment and its interpretation.

[24] Cf. e.g.: Goossens (1962: 154), Mesturini (1981: 35-56), Müller (1993: 116-121), Collard, Cropp and Lee (1995: 116), Luppe (1998: 123-126).

[25] The expression "in hexameter" used by Clement attests to the confusion between "foot" and "meter" in the scansion of iambic trimeter, which would give rise to the Byzantine dodecasyllable. Cf. Guardasole (2000: 77, n. 121).

[26] Guardasole (2000: 76-84) also relates the Euripidean passage to the beginning of the Hippocratic treatise *On Airs, Waters and Places,* with which it shares many

the idea that a good doctor must follow the practice of observation: he must not only pay attention to the symptoms of the illness, but also to the entire environment and factors which surround the patients, their age and the season of the year, as well as the land in which they live. Euripides summarizes all of these factors in two words: *diaitas* and *gēn*. The second refers to the surroundings in which the community is located (for example, the type of climate and the land, whether it is mountainous or swampy, etc.). The former encompasses the entire way of life of both the patient and the community where he lives (such aspects as type of work and diet). It is also noteworthy that the tragedian uses the term *diaitas* in connection with medicine, as this could be seen to reflect the dietary theory developed in the *Hippocratic Corpus*, in which traditional knowledge regarding diet and *dunameis* of foods was reworked.[27]

Furthermore, we have noted the conceptual and syntactic agreement between the beginning of the fragment from Euripides and various passages of the *Hippocratic Corpus*: they all start by referring to whomever seeks to practise medicine correctly.[28] The adverb "correctly" implies practising medicine according to a standard of conduct. As we have pointed out, this would be observation of the patient's environment.

This is the "new medicine" which must have been emerging during this period and must have been different from earlier medicine, where considering the environment had no place and which stood for medication and surgery above all. This medicine is reflected in fragment 403 Kannicht of *Ino*:

> What mother or father gave life to ill-named envy, as a great evil for men? Wherever does it actually have a part in the body, and live–in the hands or entrails or close by the eyes? It is in us, such that it would be a great labour for doctors to remove by surgery or potions or remedies (*pharmakois*)[29] this greatest of all mankind's diseases (*nosōn*).

Euripides metaphorically considers envy an illness that must be extirpated and he even asks what part of the body it is located in.

---

similarities. He concludes by stating that the fact that they coincide in many points proves that Euripides must have been familiar with the first part of this treatise.

[27] Vid. et fr. 213 of *Antiope,* which contains the technical term *phaulē diaita*, Cf. Guardasole (2000) 268-269, with earlier bibliography.

[28] Hp. *Aër* 1= 186, 1 Jouanna = II 12, 1 L.; Hp. *Vet. Med.* 13, 1= 133, 10 Jouanna. Cf. et Hp. *De prisca med.* 13 and *Praecept.* 4. Cf. Nestle (1938) 24 n.2.

[29] We are using the manuscript reading ποτοῖς ἢ φαρμάκοις preserved by Kannicht, in contrast with the ποτοῖσι φαρμάκοις proposed by Valckenaer and preferred by Jouan and Van Looy.

Continuing with the metaphor, he expresses his desire for the doctors to be able to get rid of it, and to this end he lists what must have been the three most common practices in medicine: surgery (*tomais*), medicines that can be drunk (*potois*) and any other type of medicine (*pharmakois*).

To conclude, it is worth pointing out fr. 1072, where Euripides echoes the new anti-surgery trend in the medicine of the period and perhaps to a certain extent reflects the Hippocratic doctrine of *kairos*:[30] "And a doctor who delays and gives the disease (*nosōi*) time has already cured it more than throug surgery."

Conclusion

In Euripides, the term *pharmaka*, which in principle refers to a herbal, animal or vegetable preparation, may be linked to both magical practices and medicine. Although on many occasions it is not at all easy to separate what is magic from what is not, as this is not our aim, we have based the differentiation between magic and folk healing or medicine primarily on the intent with which the person used the remedies.

And thus we first have several passages where the term *pharmaka* refers to poisons and magical preparations used with the intent of killing or causing some irreparable harm to another person (for example, infertility). All of these texts are closely related to magic: this is the case of Medea, the famous sorceress; Creusa, mother of Ion, who uses the magic venom of the blood of the Gorgon; and Andromache, who is falsely accused of *pharmakaeia*.

Still within the sphere of magic, *pharmaka* are also used to designate potions which seek to cure lovesickness, as in *Hippolytus*. In a passage from *Orestes,* the possibility that Helen had vanished thanks to a *pharmakon* is even considered.

In all of these texts, there is one common factor: it is a woman, sometimes with the aggravating circumstance of being a foreigner, who practises the use of both poisons and magical remedies. It could be said that in the works of Euripides, the practice of magic is closely, almost exclusively, linked to women.

In contrast, verses 646-648 of *Cyclops* contain a humorous reference to the presence of magic in society at that time: in addition to *pharmaka*, there were also spells and incantations (on many occasions even used in combination) and the figure of Orpheus is closely linked to them, serving as a sort of guarantee.

---

[30] Cf. Trédé (1992: 183-187).

Secondly, *pharmaka* also refer to the medical remedies of the period, both folk and scientific. In these cases, they are the medical preparations used to make a health problem better and antidotes against poison, such as that of a serpent.

In an important passage from *Alcestis* (vv. 965ff.), it appears that Euripides is including the two types of medicine which the people of the time had recourse to. However, in the case of Alcestis they do not good, as the power of Necessity is inexorable. On one hand we have folk medicine, represented by the figure of Orpheus, and on the other, scientific medicine, practised by the heirs of Asclepius (speaking figuratively of doctors).

Lastly, there is a series of fragments that seem to have a significant connection to Hippocratic medicine, that is to say, the scientific medicine of the period. Fragment 286 talks about observation as a necessary method of diagnosis. This is clarified further in fragment 917, which tells how a good doctor must pay attention to both the environmental factors surrounding the patient and his *diaitas*, with all that term entails. This has even been seen as a possible nod to the dietary doctrine which we begin to see expressed in the *Hippocratic Corpus*.

In fragment 403, despite speaking metaphorically of envy as an illness, in addition to an anatomical interest in locating the home of the illness, we have a list of the most common medical procedures used at the time: surgery, ingested medications and any other type of medicine (for example, fomentations, inhalants, poultices, etc.). Fragment 1072, as we have already noted, defends the anti-surgery trend that must have been new at the time, linked to the Hippocratic doctrine of *kairos*.

References

Artelt, W. (1937). *Studien zur Geschichte der Begriffe "Heilmittel" und "Gift"*, Medizinische, Leipzig: Fakultät.

Betz, H. D. (1986). *The Greek Magical Papyri in Translation*, Chicago: University of Chicago Press.

Carastro, M. (2006). *La cité des mages*, Grenoble: Editions Jerôme Millon.

Collard, C., M.J. Cropp and K.H. Lee (1995). *Euripides' selected fragmentary plays, vol. I*, Warminster: Aris & Phillips.

Derrida, J., (1981). *Dissemination*, trans. B. Johnson of *La Dissémination* (1972, Paris), Chicago: University of Chicago Press.

Dodds, E. R. (1944). *Euripides. Bacchae*, Oxford: The Clarendon Press.

Guardasole, A. (2000). *Tragedia e medicina nell' Atene del 5 secolo a.C.*, Naples: M. D'Auria editore.

Goossens, R. (1962). *Euripide et Athénes*, Brussels: Palais des Académies.

Jouan, F. and H. Van Looy (2000). Euripide. *Tragédies VIII 2ª partie*, Paris: Belles Lettres.
Kosak, J. C. (2004). *Heroic Measures. Hippocratic Medicine in the Making of Euripidean Tragedy,* Leiden-Boston: Brill.
Lloyd, G.E.R. (2003). *In the Grip of Disease*, Oxford: Oxford University Press.
Luppe, W. (1998). "Medizinische Unheilbarkeit von Göttern verursachter Krankheiten: zu Euripides fr. 292 N^2". In Ch.F Collatz, J. Dummer, J. Kollesch and M.L. Werlitz (eds.), *Dissertatiunculae criticae: Festschrift für Christian Hansen*, (pp. 123-126), Würzburg: Königshausen & Neumann.
Macías, S. M. (2008). *Orfeo y el orfismo en Eurípides*, doctoral thesis, Madrid: UCM.
Martín Hernández (2003). "La relación de Orfeo con la magia a través de los testimonios literarios", *MHNH* 3: 55-74.
—. (2006). *El orfismo y la magia*, doctoral thesis, Madrid: UCM.
Mesturini, A. M. (1981). "Magia e medicina in un frammento del Bellerofonte di Euripide", *Ann. della Fac. di Lett. e Filos. Univ. di Genova*: 35-56.
Müller, C. W. (1993). "Euripides, Bellerophontes Fr. 292 N^2", *RhMus* 136: 116-121.
Nestle, W. (1938). *Die Struktur des Eingangs in der Attichen Tragödie*, Hildesheim- Sttutgart: Olms.
Roux, J. (1972). *Euripide. Les Bacchantes*, vol. 2, Paris: Les Belles Lettres.
Seaford, R. (2001). *Euripides. Bacchae*, Warminster: Aris & Phillips.
Thomsom, G. (1979). "The problem of the *Bacchae*", *EEThess*: 424-446.
Trédé, M. (1992). *Kairós: l'à-propos et l'occasion: (Le mot et la notion, d'Homére à la fin du IV siècle avant J.-C.)*, Paris: Klinckieck.
Winkler, J. J. (1990). *The Constraints of Desire*, London: Routledge.
Winnington-Ingram, R. P. (1969). *Euripides and Dionysus. An interpretation of the Bacchae*, Bristol: Bristol Classical Press.

Galen on Body, Temperaments and Personalities

Mario Vegetti

Introduction

In order to delineate an introductory assessment of Galen's views regarding the body, it is a good idea to proceed in the same way as one of his sedulous disciples would have, or at least a careful reader of his works. Therefore one should approach his works following the systematic order that Galen himself indicates in *De ordine librorum suorum* and partially clarifies in *De libris suis* (short works that now are accessible in a more adequate version thanks to the recent edition by Véronique Boudon, based on the Vlatadon manuscript which has been lately brought to light).

In the first place, Galen prescribes to his most talented pupils, or to the readers with the highest levels of self-confidence, to start by reading the fifteen books that constitute the large treatise *De demonstratione*, which deals with issues of logic and epistemology applied to natural philosophy and medicine. All the other readers, us included as the tradition made a choice for us by preventing survival of the first work, have to start from some short writings for beginners and then continue to read through two other large sets of texts. The first group features, respectively, the vast treatise on *Anatomical procedures*, some works on physiology (among which we can include the *De naturalibus facultatibus* and probably *De usu partium* as well, which links the anatomical dimension with the physiological one) and then some writings on psycho-physiology, perhaps hinting at *De placitis* and maybe the *Quod animi mores* as well (*De ord.* II 4-6). The second group of texts comprehends *De elementis secundum Hippocratem* and *De temperamentis*; at this point, the reading sequence splits into two different paths, the one exploring the realm of pharmacology, the other leading to the field of hygiene. The second route will be object of our interest in the present analysis and revolves around the text of *De sanitate tuenda* (*De ord.* II 8-12); nonetheless, the first two texts of this group, on the elements and temperaments, are to be considered as indispensable preliminaries to the study of the therapeutic method (*De methodo medendi*) (*De libr.* VI).

Anatomy and the Bodily Machine

What image of the body is outlined in our pupil or reader's mind after the difficult voyage through the first group of writings, comprehending its anatomical and anatomo-physiological masterpieces? There are, of course, the three main organs (brain, heart and liver), regarded as the natural locations and instruments of action for the three "parts" of the soul and their own psychological functions, in keeping with Galen's interpretation of the *Timaeus*. In addition, there are other organs, such as the reproductive apparatus, the lungs, kidneys and spleen, nearly as relevant as the previous ones in Galen's description.

A most important role, however, is reserved to the three main systems, comprehending respectively nerves, veins and arteries, with the complement of sinews, muscles and bones. In his *Anatomical Procedures*, Galen vigorously complains that doctors devote more attention to the organs than to these systems and, polemically, he defers the treatment of the brain to Book 11 of his treatise. Anatomic theory, Galen argues, presents different interests and value for "a *physikos* who loves science in itself," for a philosopher who wants to demonstrate that nature does nothing in vain, for a physiologist interested in natural or psychological functions, and lastly for a doctor or surgeon dealing with wounds or lesions to be cured. The anatomists "have considered with the highest attention the part of anatomy that is completely useless for doctors [...] and with the slightest interest the most necessary and useful part," which discusses muscles, nerves, veins and arteries.

> It would have been better, actually, not knowing how many membranes are present in each orifice or the heart, nor how many veins nourish it, nor where they stem from and how [...] but to know what muscles extend the arm, the forearm and the carpus, the thigh, the leg and foot [...] and how many nerves are in each of them, where they originate from and how they are arranged... (*An. adm.* II 3 K. II 286-288).

And later he adds:

> Thus I exhort all young people to refrain momentarily from the dissection of the brain, heart, tongue, lungs, spleen, kidneys, oesophagus, larynx, foetuses and pregnant uterus, and before learn well how the arm is

> articulated […], which muscles move it, which nerves, arteries and veins are in each of its parts (*ivi* II 3 K. II 290-91).

Galen's preference for the elastic structures of these organs, and for the relative systems, is not simply induced by reasons of surgical or traumatological usefulness. The voluminous treatise *De usu partium*, which chronologically precedes the *Anatomical Procedures* and, as we will see later, is rather directed to the *physikoi* and to the philosophers rather than to medical practitioners, starts with a deeply admired description of the structure of the human hand and ends with another description of the equally wonderful trunk of the elephant, which is both its nose and hand.

The body, in Galen's view, is more than anything else a perfect machine, capable of performing precise and complex actions thanks to an assemblage of nerves, arteries, veins, sinews and bones, an assemblage made of inserts, joints, microscopic holes lodging vases within different bones, in front of which sometimes the anatomist cannot restrain his emotion. The physiologist, however, needs an extra element to set the machine in motion, something that cannot be anatomically observed. It is the psychic *pneuma*, "the first organ of the soul" according to definition given in *De placitis* (VII 3, 23, p. 444,15 D.L.): it is a fluid that is produced in the cerebral ventricles, which transmits to the brain perceptual stimuli originated from the sense organs, by flowing in the sensory nerves, and distributes movement commands by reaching the motor organs from the brain through the kinetic nerves.

At this point, the body–a perfect instrument–is ready to be given to "its soul". At the beginning of the text of *De usu*, we read that "the body is instrument of the soul"; if various animals have different bodies, that is because their souls are different. Bodies of brave and aggressive animals, such as the lions, will be suitable to attack, while those of fearful animals, such as the deer, will be suitable to escape: "all animals, hence, have a body that is appropriate to the habits (*ethe*) and faculties (*dynameis*) of the soul" (I 2, K. III 2). The renowned polemical argument against Anaxagoras, which Galen derives from Aristotle, belongs to this context: it is not true that the possession of hands is the cause of man being the most intelligent of all animals; on the contrary, it is true that "hands are appropriate instruments for an intelligent animal" (I 3, K. III 5). Men do not use hands only as instruments to defend themselves by means of arms.

> Being a peaceful and political animal, the man wrote laws and built altars and statues for the gods, and built the ship […] and invented all the arts and crafts' instruments […] and left writings to remember theories. Thanks

> to these writings and to the hands, today it is still possible for you to have a relationship with Plato, Aristotle, Hippocrates and the other ancient authors (I 2, K. III 4-5).

The admirable structure of the animal body, worthy of Polycleitus, and its perfect ability to act as instrument of the soul, in its different configurations, constitute in Galen's view the best demonstration of the truth of the Aristotelian and Stoic finalism, which he strenuously sustains in his *De usu*. In the last book of the treatise, denominated *Epode* following the manner of sacred hymns, he writes these words:

> Anyone who observes the facts with free judgment [...] observing the structure of any of the animals–as all show the sign of the wise *demoiurgos*–will understand the excellence of the celestial intelligence: and what previously seemed to you a small thing, the explanation of the usefulness of its different parts, will truly become the beginning of a rigorous theology, which is a much bigger and much more noble thing than the whole of medicine. The examination of the usefulness of the parts, therefore, will not be useful only to the doctor, but much more than to the doctor it will be useful to the philosopher, who strives to acquire knowledge of the whole nature; to these mysteries, I believe, should be initiated all the men who worship the gods, of all nations and all social classes, mysteries that are similar in nothing to those of Eleusis and Samothrace (XVII 1, K. III 360 sg.).

The anatomo-physiologist, capable of documenting scientifically the work of providence in the structure of the bodies, claims for himself the role of protagonist in the construction of a rational theology, and candidates himself to a sort of laic priesthood, which is opposed both to irrationalist cults and to the atheistic materialism *à la* Epicurus and Asclepiades.

This triumphal conclusion, to which the close or distant disciples of Galen are conducted by means of the prescribed sequence of texts regarding anatomy and physiology, nonetheless involves high prices. The difference between souls and bodies is established on the level of species; anatomo-physiology is unable to say anything on individual characters, on the relationship between the organisation of individual bodies and that of the relative souls. Moreover, the finalistic perfection of the bodily machine is unavoidably lacking any kind of information concerning the pathologies that threaten it; the only failures predictable within this theoretical framework belong to the realm of external traumas, while no conceptual space is present for the degenerative phenomena of the somatic complex.

The Completeness of Medical Knowledge

This makes even more necessary, and at the same time more abrupt, the passage to the second group of readings prescribed by Galen. The treatise *De temperamentis* begins with this surprising statement: "Animal bodies are a mixture (*krasis*) of heat, cold, wet, dry, and these *dynameis* are not mixed in the same way in all the cases" (I 1 K. I 509); similar remarks open both the *De elementis secundum Hippocratem*, and the commentary on the Hippocratic *Nature of man* as well as the treatise *De optima corporis constitutione* (K. IV 741). The beloved vascular systems, the major organs of psychophysiology, the sublime architecture of the bodily machine suddenly disappear and in their place appear the old *dynameis* of Empedoclean and Aristotelian descent, the *stoicheia* deriving from the same framework of reference, and of course the humours belonging to the Hippocratic tradition.

This abrupt transition does not lack in justifications in Galen, both on the epistemological and the strictly medical levels. The Galenic polemic carried out mainly in Book 2 of *De methodo medendi* against rationalist (or dogmatic) doctors, who followed the authoritative anatomical tradition of Herophilus and Erasistratus, plays a central role in so far as the epistemological level is concerned. The very limits of medical knowledge are at stake here. The rationalists remain faithful to Herophilus' *motto*: "let these <*sc*. the structures perceivable by means of anatomical observation> be conceived as the first things, even though they are not the first" (*M.M.* II K. X 107); in other words, they intend to "stay within the limits of the *techne*" and refuse to overcome them in order to move towards the theoretical pinnacles of natural philosophy (*akra physiologia*), which would mean to "explore the human nature as philosophers do, who reach the first *stoicheia* by means of Reason" (*ivi* 106-7), that is "the elements of homogeneous parts" (*M.M.* VI K. X 421). It is possible to say that the rationalists respect the neat division traced by Aristotle between a *physikos* and a physician (*De an.* I 1 403b9-16, *De sensu* 1 436a20 ss.). But this hierarchy does not satisfy Galen anymore. If medicine wants be conceived as a complete system of knowledge, organised in sound argumentative sequences, it cannot renounce to take control over knowledge regarding the first elements of nature, delegating this task to philosophy: this renunciation, in Galen's view, makes of the dogmatic only semi-rationalists, dimidiated *logikoi*, because their discourse is limited to the organic parts, which can be observed anatomically, and does not reach the level of the first elements that constitute the structure of the living matter.

Here it is at stake not only a (nevertheless important) question of epistemological completeness, with the connected claim on self-sufficiency of medical knowledge, which would therefore be exempt from all sectarian *diaphoniai* concerning philosophy of nature. A theory of elements is necessary, according to Galen, to face central questions of pathology, diagnostics and therapeutics regarding which anatomy and anatomo-physiology, with all their teleological elegance, remain impotent. As we have seen above, Galen actually considered the works *De elementis* and *De temperamentis* to be introductory readings to the disciplines of therapeutics and pharmacology, and this represents the second justificatory argument for the abrupt shift we have noted above.

It is not difficult to outline broadly the reasons for this shift. The approach of Galenic pathology to internal diseases remains essentially the same of the Hippocratic one based on humours; diagnostics and prognostics are still based on the semiology of bodily excretions (bile, phlegm, blood), with the addition, of course, of the pulse-theory of Herophilean descent. The therapeutic approach remains allopathic in nature (with the obvious exception of surgery), and the enormous dilation of the variety of available medications is not followed by a comparable diversification in the understanding of their *dynameis* (essentially: heating, cooling, humidifying and drying).

Another motive can be added to all the previous arguments: the need to differentiate the individual subject of cure and detect the personal somatic traits that can be useful for diagnosis and therapy, in other words exactly what anatomy was unable to do. And, as we will see, in this respect result other significant effects which affect Galen's view of the relationship between body and psychological characters.

Elements, Humours, Temperaments

Galen's strategy becomes more complex from now on. The question of interest here is how to combine the Hippocratic humoral pathology with the theory of elements-qualities (*dynameis-archai*), in order to lay the former's foundation in natural philosophy and to make the latter more readily available to medical practice. This operation is firstly performed in the *De elementis secundum Hippocratem*, which initiates the second group of works to be read by beginners (but this short treatise must be integrated with the later commentary on the Hippocratic *De natura hominis*). The difficulty of the intended operation is already revealed in the very title of the work: actually there is no proper theory of elements (*stoicheia*) in the Hippocratic texts while, right in the *De natura hominis*, there is a theory of the four humours constituting the body (blood, phlegm, black and yellow bile) and some hints at the relationship between these humours and the primary qualities of the matter (being hot and cold, dry and wet, solid and fluid). Galen's problem is to assemble conceptually the humoral theory of *De natura hominis* and the Aristotelian doctrine of elements (water, air, earth and fire) and of *dynameis* or *archai*, developed in *De generatione et corruptione.* The positive outcome of this operation shows the medical relevance of the theory of elements and its Hippocratic origin, against the Herophilean rationalists, and at the same time requires the medical re-appropriation of the principles of natural philosophy, against Aristotle.

Therefore, heat and dryness both produce and are the components of fire, which corresponds to yellow bile; cold and dryness produce earth, corresponding to black bile; heat and wetness produce air, corresponding to blood; cold and wetness produce water, corresponding to phlegm (*Hipp. Nat. hom.* I 14, K. XV 54; *De elem.* I 3-5, K. I 426-457; *De nat. fac.* I 2, K. II 4-5). So the beginning of *De temperamentis*, where we started from, now can be usefully integrated: the body is constituted by the four primary qualities of matter as well as by the elements and humours, which derive from them.

All these are mixtures (*kraseis*) that can have variable proportions. Starting from here, Galen develops his diagnostics of temperaments, from which derived potentially morbid conditions, and which determine the efficacy of action both of pathogenic factors and of medications: both of them, actually, act on bodies by means of their *dynameis*, respectively upsetting or restoring their balance.

The discussion of the perfect *krasis,* and its defective and pathogenic varieties, in *De temperamentis* is difficult and at times even painful, because Galen tries to give a theoretical structure to a collection of qualitative conceptions based on empirical, immediate evidence. He himself acknowledges the primate of tactile sensations in recognising the states of heat and cold, against Pyrrhonian sophisms (*De temp.* II 2, K. I 589-90), and firmly excludes, in agreement with Aristotle, that the categories of mean and excess could be determined by quantitative measurements, given that they are always relative and variable states. Anyway, Galen's examination accomplishes to establish the model of the *symmetros krasis* of the qualities (the language here clearly betrays a Stoic derivation, rather than a Platonic or Aristotelian one): this model is represented by the human epidermis and specifically that of the hand's palm, which in the best-constituted individual features a perfect equilibrium of heat and wetness (*De temp.* I 9). In comparison with the representation of the *eukratos physis*, which has to be considered as an utterly exceptional case (*De san. tuenda* II 4), there are eight different types of potentially pathogenic degenerations. Four derive from the excess of one single quality and other four from couples of "degenerate" qualities (hot/wet, hot/dry, cold/wet and cold/dry; clearly the couples hot/cold and dry/wet cannot exist, following the Aristotelian law of contraries) (*De temp.* I 1-2, K. I 510-8).

Personalities: The Soul and the Somatic *Kraseis*

The conception of the body as a mixture of *dynameis*, *stoicheia* and humours (a conception which we could define biophysical and biochemical) comes to substitute the architectonical and mechanic one presented by anatomo-physiology; this change in perspective is accompanied by a progressive reversal of the relationship between body and soul (however, while this connection is certainly not casual, it is difficult to go so far as to talk about causal dependency). As seen above, the body was described in *De usu* as the *organon* of the soul, with a choice of language perfectly coherent with the anatomo-physiological framework employed there.

A signal hinting at a different perspective appears in *De placitis*, in a context that, significantly, does not deal with the system comprehending psychic pneuma-nerves-muscles, but with cardiac heat and blood. Galen writes that in some cases the therapy of psychological pathologies is difficult and painful, and these are the cases of patients in which the emotional impulses (*kineseis*) became violent as a necessary (*anankaios*) consequence of their bodily constitution (*kataskeue*), while the rational part was feeble and obtuse by nature (*De plac.* V 5,28, p. 322 D.L.).

The question is presented again, but with some significant theoretical uncertainty, in an important passage of the *De sanitate tuenda*, focusing on children. In this passage is not clear whether the excessive emotional impulses (*ametrioi kineseis*) proper to the passional states produce a bad bodily *krasis*, or if these impulses are due to variations in excess and defect of the innate heat, conditions which respectively determine a bilious or phlegmatic temperament (I 8, K. VI 39-41). These passional excesses, then, are causes or signs of the mutations happened in the bodily *krasis*? Galen had previously written that "the differences between functions are consequent (*akolouthousin*) upon those of the *kraseis*" (I 5 K. VI 15), but nonetheless his uncertainty persists, as we have seen. Certainly the position held in *De sanitate* is, from this point of view, intermediate between those held in *De usu* and *De temperamentis*, even though this last work shares, as we will see, some uncertainty on the epistemological level.

To be honest, though, this treatise appears to be exempt from doubts in so far as its central theoretical line is concerned. If a man presents a good *krasis* of the *dynameis* and the consequent humours, he will present a well-tempered soul: generous, wise, with good *thymos*, intermediate between excessive courage and cowardice, sympathy and envy, capable of impeccable mental activity (II 1, K. I 576-7). The excess of wetness and cold (but in some cases of dryness and cold as well) determines coward and irresolute characters; on the contrary, given that in Nordic populations, such as the Celts, Germans and Scythians, the blood and bodily heat retreat into the inner organs and are compressed until they boil, they are courageous, with abundant *thymos* and quick in tacking decisions (II 6, K. I 626-8). It is worth noting that in these passages the language relative to the psychological characters connected to humoral temperaments starts to be developed: the "melancholic" type and the "phlegmatic" type are mentioned, the former deriving from a *krasis* of hot and dry, the latter featuring a *krasis* of cold and wet (II 6 641-3); but this is only an embryonic manifestation, given that Galen never formulated a complete system of characters and related temperaments, which will be completely developed in the mediaeval tradition.

The comparison with physiognomy is relevant, as this form of knowledge interprets physical appearances, by means of similarity with animals, as a *sign* of the soul's condition. For example having a hairy chest, similarly to a lion, indicates a courageous disposition. Galen writes that the *physikos*, instead, tracks down the causes both of the exterior *signal* and of the psychological disposition that is signalled. Therefore, the hairy chest is due to the high thermic gradient of the heart, which is as well the cause of the excess of *thymos* of the subject in question (II 6, K. I 624-5). Up to now, then, it seems that the somatic *krasis* is actually the cause of psychological configurations rather than its organ, as was stated in *De usu.*

At this point of *De temperamentis*, though, Galen formulates a doubt, according to him shared by Aristotle as well, which demonstrates how aware he was of the deep consequences and issues involved with this shift. Is it not an error, he wonders, to give up considering the natural cause of our bodies' construction as some sort of demiurgic power, which adapts the different material parts to the soul's characters, therefore re-establishing the soul's primate? And this power should not be attributed to "a holier cause than heat, cold, dryness and wetness?" (*De temp.* II 6, K. I 635-6). This doubt was already briefly formulated in a previous passage: there the best constitution resulted (*hepomene*) from the *eukrasia* of the four elements, but the possibility of the existence of "some other cause, more divine and superior" was mentioned as well (I 9, K. I 567). Nevertheless, adds Galen in a reassuring fashion, these questions are so complex that they continue to challenge the best philosophers, and are not supposed to obstacle the examination of *kraseis* and their somatic and psychological effects.

It is not easy to clarify unequivocally the reasons that drove Galen to resolve the doubts and uncertainties expressed both in *De sanitate tuenda* and, more explicitly, in *De temperamentis*, in order to move resolutely towards a complete subordination of the soul to the somatic complexion, which he elaborates in one of his last works, the *Quod animi mores* (the Greek title actually says that the soul's *dynameis* follow–*hepesthai*–the somatic *kraseis*). In this work, moreover, Galen invoked in support of his theory exactly the traditional "best philosophers", from Plato (rather reluctantly) to Aristotle and Posidonius. According to Moraux, it is necessary to look at this question assuming a chronological development which goes from the instrumental approach of *De usu* to the naturalist one featured in the last work, passing by the intermediate stage of *De temperamentis*. Lloyd and Donini developed a different perspective, hypothesising that the diversification in argumentative strategies is due to the different audience of each treatise (intellectual and philosophers on the one hand, medical practitioners on the other). Even though these hypotheses, especially the second one, have many good points, their weakness lies in the fact that, as we have seen, shifts in perspective are present within the same work; in addition, it is necessary to remember that Galen conceived his writings as a coherent and systematic *corpus*, independently from the addressees of the individual works. I will return to this point in my conclusions.

Now, instead, it is worth to outline the argumentative line of the *Quod animi*. The starting point is represented by the Platonic *Timaeus*, according to Galen's interpretation of the text provided in his *De placitis*: the tripartition of the soul in *logistikon* and two other "mortal" parts, *thymoeides* and *epithymetikon* is presented there, as well as the localisation of these parts in the somatic organs, respectively in the brain, heart and liver (with the interesting omission of the sexual organs so far as the *epithymetikon* is concerned). But it is necessary to ask: what does it mean to say that the parts of the soul are located in specific somatic "locations"? Clearly, they do not simply inhabit them as tenants in a flat (*Quod animi* 2, K. IV 769). An initial, approximate answer comes from the Aristotelian *De anima*, which considered the soul as the form (*eidos*) of an organic body. Galen, though, corrects Aristotle on the basis of Plato: there is not a whole soul that represents the form of the body, but its different parts represent an equal number of forms of the relative organs. But, still, what does "form" mean in this context? Galen apparently finds an answer to this question in the Aristotelism of the first and second century AD, perhaps specifically in Andronicus, who suggested to interpret the concept of "form" as somatic *krasis* or *dynamis* (4, K. IV 782); but a similar theory could be traced back as far as to the Pythagorean Simmias of the Platonic *Phaedo* (86b-c).

Actually, to talk about *dynamis* means to imply that the parts of the soul stand to their organs in the same way as the faculty of sight stands to the eye, and this is still within Aristotelian grounds. But to talk about *krasis* means to say that the parts of the soul consist in a specific mixture of elements, material properties (3, K. IV 774; 4, K. IV 780) and humours (3, K. IV 777), which allows the body to function. Therefore, it is not difficult for Galen to argue that at least the "mortal" parts of the soul do not follow the *krasis* of heart and liver, i.e. are not enslaved (*douleuein*) to it, but actually are the same thing of this *krasis*, consequently being entirely corporeal and, therefore, mortal (3, K. IV 773; 4, K. IV 782). Thus the *thymos* will consist in the excess or defect of innate cardiac heat, and the *epithymetikon* in the variations of the hepatic *krasis*.

The question of the *logistikon* is made more complex by the Platonic assumption of its immortality, which would turn it into an *ousia* independent from the body, and specifically independent from the brain. With regard to this question, Galen sustains a cautious agnosticism (3, K. IV 773), which seems to derive more from formal acquiescence than from genuine uncertainty. Plato himself was unable to explain why the rational soul departs from its organ when the latter is severely impaired; and it is clear that, also during lifetime, the behaviour of *logistikon* is compromised and drawn to folly by the excesses of cold or heat, bile or phlegm, by drugs such as hemlock or by wine (3, K. IV 776-7). Given that the pathological alterations of the brain force the rational soul to depart from the body, or at any rate drastically alter their functions, even those who recognise the soul's autonomous substance will have to admit its subjection (*douleia*) to the bodily organ (*ivi*, 779). And this is what Galen claims that Plato admitted as well in the end, quoting extensively passages from the *Timaeus* (86c-e) and the *Laws* (II 666a-b, 674a) where the damages inflicted to the soul by means of noxious humours, wine or superabundance of sperm are described in order to support the theory of the involuntary nature of evil deeds, which will be relevant to Galen as well (6, 10).

The main anthropological argument that Galen states several times in order to support his thesis is based on the observation of the natural differences, concerning characters and intellectual capacities, presented by children before any form of educational and environmental conditioning or after equal amounts of external interventions (2, K. IV 768-9; 11, K. IV 816-9). This proves that the principle of the intellectual and moral decadence does not come from external sources but is endogenous, and thus could consist only of those imperfect *kraseis* of the organs that determine the nature of psychological activities.

Just as myopia depends on a bad conformation of the eye, stupidity, evilness, intellectual and moral deviance depend on a defective *krasis* of the brain, which could have developed during embryonic evolution or supervened afterwards as a result of illnesses or dietetic imbalances. Galen derives from all this two important consequences. The first consists in a sort of moral determinism on an organicistic basis: nobody decides freely and voluntarily to be good or bad, because this is determined by his corporeal temperament; nobody, therefore, can be considered to be responsible for his moral conduct, either good or bad. This does not mean, however, that it is not possible to press charges and punish criminal conducts, although involuntary: here the social body's self-preservation is at stake, and capital punishment of the irredeemable wicked individuals protects the good ones without harming the wicked themselves, given their depravity (11, K. IV 814-6).

The second consequence of the organicistic determinism is doubtless the most significant in Galen's view. If the moral and intellectual quality of self-conduct depends on corporeal conditions, then it is a task for the dietetic practitioner, not the philosopher or the politician, to develop a standard of social hygiene that aims at preventing, or correcting, deviant forms of behaviour. As Galen puts it:

> At least now those who do not accept the fact that food can make some people more temperate, others immoderate, some more self-controlled, others less so, and courageous or coward, gentle or quarrelsome, will return to their senses and come to me in order to learn what they have to eat and drink. They will gain much profit concerning moral philosophy and in addition to this, as they would become more intelligent and with a better memory, they will enhance their virtue by means of the faculties of the rational soul. In addition to different types of nourishment and drinks, I will teach them about the winds, the environmental temperaments and what regions should be chosen or avoided as well (9, K. IV 807-8).

In Search of a Unified Theory

The triumph of the physician is probably one of the objectives of Galen's argument in the *Quod animi* but, as we have seen, it is part of a more complex epistemological framework, which we can recapitulate now. Following the order of the readings prescribed by the author, we have first encountered an image of the body on an anatomo-physiological basis, with the prevalent position of the system made of brain-nerves-motor organs. In so far as the relationship body-soul is concerned, to this image corresponded a conception of the body as an instrument, (*organon*) available to the faculties of the soul. Later, instead, we have met a description of the body as a mixture of elements, material qualities and humours; and this corresponds to the thesis which sustains the dependence of the soul from the material composition of the body, and specifically of the internal organs associated with the soul's functions.

Is there an insuperable opposition between these two different conceptions of the body? And is there a causal nexus between these and the different conception of the relationship between body and soul? Giving an answer to the first question will at least help in developing a better formulation of the second problem.

In Book 2 of *De partibus*, Aristotle clearly defined a sequence of three levels of organisation (*systasis*) of living matter. The first presented elementary qualities; the second was constituted of the homeomeric parts composed from the first qualities, i.e. tissues such as flesh, bone, blood and so forth; the third level comprehended the organs made of tissues, such as the hand, eye, etc. (*De part.* II 1). On this last level we find the scientific area proper to anatomo-physiology and here arises the problem of the function of each organ, and also of the relationship between the soul and an "organic" body (*De an.* II 1 412a28). In Aristotle, the organic body as a whole is conceived as the instrument of the soul, which means that each body is fitting to the qualities and functions of his species: the body of a lion will be vigorous because this species is aggressive, the human body will be erect and with hands to be suitable to the "divine" and intelligent nature of this species. But, as we have seen, it is not possible to overcome the specific level of this instrumental connection between body and soul.[1] Beneath this level, the differences between individuals depend

[1] Even when Aristotle stops at the level of tissues, as in the exam of the properties of the blood in the *De part. Anim.* II 2-4, the relationship between the blood's composition and psychological qualities remains always on the level of the species. However there are some passages, like *Eth. Nicom.* VI 13, 1144b6 ff. and *Categ.* 9b33 ff., where *ethe* are said to be acquired "by nature" and "since birth", which

on *ethos*, not on the soul, and in human beings the *ethos* derives from education given by family and society; control, correction and development of moral conduct will be, therefore, competence of educators with different roles, from the father to the politician and the philosopher.

Galen is not willing to, and neither can, go down the path described by Aristotle, moving between different levels of organisation of the bodies. Rather than a linear path, he describes a crossroad: on the one hand, there are the organs and their functions; on the other, the *krasis* of elements and qualities, with the addition of humours. The first way leads towards anatomy, a prestigious type of knowledge with an illustrious traditional background that produces an image of the body as a perfect machine, which is the most evident proof of the immanent teleology of nature, the divine architect who designed it. The second way deals with different types of mixtures, in most cases quite defective and imbalanced, which seem more similar to the product of a clumsy pharmacist than to the work of the providential *artifex*, in such a way to make Galen very hesitant in connecting them to some "superior cause", as we have seen. Nonetheless, this direction is the only useful one in so far as the diagnosis and therapy of pathological individual predispositions are concerned, as well as for the interpretation of the humour's imbalances, the analysis of the medication's effectiveness, the hygienic and dietetic prevention: all these are crucial questions for the practice of medicine and the anatomo-physiological knowledge has little to say about them (with the obvious exception of surgery).

How does this influence the question of the relationship between body and soul? The language of the elementary *krasis* extends to that of humoral temperaments (melancholic, choleric, phlegmatic); this language reaches the sphere of individual psychological dispositions, also by means of physiognomics (to which, as we have highlighted above, the theory of temperaments is not considered as an alternative but, rather, as its scientific foundation). Moreover, the individual psychological sphere is parallel to that of the pathological dispositions and therapeutic interventions, which are differentiated on an individual basis as well. Out of the dimension comprising the organs and their functions, it is now impossible to conceive the body as the organ of the soul, a notion which was linked to the level of the species. The somatic complex of elements and humours seems to be able to explain individual psychological

---

could refer as well to some sort of conditioning of the individual character by the conformation of the somatic complex, even though Aristotle does not explicitly make this conclusion. I am grateful to Cristina Viano for having pointed these passages to my attention.

organisations, the structure of each singular personality; and, in Galen, "explaining" implies a causal connection. What we would call *personal temperament* "follows", using Galen's terminology, the somatic *krasis* and to "follow" means to depend on it, as in the enslavement mentioned in the *Quod animi*, thus abandoning Plato's views.

This work, then, does not only aim at professional propaganda. Galen's oscillations between different theoretical conceptions of the relationship between body and soul seems to me to be connected to a real theoretical obstacle due to the coincident presence of intellectual and epistemological frameworks that are difficult to organise coherently. The missing attempt to unify them all, substituted by a cumulative programme of readings, doesn't necessarily imply an absence, in Galen, of *esprit de système*. On the contrary, perhaps, it betrays an excessively systematic ambition, which aimed at comprehending in a unitary form Aristotelian natural philosophy, Aristotelian-Stoic teleology, Aristotelian and Alexandrian anatomo-physiology and the nonabandoned physiopathology of humours, of Hippocratic origin. All these elements are present in the order provided by Galen's works. But their utility inevitably becomes appealing to different types of readers: on the one hand intellectuals and natural philosophers, on the other medical practitioners, physiognomists and several other types of psychologists *ante litteram*. Later on, instead, the division will be between teleological and providentialist explanations on the one hand, materialistic and mechanistic ones on the other. Keeping all these aspects together in a unified theory was a task superior to Galen's capacities, if not to his ambitions.

Bibliographical note

With regard to *De ordine* and *De libris* cf. the edition, with extensive introduction, by V. Boudon-Millot, in *Galien.* Tome I, Les Belles Lettres, Paris 2007. On Galen's philosophy of nature, see P. Moraux, *L'aristotelismo presso i Greci*, vol. II, tomo 2, pp. 293-351, trad. ital. Vita e Pensiero, Milano 2000; R. J. Hankinson, *Philosophy of Nature*, in Id. (ed.), *The Cambridge Companion to Galen*, CUP, 2008, pp. 210-241. On *De sanitate tuenda,* see S. Grimaudo, *Difendere la salute*, Bibliopolis, Napoli 2008. On psychology and the *Quod animi* P.L. Donini, *Tre studi sull'aristotelismo nel II secolo d.C.*, Paravia, Torino 1974, pp. 127-185; Id., *Psychology*, in *Cambridge Companion* cit., pp. 184-209; R.J. Hankinson, *Actions and passions: affection, emotion, and moral self-management in Galen's philosophical psychology*, in J. Brunschwig-M. Nussbaum (eds.), *Passions and Perceptions*, CUP, 1993, pp. 184-222;

G.E.R. Lloyd, *Scholarship, authority and argument in Galen's* Quod animi mores, in P.Manuli-M. Vegetti (eds.), *Le opere psicologiche di Galeno*, Bibliopolis, Napoli 1988, pp. 11-22. On Galen's epistemology, cf. M. Vegetti, *Modelli di medicina in Galeno*, in V. Nutton (ed.), *Galen: Problems and Prospects*, Wellcome Inst., London, pp. 47-63; Id., *L'immagine del medico e lo statuto epistemologico della medicina in Galeno*, in W.Haase-H. Temporini (eds.), *ANRW*, II, 37.2, 1994, pp. 1672-1717 (= *Dialoghi con gli antichi*, Academia Verlag, Sankt Augustin 2007, pp. 227-278. On Aristotle, see C. Viano, *Aristotle and the starting point of moral development: the notion of natural virtue*, in S. Stern Gillet-K. Corrigan (eds.), *Reading Ancient Texts. Essays in Honour of Denis O'Brien*, vol. II, Brill, Leiden-Boston 2007, pp. 23-42

KNOWING THE BODY: RENAISSANCE MEDICINE AND THE CLASSICS

HELEN KING

In this paper I would like to address the themes of the volume through three, related, areas. The first of these concerns identifying the fundamentals of Greek medicine, including gynaecology. In the second section, I will illustrate how the Renaissance represented the classical medicine on which it drew. Finally, by looking at the transmission of a single Hippocratic case history, I will give a few examples of the multiple ways in which the medicine of the sixteenth century–and indeed the seventeenth century, as there is no clear break in this topic at 1600–worked to retain Hippocratic *content*, even while developing from it different *theory*. I will thus demonstrate the continuing versatility of Greek medical science in the age of observation.

What is Greek medicine?

First, what do we mean by Greek medicine? Many claims have been made for the "Greek miracle" as a shift in thinking from the mid-sixth century BCE onwards; for the origin of rationality in medicine, the retreat from superstition and magic, and the emergence of the belief that natural causes lie behind disease, so that natural substances can cure it. Much of this is overstated. For example, with regard to the alleged retreat from superstition and magic, the Hippocratic writers rarely attack religion and, where they do, they attack individual wandering healers rather than the religion of the city-state. This can be seen in the treatise *On the Sacred Disease* 3-4, where purifiers and magicians are attacked for their deception of sufferers, but the religious healing of the sanctuaries is not condemned (van der Eijk, 1990; Laskaris, 2002: 71; Lloyd, 2003: 46).

Historically, as the needs of those seeking the authority of a connection to the classical Greek past have shifted, "Greek medicine" has often become shorthand for "rational medicine" or for "medicine based on experience", for "humoral medicine" or for "scientific medicine". As Andrew Cunningham pointed out, "Everyone is on the side of both reason and experience, and reason and experience are on the side of everyone, but they mean different things to different people" (2001: 107). Over the past two millennia, as medical theories have changed, so different treatises of the Hippocratic corpus have been moved into the foreground, and back into the shadows; Thomas Sydenham liked *Epidemics* and *Airs, Waters, Places* because of the importance of observation there; the medical meteorologists of the first half of the eighteenth century agreed with his choice of texts, but used them to discuss the weather as a cause of disease. In the period from the seventeenth to the nineteenth centuries there could be a "chemical Hippocrates", an "experimental Hippocrates" as well as a "vitalist Hippocrates" (Cantor, 2001).

In his commentary on the Hippocratic treatise, which he identified as bringing together the texts *On generation/Nature of the child/Diseases IV*, the late Iain Lonie placed the doctrine of the four humours as the centre of the Greek miracle, claiming that

> It is important that the search for constituent humours marks a revolution in medical science, a new way of looking at the human body, and this revolution was the result of philosophic development in the fifth century (1981: 60; cf. Jouanna 1999: 210).

Jacques Jouanna, in his monumental work *Hippocrates* (1992 Fr.), noted:

> The one domain of Hippocratic physiology where the imagination of certain physicians succeeded in creating a system of perfect clarity, in order to account for an entirely obscure world, is that of the humors (1999: 314).

While the relationship between obscurity and clarity is I think entirely correct, there are two problems with this sentence; first, it places the focus on the humours, and second, it seems to imply that there was a single "Hippocratic physiology", which is not really the case.

Certainly, the main way in which this "obscure world" of the body's interior became "clear" to the physician was through observing what came out of the body; fluids, not just the humours, but also non-humoral fluids including sweat, semen, urine and tears. It is now a commonplace that the pre-modern body–whether it was male, or female–was "a body of fluids" rather than a "body of organs", but the dominance of the four humours in western medicine has led to these other fluids being sidelined (Paster,

1993; Schoenfeldt, 1999: 36-7).[1] In the ancient Greek "body of fluids", fluids moved through channels, and disease could arise because of blockages in these. Fluids were attracted to "structures" (Gk. *schēmata*), which were seen as having different shapes and textures that affected their ability "to attract and absorb moisture from the rest of the body," in the words of the Hippocratic treatise *On Ancient Medicine* 22.

What precisely are these fluids within the body? While it is still widely assumed by those outside the field of the history of ancient medicine that "the four humours"–blood, phlegm, yellow bile and black bile–represent *the* Hippocratic model of the body, it is now acknowledged by scholars working on ancient medicine that the Hippocratic corpus has *no* single dominant model of the fluids that constitute the body. The Greek *ikmas*, "humour", simply means "moisture", and is something plants need in order to sprout; like the rest of the language taken from ancient Greek and used in later medicine, it is not, in origin, a "medical" term (Thomas, 2000: 50-52).

Ancient writers who do operate with a largely humoral model appear to differ as to whether humours are naturally present in the body, only becoming dangerous if there is too much or too little of them, or whether they are considered to be intrinsically harmful substances made from blood. Furthermore, even in those treatises historically imagined to be closest to the ideas of the historical Hippocrates, such as *Epidemics* 1 and 3, humours are not always at the centre of the picture of disease causation. Caterina Licciardi has noted that, in these two treatises, while the humours feature in descriptions of patients' symptoms, with bilious or phlegmatic matter being present in discharges, nevertheless "les humeurs ... ne jouent jamais le rôle de facteurs causaux" (1990: 328). In *Epidemics* 2, 4, and 6, regarded traditionally either as the work of Hippocrates' son Thessalus or as notes by the great man himself, phlegm and bile do appear as causes of disease, but blood does not, while in books 5 and 7 of this collection, which appear to be the most recent and for whose Hippocratic authorship nobody has argued, bile (both yellow and black) features as a cause, but only in the first part of book 5 (Licciardi, 1990: 331-32; 334-35). The main concern of these texts is to describe the symptoms, not to speculate on causes. Looking outside the *Epidemics* collections, *Regimen* operates with two elements, fire and water, with the human body in flux between

[1] Cunningham (2010: 158) cites Winslow's (1733) comment that "The history of the fluid parts ... properly belongs to what is called Physiology or the Animal Oeconomy." H.F.J. Horstmanshoff, H. King, and C. Zittel, eds. (forthcoming), aims to redress the balance in favour of the non-humoral fluids in western medicine.

the two, while in *Affections* and *Diseases 1*, while fluids are the constituent parts, only two fluids are named: bile and phlegm. In the texts devoted to the female body, although there are a few occasions on which humours are mentioned, such as "phlegmatic menses", in general blood is so dominant that humoral theory appears to be ignored (DW 1.11, L 8. 44-46; King, 1998: 69).

Those working on ancient medicine have long been aware that there was ancient Greek medical writing from the time of Hippocrates that was not included in the *Hippocratic corpus*, and scholars have recently started to question even the utility of the category "Hippocratic medicine".[2] The lost works of some other writers of the Hippocratic period are only known to us from summaries in the papyrus known as *Anonymus Londinensis*, dating from the first century CE and discovered at the end of the nineteenth century, and assumed to have been written originally by a pupil of Aristotle (Jones, 1947; Manetti, 1990; Manetti, 2011). Where did these writers stand on humoral medicine? Thanks to *Anonymus Londinensis*, we know for example about the 3-humour theory of Philolaus of Croton (blood, bile and phlegm) and another 3-humour theory linked with Thrasymachus of Sardis (bile, phlegm and pus) (Nutton, 2004: 80). In the Hippocratic *Generation/Nature of the Child* 3 and the related *Diseases* IV 32, an alternative 4-humour theory is found, with the four being blood, phlegm, bile and water.

The famous account by Thucydides of the plague of Athens, which affected Greece in 431/430 BCE, is interesting here in providing a lay view of the humours. In his extensive list of the symptoms that were experienced in all parts of the body (*Peloponnesian War*, 2.49.3), Thucydides mentions the production of "every kind of bile named by the medical profession."[3] It is worth noting that he does not mention "both kinds of bile"–which would have evoked the 4-humour model with its two "biles"–but "every kind", suggesting perhaps that the historian was most familiar with a multi-bile model, or alternatively that he was aware that different medical practitioners functioned with different lists of the categories of bile. For example, the Hippocratic treatise *Regimen in Acute Diseases* distinguishes between "black bile" and "bitter bile":

---

[2] The 13th Colloque Hippocratique, held in Texas in 2008, had as its theme "What's Hippocratic about the Hippocratics?" The proceedings, edited by L. Dean-Jones and R.J. Hankinson, are in press (Leiden: Brill).

[3] Parry (1969) used this passage to argue that Thucydides knew the Hippocratic texts.

> ... the sharpness obtained from vinegar is more beneficial to those with bitter bile than those with black bile because it dissolves bitter substances, turns them into phlegm and fetches them up. Black bile is lightened, brought up and diluted, for vinegar brings up black bile (61: tr. Lloyd).[4]

But it is in only one of the 60 or so "Hippocratic" texts–that is, in *On the Nature of Man*, thought to date to the late fifth century BCE–that the specific "canonical four" of blood, phlegm, black bile and yellow bile are named, and a fairly-developed picture of the humours is given. For a modern audience trying to link the humours to what we can now observe, the most worrying of these fluids is "black bile", which does not seem to correspond to any normal body fluids. Langholf (1990: 47) argued that the author of *On the Nature of Man* had worked in an area where he observed the black urine of malaria sufferers, and so created the concept of black bile to account for this. Even in this treatise, where the humours are intrinsic to the body, health is not seen as a simple matter of balance; the levels depend on the seasons of the year, so that what is a normal amount of blood in the spring would be regarded as an excess of blood in the winter.

> So too in man sometimes phlegm is powerful, sometimes blood, sometimes bile, first yellow, and then what is called black bile. The clearest proof is that if you will give the same man to drink the same drug four times in the year, he will vomit, you will find, the most phlegmatic matter in the winter, the moistest in the spring, the most bilious in the summer, and the blackest in autumn (*Nature of Man*, 7).

So if the four humours did not form the basis of the physiology of the Hippocratic corpus, why do we think that they did? The reason why the four humours became so powerful in western medicine was that Galen, writing around 190 CE, chose to elevate the section on humours in *On the Nature of Man* to the status of a "genuine work of Hippocrates" when writing his commentary on it, and not only used, but also developed in far greater detail, the 4-humour model. For Galen, the first part of this treatise, where the four humours are described, "contains, as it were, the foundation of the whole Hippocratic science" (CMG 5.9.1; Smith, 1979: 170). But, as we have seen, in fifth-century BCE Greece, contrary to what Galen would have us believe, it is clear that there were multiple models of the humours, even if we only look at the medical texts of the Hippocratic corpus.

---

[4] "Bitter bile" is also mentioned in *Regimen in Acute Diseases* 53, and in *Epidemics* 3.12.14.

For a contribution to a volume on the *long* tradition of Greek Science (4th c. BCE-17th c. CE), the situation is complicated further by the point that this most enduring figure of "Greek medicine" was not just Greek, but Roman. The most prolific prose writer of antiquity, Galen was born in the Greek half of the Roman Empire, wrote in Greek but spent much of his life in Rome, where he was doctor not only to the Roman emperor Marcus Aurelius and his family, but also to many of the elite of the city. As Wesley Smith summarised the situation, Galen "held to a four-humor theory and looked for a Hippocrates in his own image" (Smith, 1979: 50). Rather than finding such a Hippocrates, he created one; he decided that those Hippocratic treatises that most resonated with the views he himself held were by the historical Hippocrates, those that he found less convincing then being attributed to other members of Hippocrates' family.

Galen worked in the second century CE, the period known as the "Second Sophistic", where writers idealised the fifth-century Greek miracle, wrote in Attic Greek, and idolised the figures of classical Athens. There was a great interest in the city of Athens itself–it was visited by the emperor Hadrian three times. It was precisely in Asia Minor, Galen's part of the Empire, that the Second Sophistic as a cultural movement was strongest. Graham Anderson took at face value Galen's comments deriding "sophists", arguing that medicine was not taken up in this philosophical and cultural movement because:

> The professional and experimental values of a doctor had little time for the world of approximation, equivocation, showmanship and self-proclaimed authority so typical of successful public speakers (Anderson, 1993: 10)

Heinrich von Staden rightly took issue with this view, and has shown how Galen's interest in spectacular public displays of his prowess in both animal dissection and animal vivisection to entertain and amaze his audiences can be seen as epideictic in precisely the sophistic sense. They are showmanship; they are about proclaiming his personal authority (von Staden, 1994; 1997: 36). As Maud Gleason noted, while "Reverence for Galen as a pioneer of scientific rationalism makes it difficult to concede that his activities might have anything in common with wonder-working competitions and the popular performances of mountebanks" (2009: 100), he should be seen as using "a rhetoric traditional to Greek science, a way of arguing that involved a unitary view of nature and an emphasis on homology between animals and man" within a more Roman rhetoric of violence, power and control (2009: 113). Galen is in some senses a Roman doctor, but in more senses a Greek doctor, not just because of his origins in Pergamum but also because of his cultural position in the Second Sophistic (Kollesch, 1981).

Even before Galen in the second century CE, Romans constructed the medical past so that–for better or for worse–it was not only Greek, but was also often centred on Hippocrates. A century or so before Galen was writing, in 47 or 48 CE, the doctor and drug expert Scribonius Largus had called Hippocrates "the founder of our profession" (*Medical Recipes*, preface 5). Scribonius' Latin is such that it has been suggested that, like Galen, he may have been a doctor with Greek origins, possibly a freedman; he accompanied the emperor Claudius on his British expedition, perhaps as an army physician, or perhaps to serve one of the elite Romans on this mission (Hamilton, 1986: Baldwin, 1992). When Scribonius Largus looked back at the medical past, however, he did not offer a simple view of a Hippocratic peak followed by a steady decline. Herophilos, the Greek physician working in Alexandria in the third century BCE, was once considered, Scribonius Largus tells us, "among the greatest"; Hippocrates was the founder of the profession, and of the *Oath*; but it is yet another, more recent, Greek doctor, Asclepiades of Bithynia, who is singled out by Scribonius as "the greatest proponent of medicine" and as "that great master of medicine" (*Medical Recipes*, preface 7). However, Scribonius did go on to criticise Asclepiades because he did not recommend drugs for sufferers from fever or acute conditions, preferring instead only dietary measures. Interestingly, Scribonius criticises those who have "obtained the full title of doctor from a knowledge of [only] one branch of healing"; as we will see shortly, something very similar was written by Vesalius in 1543. The ideal of a unified medicine, rather than a medicine divided by specialism, is an enduring one in the history of Greek medicine. It is found in the Greek historian Herodotus (*Histories*, 2.84), who describes Egyptian medicine as so specialised that each doctor treats only one part of the body; there is an implied contrast here with Greek medicine, as there would be no reason to find this an interesting feature of Egyptian culture if the Greeks shared this approach, and indeed there is little sense of specialism within the ancient Greek medical writings.[5] Herodotus also comments that the Egyptians are the healthiest of all peoples, but this is because of the equanimity of their climate, not because of their medicine (*Histories*, 2.77.3).

Not all those writing at the time of Scribonius saw Greek medicine in entirely positive terms; famously, Pliny complained that "Medicine changes on a daily basis, being tarted up again and again, and we are swept along on a blast from the clever brains of Greece" (*Natural History* 29.5.1). Here what is being criticised is partly the professionalization of medicine, when the

---

[5] In terms of specialism, the nearest Hippocratic medicine comes to this is the statement that the treatment of the diseases of women differs greatly from that of men, *Diseases of Women* 1.62 (Littré 8. 162).

Romans still held the ideal of the *paterfamilias* whose care for his household includes treating them when they are sick, and partly a dislike of change of a superficial nature. For Pliny, Greek doctors keep changing their theories and their remedies, and end up confusing their (Roman) patients (Nijhuis, 1995). This is very different from the way that Greek doctors themselves present their past: where Pliny sees change, they see continuity, even if this is manufactured continuity with a figure of the distant past, Hippocrates, which allows them to rise above the acrimonious disputes of their own day.

So far, then, I have suggested two ways in which Greek medicine saw itself and was seen by others, even in the earliest days of its history. First, it presented itself as having a history, rather than being based on novelty. This history was focused on the person of Hippocrates and, following Galen, it centred on a 4-humour model of the body. Second, it saw itself as a unified medicine in which one practitioner knew about different medical areas–diet, surgery, drugs–rather than having only limited knowledge of just one part of the body.

Greek medicine in the Renaissance

I have described elsewhere (King, 2006) how, in Galen's period, under the influence of the Second Sophistic, the Greek medical past up to the second century CE was constructed as a period of decline from a fifth-century BCE (Hippocratic) peak. Galen's contemporary Aelius Aristides, perhaps the most famous patient of antiquity because of his accounts of his own symptoms in his *Sacred Tales*, provides a further example of a second-century CE writer considering the period from fifth-century Greece to his own day as one of "decline". We could set this decline in an even wider context of ancient Greek ideas about medicine because, in medical terms, the Hippocratic "peak" was necessary because of an even earlier "decline" that was envisaged from the Hesiodic Golden Age in which diseases were not known; they only emerged from Pandora's jar as part of the establishment of the conditions of the present world (Hesiod, *Works and Days* 102-4). There was an opposite model, one of "ascent": this held that medicine was part of the raising up of mortals from a bestial condition towards something nearer to the gods. Its key texts were Prometheus' speech in Aeschylus' *Prometheus Bound*, about the benefits he brought to the mortal world (ll. 478-483), and the Hippocratic *On Ancient Medicine*, which talks about the primitive, indigestible, disease-inducing diet of earlier generations and looks to the *iatros* as the hero-figure responsible, over a considerable period of time, for creating diets more appropriate to the human condition. The *iatros* is the "discoverer" of the best diets for both the healthy and the sick. So we have a first "peak" of medicine in the distant Golden Age (no disease, so no need for medicine); then decline from the Golden Age as diseases run riot in the world; followed by a new peak as Prometheus–or the *iatros* of *On ancient medicine*–discovers the cures for diseases. Following the fifth-century Greek "peak", Galen would argue, there was then decline again, until Galen recovered the true Hippocratic medicine.

In the Middle Ages, Hippocrates and Galen were represented sitting with a third figure, the Arab Avicenna, forming a medical triumvirate; each held his own book, but it was Avicenna, interpreter of Galen, sitting in the centre, who carried the orb and sceptre, and wore the crown.

But by the sixteenth century there was a strong reaction against Arabic medicine as having mangled the true Greek heritage, both in translating the texts and in introducing barbarous methods of cure. Of course, we now see things rather differently–if it were not for the Arabic translators of Galen, our knowledge of Galenic medicine would be far more limited. One focus for disagreement in the sixteenth century concerned bloodletting; most sixteenth-century writers, following Galen, agreed that bloodletting is generally a good idea in many conditions, but argued over which site should be used in which condition: a vein on the same side as the problem, or a vein on the opposite side (King, 1999)? The debate centred on a passage of Galen concerning revulsion, the practice of shifting a humour from the affected part to a healthy part in order to release it (Galen, *To Glaucon* II.4). The Parisian Pierre Brissot (1478-1522) condemned as an Arab misinterpretation the practice of bleeding from the site furthest away from the affected part. Instead, he said, one should remove a good quantity, and from the side affected by the disease; so, for chest pain, one should take blood from the arm on the affected side (O'Malley, 1964: 67-68).

In the powerful reaction against Brissot, Hippocrates was cited by the traditionalists. *Nature of Man* 11, which describes the main blood vessels of the body, was used to support bloodletting from the opposite side; it instructs the reader to cut "as far as possible from the places where the pains are wont to occur and the blood to collect" *(Nature of Man* 11, p.33 Loeb). Hippocrates apparently supported the Arabs! But Hippocrates was also cited by supporters of Brissot: in particular, *Aphorisms* 5.68, which says that the forehead should be cut for pain at the back of the head, thus recommending a site *near to* the affected part. It is significant that both sides of the great blood-letting debate went back beyond Galen to the Hippocratic corpus, but the conclusions they drew from the Hippocratic treatises were opposed. This process, of going back to Greek medicine in search of ancient authorities to support one's position, has been discussed in detail by Andrew Cunningham (1997). Cunningham showed how Vesalius looked to Galen in this role, while Realdo Colombo looked to Herophilus and Erasistratus, and Girolamo Fabrizi to Aristotle. The Greek tradition provided different "forefathers" depending on what it was that needed to be "fathered".

The "return to [various different texts associated with] Hippocrates" has, I would argue, even more benefits than those gained by simply moving to a different part of the Greek tradition. Sixteenth-century medicine generally retained Galen's model of twin peaks, Hippocrates and Galen himself, with Galen as the higher peak. But looking back to Hippocrates was, historically, an extremely effective strategy precisely because there is so much variation within the texts preserved as the *Hippocratic corpus* that entirely opposite positions were equally able to find support. Greek medicine has therefore been able to preserve the impression of unity, even as change occurs, by focusing on a different part of its Hippocratic heritage.

I would now like to look briefly at the model of the past in Vesalius, and its implications for what was perceived as "Greek medicine". In the *Preface* to the *De fabrica corporis humani* (1543) Vesalius produced a polemic against the perceived inadequacies of the medicine of his own day. Just as everyone else had done, he played the game of picking a genuine work of Hippocrates. For him, the top contenders as what he called the "most perfect" works were not the usual suspects–*Epidemics*, *Nature of Man*–but *The Doctor*, *On Fractures* and *On Joints*. The last two fit with his (Galenic) concern with the bones as the structure of the body; the first volume of *De fabrica* is devoted to the bones, and Vesalius is of course depicted on the title page of the volume with a skeleton behind the dissection table.

Vesalius also tells a story of the decline of medicine from its ancient beginnings, a decline starting for him well after Galen, at the fall of the Roman Empire. Interestingly, in a text on dissection, on "ana-tomy", literally the cutting apart of the body, the great evil is "fragmentation": he talks about how "the art of medicine began to be torn in pieces long ago" (Vesalius, 1998: xlvii). "All branches of knowledge, which had previously flourished gloriously and been practiced in the proper manner, began to deteriorate" (Vesalius, 1998: xlviii). He regards the lost ideal as being the bringing together of control of the three branches of diet, drugs and surgery in a single person, in contrast to the medicine of his own day when nurses supervise diet, apothecaries drugs, and barbers all manual operations; what he presents as the traditional "threefold approach to medicine cannot be broken up" (Vesalius, 1998: xlix). In the *Preface*, he did not come down firmly in favour of any single period as the Golden Age of medicine; he regarded as valuable Hippocratic medicine, the Alexandrian anatomists of the third century BCE, and Galen, even though he had been "deceived by his apes" (Vesalius, 1998: liv) into thinking that the human body had features which in fact are not found there. Of course, this is very far from being a naïve move. Herophilus and Erasistratus must be in the frame, because what they were known for was human dissection, and it was precisely that which was the subject of Vesalius' title page. I would identify the figures in "antique costume" on the title page as Herophilus and Erasistratus looking over Vesalius' shoulder, and Galen on the other side of the dissection table, with Hippocrates not being present: I would argue that his absence is not due to him losing his position in the hierarchy of the medical past, but because he is the hero of a particular ideal of Greek medicine, unfragmented medicine, for which he is the ethical master of the bedside manner, ministering to the "whole" and living patient.

Vesalius' own insistence on human dissection as the centre of the medical curriculum very firmly looked to the Greek past for its basis. As well as paying homage to Herophilus and Erasistratus, it managed to draw authority from Galen–even though he was not able to dissect humans–because he dissected every day, either to practice or to display his skill to an audience, meaning that he was "easily first among the professors of dissection" (Vesalius, 1998: liv). Even when noting the errors in Galen due to his reliance on apes, Vesalius was in fact doing precisely what Galen had urged his followers to do: to look at the body itself. Galen's influence on sixteenth-century medicine was to show that "anatomy was fundamental to medical knowledge and practice and that dissection was fundamental to anatomy" (Park, 2006: 166). Vesalius has rightly been described by Andrew Cunningham in terms of his activities in dissection and vivisection as "simply Galen restored to life" (Cunningham, 1997: 115) and by Hilary Nunn as "a star" of the dissecting theatre (2005: 11).

When Vesalius looked at the period after Galen, he argued that there had been a further decline from what he presented as Galen's ideal of one person combining intellectual knowledge (the sphere of the physician), manual dexterity (the sphere of the surgeon) and ability with drugs (the role of the apothecary). In Vesalius himself, according to Vesalius, the three spheres had been reunited; the ideal of classical Greek medicine had been realised afresh.

The ideal of a unified medicine, performed by one person for all patients and for all branches of healing, rather than a medicine divided by specialism, is an enduring one in the history of Greek and Roman medicine; but how does medicine for the female body fit into this model? Despite an insistence in the Hippocratic *Gynaikeia* treatises (the *Diseases of Women*) that the healing of the diseases of women differs greatly from that of men, and that women's bodies are different from those of men because of their wet and spongy flesh (King, 1998), there is no evidence for "gynaecologists" in the ancient world. While we have works devoted to the female body (the Hippocratic *Gynaikeia*, Soranus' *Gynaecology*) these are simply evidence of a debate over just how different that body was thought to be, rather than of medical personnel who only treated women (Flemming, 2007).

In the sixteenth century, a further form of fragmentation occurred, with the first printing of Latin translations of the Hippocratic *Gynaikeia* treatises; Marco Fabio Calvi's translation was published in 1525. As Galen had not left any treatises devoted to women's diseases, these treatises filled a gap, and meant that Hippocrates came to be seen as the "father of gynaecology", the best guide to the complex creature that was woman. Compilations of gynaecological materials were printed from 1566, and Hippocrates' insights into women were praised even by those who had not in fact read the treatises (King, 2007). So, even when a new area of medicine was coming into being, it was to Greek medicine that its supporters looked.

The transmission of a Hippocratic case history

In this final section, I want to look briefly at how just one case history of the Hippocratic *Epidemics* collection was re-read in the sixteenth century and into the seventeenth century.[6] This is the story of Phaethousa of Abdera, who became masculine in appearance, even growing a beard, when she stopped menstruating; despite attempts at medical treatment, she died. Due to the sixteenth-century interest in Hippocrates as an expert on the diseases of women, the story was told and re-told for several hundred years, and examining its retellings illustrates how Greek medicine retained its central place even as medical theories and therapies shifted at the end of our two-thousand-year period.

The case history reads as follows:

> In Abdera, Phaethousa the wife of Pytheas, *oikouros*, having borne children in the preceding time, when her husband was exiled stopped menstruating for a long time. Afterwards pains and reddening in the joints. When that happened her body was masculinized and grew hairy all over, she grew a beard, her voice became harsh, and though we did everything we could to bring forth menses they did not come, but she died after surviving a short time. The same thing happened to Nanno, Gorgippos' wife, in Thasos. All the physicians I met thought that there was one hope of feminising her, if normal menstruation occurred. But in her case too it

[6] This section will be developed in my forthcoming monograph on the Renaissance and early modern uses of the stories of Agnodike (the "first midwife") and Phaethousa.

was not possible, though we did everything, but she died quickly (*Epidemics* 6.8.32, tr. Smith).

The description of Phaethousa as *oikouros* could mean simply that she kept at home (in the *oikos*, or house), that she was the mistress of the house, or–in some versions of the case history preserved in manuscripts–that she was not Pytheas' wife, but his housekeeper. In another manuscript variant, she is not even *oikouros*, but *hê kouros* or "the servant"; so, some translations make the central character "Phaethousa, the maidservant of Pytheas" (Manetti and Roselli, 1982: 194-5).

How should we understand this story in its original context? It concerns the risks of menstrual suppression, and the hairiness which, in later medical sources from the Greco-Roman world, is the "mark of a man" (Gleason, 1990: 400), indicating the greater male body heat that also enables a man to "concoct" his blood into semen. Women are normally too cold to do this: Phaethousa's menstrual blood may be absent because it is turning into her beard and body hair. The selection of incidents–her previous child-bearing, the absence of her husband–indicates what is seen as diagnostically significant for the writer of this case history. She does not in any sense change her sex, nor is this a previously concealed "true sex" being revealed late in life; as she has given birth, she is clearly a woman, and indeed the use of feminine parts of speech, at the end of both her story and that of Nanno, underlines this.

Despite this not being a story of sex change, however, in sixteenth-century readings it was most commonly placed within the context of collections of stories of this kind, typically alongside accounts from ancient writers such as Pliny, Phlegon and Diodorus Siculus as well as more recent infamous cases. Diodorus Siculus (32.11) tells of an Epidaurian girl called Callo who was born without an "opening" in "the orifice with which women are naturally provided," so when she married she could only have "unnatural" intercourse. She then developed a tumour on her genitals, which none of the physicians summoned to see her could treat, but "a certain apothecary" cut into it and testicles and an imperforate penis were found inside. She "laid aside her loom-shuttles and all other instruments of women's work" and dressed as a man, now known as Callon. In the preceding story, Diodorus tells us of Heraïs (32.10); as with Callo, a tumour appeared, this time at the base of her abdomen, and the area continued to swell, with high fevers occurring. This is an interesting feature, as it suggests that her body was "hotter" than normal for a woman, thus assimilating her to the male, and makes me think of the reddening of Phaethousa's joints. Heraïs' physicians applied remedies for what they assumed to be the problem, which was an apparent tumour at the mouth of the womb. This tumour burst, "and projecting from her groin there appeared a male genital organ with testicles attached."

Heraïs, like Phaethousa, was married when she became ill; to "a man named Samiades. He, after living in wedlock with his wife for the space of a year, went off on a long journey." In both cases, then, we could suggest that it was the absence of a husband that triggered the changes; it is as if femininity is so fragile that it depends on a husband to stabilise it. With Heraïs, however, in contrast to Phaethousa, there is nothing said about previous childbearing. After her transformation, Heraïs tried to cover up what had happened by continuing to dress and to act as a woman; in fact, perhaps, to over-act. On her husband's return, she refuses to sleep with him, and so he takes her to court, where her transformation is revealed. I see a further echo of the Phaethousa story in Diodorus' statement that, before the final revelation, Heraïs "continued to conduct herself as *oikouros* and as one subject to a husband." Phaethousa too is labelled as *oikouros*, a good stay-at-home wife, before her husband leaves and her beard grows. But Heraïs seems to be adopting the *oikouros* model in order to make her behaviour seem more "womanly" and thus to cover up the terrible secret beneath her clothes.

These, then, were the sorts of story with which Phaethousa kept company in the sixteenth century and at the beginning of the seventeenth century. By about 1600, her story was very widely known, to the point that authors assumed that their readers would not need the details. The 1607 English translation of Simon Goulart's *Admirable and Memorable Histories* simply assumed that the reader would know it, using it as an introduction to a story of a woman in Spain who changed sex, saying that the author was "emboldened" to tell what would have otherwise seemed only a "tale made for pleasure" by "That which I haue read in HYPOCRATES in his sixt Booke of *the popular diseases*[7] of PITHVLIA the wife of PYTHEVS" (1607: 275). The authority of the Greek medical tradition is clear; it enables the reader to tell truth from fiction. The story Goulart told did not involve the death of Phaethousa. In other versions, the story is compressed, but also undergoes significant changes. For example, in *Wits Theater of the Little World* (a commonplace book of prose sayings arranged under headings with innumerable historical examples collected under each), it is given as follows: "Phaethusa, the wife of Pytheus, thought so earnestly vpon her husbands absence, that at his returne, she had a beard growne vpon her chinne. *Hier. Merc.*" (Allott, 1599: 110)

[7] "The Popular Diseases" was a common English translation of the title "Epidemics".

Here, the cause of the symptoms is her thinking about her husband, and it seems to be suggested that by thinking, she comes to resemble him, apparently a variation on the common early modern theme of the maternal imagination, in which thinking about something can influence the appearance of the unborn child. None of this comes from the stated source, Mercurialis, who instead explains that the story shows how, in menstrual suppression, the humours are spread around the body, causing changes to the hair, beard and voice (1587: 110). Furthermore, in this 1599 version, the husband returns; I wonder whether this is a result of fifty years of proximity in sex change story lists leading to a merger of Phaethousa with Heraïs.

For those writers who included Phaethousa in a list of sex change stories, there were two major difficulties: her previous childbearing, and her death. For some sixteenth-century writers, the first of these difficulties was so important that it meant she must be placed somewhere other than in a sex change list; they underlined that she must have had a womb, and indeed all other female organs of generation. For example, André du Laurens, whose *Historia anatomica humani corporis* was first published in Latin in 1595, with its first French edition in 1610 (Schleiner, 2000: 184), told the story of Phaethousa in response to a Quaestio concerning whether the genital parts of women differ from those of men solely in location–as "the ancients" thought–and whether a woman can turn into a man; it is worth noting that these two issues were considered closely linked (du Laurens, ?1600: 275; 1613: 225r-v). While the ancients believed that these were the same organs, but located inside or outside according to sex, and "today almost all doctors think the same," du Laurens insisted that the organs differ not only in location but also in number, form and structure. Like his contemporaries, he believed that sex "change" from female to male was in fact due to the availability of sufficient heat revealing previously hidden organs, but for him Phaethousa did not fall into the sex-change category. Instead it was grief at her husband's exile that precipitated her symptoms, but he notes that we read nothing of a change in the position of her sex organs. As evidence, he discussed the end of the story: her death (Schleiner, 2000: 187). "Hippocrates" says that everything possible was done to "provoke the flowers," but with no success, and she died. The point, then, is that she remains a "she"; du Laurens specifies that she retains female genitalia, a womb, and all the parts necessary for the expulsion of menstrual blood. Du Laurens's work was followed closely in 1615 by Helkiah Crooke, whose *Microcosmographia* was reprinted in 1651 and proved a very influential text in early modern England.

As for the second difficulty, her death, one solution was simply not to mention it. For example, in Amatus Lusitanus' *Curationum Medicinalium Centuriæ Septem* (1552: Cent.2, Cur.39), the sentence about her unsuccessful treatment and eventual death is omitted, the text instead moving straight to the comparable case of Nanno.

A wide range of alternative diagnoses were also built on the story of Phaethousa. For example, in 1610 Jacques Ferrand, in a chapter on "Melancholy in Married Persons", suggested that women whose husbands are absent for long periods of time can close up, becoming "unperforated"; surgical intervention is needed to cure them. Ferrand disagreed explicitly with Hippocrates; or, rather, he disagreed with what he thought Hippocrates had said. He wrote:

> Although our divine ancient [Hippocrates] claimed that Namysia and Phaëtusa were changed into men, I suspect, rather, that they suffered from this ailment–one more rarely seen in women than its counterpart [in men] which is often responsible for the breakdown in marital relations (Beecher and Ciavolella, 1990: 340).

Many writers read the Hippocratic Phaethousa within a more Galenic model of the body, with the cause being not simply diverted menstrual blood, but the absence of "female seed", a Galenic construct not found in the Hippocratic *Epidemics*. Galen describes the effects of retained "female seed" in the treatise *On the Affected Places* (6.5, Kühn, 8.413 ff). This diagnostic treatise is arranged from head to foot, and here Galen considers both men and women in whom there is a natural "polyspermacy" (Gk *polyspermoi*); that is, they tend to produce more seed than other people do (Flemming, 2000: 333-335). As Monica Green (2001: 24) has pointed out, what is new here in Galen, when his work is compared to the Hippocratic corpus and the other works on women's diseases from Greco-Roman antiquity, is that it is the excess of the woman's own seed that causes her symptoms, rather than her illness being due to the absence from her womb of a *man's* seed. In his treatise on the physiology of the beard, Ulmus (1603: 309) noted that Phaethousa's previous pregnancy shows not only that she had a womb, but also that it was of the right temperature and humidity, and that she was producing "seed".

Conclusion

Historically, the Greek medical tradition has been characterised by its flexibility. It can be seen as empirical or rational, depending on what is fashionable in the later medicine that makes use of the ancient texts. The sheer range of different models of the body in the *Hippocratic corpus* makes Hippocrates potentially into the "father" of almost any medical system. While Galen narrowed down "Hippocratic medicine" to the four humours, other medical theories have been able to mix and match from the many treatises in the Hippocratic corpus to create their own Hippocratic genealogies, and thus to attract the authority that still comes from demonstrating (or creating) a link to Greek medicine. While Roman writers could dismiss Greek medicine as all about fashion and novelty for its own sake, even in the early Roman empire there was already a sense of a long past, in which each writer built on the work of his predecessors, and a strong admiration for Hippocrates. Just as the ancient Greeks had seen medicine as a unified field in which specialism was dangerous, so Vesalius argued for a unified medicine, but one in which all medical roles came together in his own person. Here he was representing himself as a true successor to Galen, who in turn presented himself as the true heir of Hippocrates.

In the area of the diseases of women, the Hippocratic corpus contained dedicated treatises on women's diseases. As a result, in the sixteenth century Hippocrates came to be seen as the expert on women, meaning that significant content from the Hippocratic texts–such as Phaethousa's story–became important to a range of theories on the possibility of sex change and the causes of disease. In the process, the story was abbreviated, with her death being omitted, and with new theories of the cause of her symptoms being added. Mergers could also take place between Hippocratic stories and those told by the non-medical writers of antiquity, as well as between Hippocratic and Galenic theories about the processes of generation. Those who distanced themselves from Hippocrates still needed to account for Phaethousa's symptoms. Even in the age of observation, Greek medical science continued to thrive.

References

Allott, Robert (1599). *Wits Theater of the Little World*, London: J. Roberts for R. Ling

Amatus Lusitanus (1552). *Curationum Medicinalium Centuriæ Septem*, Venice.

Anderson, G. (1993). *The Second Sophistic: A Cultural Phenomenon in the Roman Empire*, London: Routledge.

Baldwin, B. (1992). "The Career and Works of Scribonius Largus", *Rheinisches Museum für Philologie* 135: 74-82

Beecher, D.A. and M. Ciavolella, eds. (1990). *Jacques Ferrand; A Treatise on Lovesickness*, Syracuse, NY: Syracuse University Press.

Cunningham, A. (2001). "The Transformation of Hippocrates in Seventeenth-century Britain." In D. Cantor (ed.), *Reinventing Hippocrates* (pp. 91-115). Aldershot: Ashgate.

—. (1997). *The Anatomical Renaissance: The Resurrection of the Anatomical Projects of the Renaissance*, Aldershot: Scolar Press

—. (2010). *The Anatomist Anatomis'd. An Experimental Discipline in Enlightenment Europe*, Aldershot: Ashgate

du Laurens, André (?1600). *Historia Anatomica Humani Corporis*, Frankfurt: Matthaeus Becker.

—. (1613). *Les Œuvres*, Paris: P. Mettayer

Flemming, R. (2000). *Medicine and the Making of Roman Women. Gender, Nature and Authority from Celsus to Galen*, Oxford: Oxford University Press

—. (2007). "Women, Writing and Medicine in the Classical World", *Classical Quarterly* 57: 257-279

Gleason, M.W. (1990). "The semiotics of gender." In D.M. Halperin et al. (eds.), *Before Sexuality: The Construction of Erotic Experience in the Ancient Greek World* (pp. 389-416), Princeton, NJ: Princeton University Press.

—. (2009). "Shock and Awe: The Performance Dimension of Galen's Anatomy Demonstrations." In Gill, C., T. Whitmarsh and J. Wilkins (eds.), *Galen and the World of Knowledge* (pp. 85-114), Cambridge: Cambridge University Press.

Goulart, I. (1607). *Admirable and Memorable Histories Containing the Wonders of our Time. Collected into French out of the best authors. By I. [sic] Goulart. And out of French into English. By Ed. Grimeston*, London: George Eld.

Green, M.H. (2001). *The* Trotula*: a medieval compendium of women's medicine*, Philadelphia, PA: University of Pennsylvania Press.

Hamilton, J.S. (1986). "Scribonius Largus on the Medical Profession", *Bulletin of the History of Medicine* 60: 209-216.

Horstmanshoff, H.F.J., H. King and C. Zittel, eds. (forthcoming). *Blood, Sweat and Tears: The Changing Concepts of Physiology from Antiquity into Early Modern Europe*, Leiden: Brill.

Jones, W.H.S. (1947). *The Medical Writings of Anonymus Londinensis*, Cambridge: Cambridge University Press.

Jouanna, J. (1999). *Hippocrates* (tr. M.B. DeBevoise), Baltimore: Johns Hopkins University Press.

King, H. (1998) *Hippocrates' Woman: Reading the Female Body in Ancient Greece*, London: Routledge.

—. (1999). "Hippocratic Gynaecological Therapy in the Sixteenth and Seventeenth Centuries." In I. Garofolo et al. (eds.), *Aspetti della Terapia nel Corpus Hippocraticum, Atti del IX[e] Colloque hippocratique, Pisa, 25-29 settembre 1996* (pp. 499-515), Firenze: Leo Olschki.

—. (2006). "Where Did Medicine Come From? The Origins of Medicine in the Second Century AD." In S. Goldhill and R. Osborne (eds.), *Rethinking Revolutions Through Ancient Greece*, (pp. 246-263), Cambridge: Cambridge University Press.

—. (2007). *Midwifery, Obstetrics and the Rise of Gynaecology: Users of a Sixteenth-Century Compendium*, Aldershot: Ashgate

Kollesch, J. (1981). "Galen und die zweite Sophistik." In V. Nutton (ed.), *Galen: Problems and Prospects* (pp. 1-11), London: The Wellcome Institute for the History of Medicine

Langholf, V. (1990). *Medical Theories in Hippocrates. Early Texts and the 'Epidemics'*, Berlin and New York: de Gruyter,

Laskaris, J. (2002). *The Art is Long: On the Sacred Disease and the Scientific Tradition*, Leiden: Brill

Licciardi, C. (1990). "Les causes des maladies dans les sept livres *Epidémies*." In Potter, P., G. Maloney and J. Desautels (eds.), *La Maladie et les maladies dans la Collection hippocratique, Actes du Ve Colloque international hippocratique* (pp. 323-337), Québec: Éditions du Sphinx

Lloyd, G.E.R. (2003). *In the Grip of Disease: Studies in the Greek Imagination*, Oxford: Oxford University Press

Lonie, I.M. (1981). *The Hippocratic Treatises 'On Generation', 'On the Nature of the Child', 'Diseases IV'*, Berlin and New York: De Gruyter.

Manetti, D. (1990). "Doxographical Deformation of a Medical Tradition in the Report of the Anonymus Londinensis on Philolaus", *Zeitschrift für Papyrologie und Epigraphik* 83: 219-233.

Manetti, D. (2011). *Anonymus Londinensis*, Leipzig: Teubner.

Manetti, D. and A. Roselli (1982). *Ippocrate. Epidemie Libro Sesto* (Biblioteca di studi superiori, 66), Firenze: La Nuova Italia.

Mercuriale, Girolamo (1587). *De morbis muliebribus praelectiones* (in Caspar Bauhin (ed.), *Gynaeciorum Sive de Mulierum Affectibus*

Commentarii Graecorum, Latinorum, Barbarorum..., Basle: Conrad Waldkirch, 1586-88)

Nijhuis, Karin (1995). "Greek Doctors and Roman Patients: A Medical-Anthropological Perspective." In van der Eijk, P., H.F.J. Horstmanshoff and P.H. Schrijvers (eds.), *Ancient Medicine in its Socio-cultural Context* (pp. 49-68), Amsterdam; Atlanta, GA: Rodopi.

Nunn, H.M. (2005). *Staging Anatomies: Dissection and Spectacle in Early Stuart Tragedy*, Aldershot: Ashgate.

Nutton, V. (2004). *Ancient Medicine*, London: Routledge.

O'Malley, C.D. (1964). *Andreas Vesalius of Brussels, 1514-1564*, Berkeley, CA: University of California Press.

Park, K. (2006). *Secrets of Women*, NY: Zone Books.

Parry, A.M. (1969). "The Language of Thucydides' Description of the Plague", *BICS* 16: 106-18.

Paster, G. Kern (1993). *The Body Embarrassed: Drama and the Disciplines of Shame in Early Modern England*, Ithaca, NY: Cornell University Press.

Schleiner, W. (2000). "Early Modern Controversies about the One-sex Model", *Renaissance Quarterly*, 53: 180-191.

Schoenfeldt, M.C. (1999). *Bodies and Selves in Early Modern England: Physiology and Inwardness in Spenser, Shakespeare, Herbert, and Milton*, Cambridge: Cambridge University Press.

Smith, W.D. (1979). *The Hippocratic Tradition*, Ithaca, NY: Cornell University Press.

Thomas, R. (2000). *Herodotus in Context: Ethnography, Science, and the Art of Persuasion*, Cambridge: Cambridge University Press.

Ulmus, Marcus Antonius (1603). *Physiologia Barbae Humanae*, Bologna: Giovanni Battista Bellagamba.

van der Eijk, P. (1990). "The 'Theology' of the Hippocratic Treatise 'On the Sacred Disease'", *Apeiron* 23: 87-119.

Vesalius, Andreas (1998). *Andreas Vesalius On the Fabric of the Human Body. Book 1: The Bones and Cartilages*. Tr. W.F. Richardson, San Francisco, CA: Norman Publishing.

von Staden, H. (1994). "Anatomy as rhetoric: Galen on dissection and persuasion", *Journal of the History of Medicine and Allied Sciences* 50: 47–66.

—. (1997). "Galen and the 'Second Sophistic'." In R. Sorabji (ed.), *Aristotle and After*, London: Bulletin of the Institute of Classical Studies, Supplement 68: 33-54.

THE *FORUM VULCANI* IN THE WORK OF JUAN HUARTE: GEOGRAPHICAL ARGUMENT AND RENAISSANCE MEDICINE

ANDRÉS VÉLEZ

Pozzuoli, an emblematic place

Decisive implications are always derived from the way in which the representation of a place is used within a given discourse. The occurrence of the *Forum Vulcani* in Pozzuoli in Juan Huarte's *Examen de Ingenios para las ciencias* (1575-1594) motivates this case study for that very reason.

Among the geographical representations circulating in Europe in the sixteenth and seventeenth century, Pozzuoli was an iterative appearing reference in the manner of engravings, maps, panoramic views, and travel descriptions. It was regarded as an icon, an emblematic place.[1] Located to the north of the Gulf of Naples, Pozzuoli was widely known from Antiquity (as Dicearchia or Puteoli), throughout the Middle Ages up to the nineteenth century. Much of its subsequent fame still appeared in the vast series of descriptions produced by European travellers in the so-called *Le Grand Tour* and right before the marvellous discovery of the Herculaneum and the Pompeii's site. What draws attention in the references transmitted is the variety of values and attributes by which Pozzuoli is known: it is as if Pozzuoli was a multivalent field full of meanings. All the descriptions of this *Orbis Terrarum deliciae*[2] praise the beauty of its landscape, the

[1] See the useful compendium edited by P. Amalfitano ([ed.], 1986). Cf.: Also M. Sirpettino (1999).

[2] This is one of the phrases chosen by Joris Hoefnagel and Abraham Ortelius in their representation of the Gulf of Baiae, in the Campania region, which is the emplacement of Pozzuoli. The landscape appeared in the Georg Braun and Frans Hogenberg Atlas *Civitates Orbis Terrarum* (1593 Vol. 3). p. 52. The *locus* of this *Campania Felix* recalls Virgil's and Horace's verses.

fertility of its lands, the medicinal power of its baths and hot springs, its constructions and the marvellous tunnels and underground galleries that charm the visitors' curiosity.[3] Furthermore, mythology and history made it a place of memory: the gates of hell and through which Ulysses and Aeneas descended into Hades are located there; at the east of the Sinus of Pozzuoli was supposed to be the Lake Avernus, the Sibyl's Cave, the Mount Cumae as well as the surrounding Phlegraean Fields, the blazing lands, where the Gigantomachia took place.[4] Even in this non-exhaustive enumeration, one becomes aware of the diversity that such a place involved. Aiming to classify, one might say that Pozzuoli's descriptions oscillate between two extremes: on the one hand, those typical representations of the *locus amoenus*, with the beauty and delight produced by Nature and the works of men. On the other hand, the infernal and volcanic imagery: mephitic waters, sulphur quarries, fumaroles, unexpected underground eruptions, and the underworld life. During the humanist culture of the Renaissance this contrast intensely enhanced the place's charm amongst travellers, artists, antiquarians and scholars and, needless to say, cosmographers and geographers.

Strabo presents the best description and overview of the region:

> Next in order come the headlands that are in the neighbourhood of Dicaearchia, and then the city itself. In earlier times it was only a port-town of the Cumaeans, situated on the brow of a hill, but at the time of Hannibal's expedition the Romans settled a colony there, and changed its name to Puteoli from the wells there–though some say that it was from the foul smell of the waters, since the whole district, as far as Baiae and Cumae, has a foul smell, because it is full of sulphur and fire and hot waters. And some believe that it is for this reason that the Cumaean country was called "Phlegra" [*Blazing-land*], and that it is the wounds of fallen giants, inflicted by the thunderbolts, that pour forth those streams of fire and water. And the city has become a very great emporium, since it has havens that have been made by the hand of man–a thing made possible by the natural qualities of the sand, for it is in proper proportion to the lime, and takes a firm set and solidity. And therefore, by mixing the sand-ash with the lime, they can run jetties out into the sea and thus make the wide-

[3] The Peter of Eboli poem on the Pozzuoli's baths *De balneis Puteoli* gave to the city much of its celebrity since the thirteenth century. Beside the classical literary references to Pozzuoli, Baiae, and the entire Campania Felix, there are the two major studies by J. D'Amato on the antiquarian tradition that surged through this area from the Middle Age (D'Amato, 1993; 1996).

[4] Virgil's works, particularly Aeneid VI, provides the major part of canonical references to this region. With Virgil's guide, the territory unfolds itself like an open book. F. Zevi (1999) offers a summary.

open shores curve into the form of bays, so that the greatest merchant-ships can moor therein with safety. Immediately above the city lies the Forum of Hephaestus [*Forum Vulcani*], a plain shut in all round by exceedingly hot ridges, which in numerous places have fumaroles that are like chimneys and that a rather noisome smell; and the plain is full of drifted sulphur.[5]

This description would become one of the most influential texts for subsequent representations. What is determinant in Strabo's description is its point of view. In the text, the outline of Pozzuoli and its harbour environs is made from above, that is, from the *Forum Vulcani* that hovers, albeit menacingly, over the amenities of the bay. From that vantage point the description highlights and praises both the volcanic and mineral properties of the region and its suitability for crops, construction and industry. This emotional content in Strabo's description was to find a coming to life with the 1538 volcanic eruption and earthquake that shook and transformed the entire region to the extent that a new mount–*Monte Nuovo*–was upheaved thereby. In a contemporary engraving of the time we can see an image of how Pozzuoli looked like at the time of the cataclysm. The cartographic document cannot avoid showing the pathos in representing the victims scattered along the coast or flying over the crater as a consequence of the blasting wave's force (see detail, Fig. 1).[6]

---

[5] Strabo, *Geography*. L. V, 4. 6.

[6] The map shows the entire Gulf of Naples, the Vesuvius, Pozzuoli and the so-called Monte Novo in eruption. The legend that accompanies and guides us in understanding the map speaks about the threat of fire: "Il vero disegnio in sul propio luogho ritratto del infelice paese di Posuolo quale, E,M, 60; et del monte di novo nato in mare et in terra, condanno et morte di molti habitatori et spavento di chiunque lo vede; comincio a buttare fuogo, pietre e cenere ali 29 de Setembre 1538 et ancor' seghuita, con minaci horribili." This document, edited perhaps by Antoine Lafreri (1512-1577) and signed by G.A. Vico, is part of the cartographical collection *Geographia, Tavole moderne di geografia de la maggior parte del mondo di diversi autori raccolte et messe secondo l'ordine di Tolomeo con i disegni di molte citta et forterezza di diverse provintie stampate in rame con studio et diligenza in Roma.* This *recueil*, with more than 600 engravings, was reproduced throughout Europe since 1570 with an important influence. See, R.V. Tooley (1939). The reference of the document in the *Bibliothèque National de France* (BNF) is Res. Ge. DD. 1140 (98 RES).

Fig. 1

In addition to news and circulation of Strabo's text, the chronicles and descriptions of European travellers, who dared to go beyond Rome, propelled the iconographic strength of this region in the sixteenth century and revived curiosity for this fire and sulphur laden earth.[7] This event had its obvious impact on the visual and literary culture of the period, resulting in a considerable production of engravings, maps and chorographical descriptions and, very especially, in the updating of the Classic testimonies and poetic texts of Cicero, Virgil, Horace, Strabo, Pliny and Suetonius. A

[7] On the 1538 eruption, there is a letter brought by Marcantonio delli Falconi, a report from Simone Pozio *De conflagratione agri Puteolani,* and the Girolamo Borgia's poem *Incendium ad Avernum locum*. As an example of some travellers visiting Puzol see: Joseph Catin. *Voyage aux Champs Phlégréens (1568)* (ed. 1997); also, M. Letts (1918). In order to study how Italy was seen in French culture, there is the bibliographical review made by V. Castiglione Minischetti *et al* (2006). We mention also the texts and drawings of Pozzuoli and the *Sinus Baianus* made by Francisco de Olanda during his Italian journey between 1539 and 1540, *Os desenhos das antiguallas que vio Francisco d'Ollanda, pintor portugues.*

good example of this being the *Descrittione di Tutta l'Italia* (1550) by Leandro Alberti engaged in a detailed description of Pozzuoli and based on the Classical sources. All these documents circulated in Europe enriching libraries and collections. It is more than legitimate to suppose, therefore, that Pozzuoli and its volcanic environment was to become a real landmark, and remarkably, a cartographic and geographic emblem such as can be seen in the Abraham Ortelius' Atlas *Theatrum Orbis Terrarum* or the Georg Braun and Frans Hogenberg's *Civitates Orbis terrarum* which shows and praises the World and its places for remembrance in the form of maps, views and descriptions.

In this line of meaning, the best view of the Pozzuoli region is rendered by an engraving drawn by Jan Van Der Straet and edited by Philip Galle in 1583 (Fig. 2).[8] This chorographic image depicts the environs of the Pozzuoli harbour and the plains stretching out eastward from it. Noteworthy in Van Der Straet's landscape is the relevant parallelism with Strabo's description, in so far as the landscape is seen and oriented from the *Forum Vulcani* or *Solfatara*. Peculiar to this image is the presence of human beings on land and sea. The foreground of the engraving shows how the fumaroles, the sulphur and sand quarries as well as the hills are traversed by journey men, traders, workers, peasants, locals, and artists, some transported in *seggioli* or sedan chairs most surely going to the hot springs and baths, some on foot and others on donkeys and horses. The *Forum Vulcani* is not portrayed then as a place of terror but is rather humanized through the roads that cross it, the passengers, the workers extracting sulphur and the pack animals. Human activity dominates the site looking down on the bay filled with ships and galleons: work, trade, transportation, medicinal practices, and the contemplation of the artist finally shape the entire landscape, naturalizing men and humanizing space.

---

[8] The map shows *ad vivum* the activity in the *Forum Vulcani*. The map is dated in 1587, and it was a part of the cartographic collection *Tavole moderne di geografia…* . R.V. Tooley did not include it in his study (1939). The text accompanying the landscape says: "SPECTATORIBUS PHILP. GALLAEUS. Forum Vulcani vocavit antiquitas, locum in agro Puteolano, quem hodie vulgo La Solfatara appellant. Campum Phlegreum nominant qui fabulis delectantur Poeticis Hunc Vobis rerum naturalium admiratoribus exhibem ad viuum manu excellentissimi pictoris Joan. Strad. Brugensis depictum boni consulite 1587." BNF: GE DD -655 (106RES).

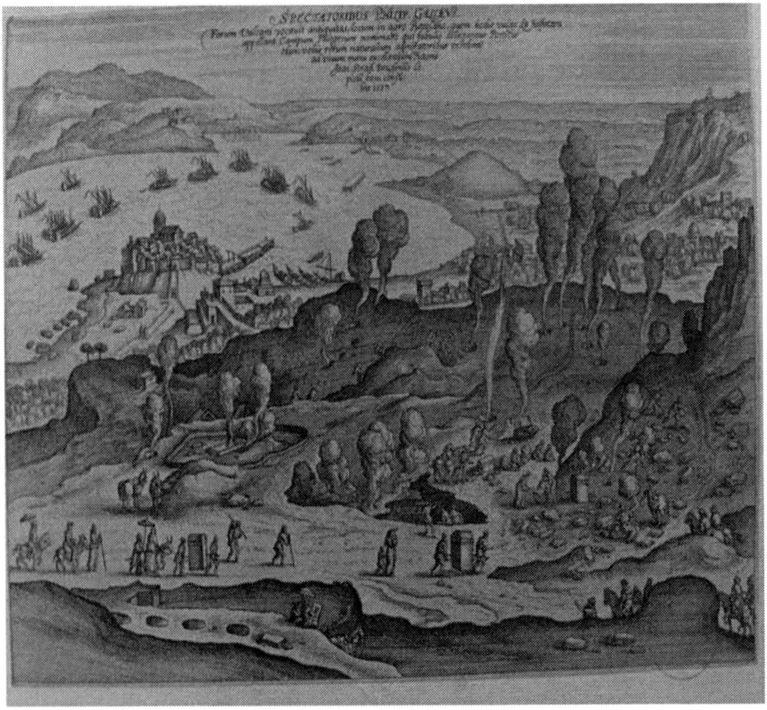

Fig. 2

By the end of the sixteenth century, news, images, praises and descriptions of Pozzuoli were known to almost every scholar and person with a humanistic education or with a modicum of intellectual and artistic curiosity. This is indeed the case of Juan Huarte. We can find a reference to the *Forum Vulcani* in the edition expurgated by the Inquisition, corrected by Huarte, and published in 1594. In his work the Navarre physician refers to it as *The Forges of Vulcan in Pozzuoli* (*Las Herrerías de Vulcano en Puzol*) and in the context of an interesting disquisition seeking to prove the paradoxical cosmological argument that fire is the heaviest element whose natural place is the depths of the Earth. At first sight the appearance of Pozzuoli in the *Examen de Ingenios*, a text known for its speculative, medical and philosophical nature, could be nothing but surprising. Nonetheless, one cannot underestimate such a manifestation and disregard it as mere vain ornament. As will be discussed further ahead, the mentioning of *The Forges of Vulcan in Pozzuoli* within the framework of an argumentation on Natural Philosophy must be seriously taken into consideration in order to draw fair conclusions about the role that an *argumentum a loco* has within a discourse which is not eminently geographical. In the lines below I intend to analyze how geographical knowledge is operational within the Huarte's text. This question leads to another of a more general order: How do the traditional mechanisms of geographical knowledge work and shape humanist scientific discourse?

Tradition and Geographical Humanism

The *Examen de Ingenios* itself is the best evidence of Huarte's humanistic formation. Although his proposals on psychology were innovative and curious at the time, his work and method reiterate the gesture akin to the Renaissance of turning to Antiquity for sources. In this sense, Huarte prepared and conceived his arguments from an erudite dialogue and a reading of Classical tradition. It must be pointed out that his relationship to the Greek texts is achieved and matured through the *passeurs* of the time, i.e. the Latin translations, and all the collected texts in encyclopaedias, *Miscellanea*, and *Florilegia.* His studies in Medicine in the University of Alcala with Francisco Vallés and Cristobal Vega, Spanish introducers of the practice of dissection and of Vesalius's anatomy from 1551 and translators of Hippocrates and Galen, allowed him to access the critical study of ancient medical humanistic knowledge.

There are many references and ways in which Huarte incorporates and updates the classical knowledge to demonstrate and develop his ideas on the cognitive and practical skills of human beings.[9] In his theory of *Ingenio*[10] or human nature we can find the artistic, philosophical, medical and religious concerns that the *cultural imagination*[11] has produced

[9] Also in this sense, as in almost all the historical interpretation on Huarte, see M. Iriarte (1939), as well the study of J. Pigeaud (2008), especially the chapter "Fatalisme des tempéraments et Liberté spirituelle dans l'*Examen des Esprits* de Huarte de San Juan" (2008: 425-469).

[10] We do not translate the word *Ingenio* for Wit as is customary. We follow the opinion of John Steven in his *New Spanish and English Dictionary* (*1726*). When he tries to define the word *Ingenio* he writes: "INGENIO: wit and ingenuity. It is not that light sort of wit which we commonly mean by this word, but a solid capacity and ingenuity, as is signified by the Latin *ingenium*; Whence the Spanish. It is also any manner of engine or invention to perform that with ease, which before was difficult."

[11] On this very useful concept for the historiography of the transmission of Western *medico-philosophical* thought I refer to the masterful works of J. Pigeaud (1989; 2008).

throughout the centuries: the relation between body and soul, the idea of a man of genius, the artistic creation and the melancholic character, the inquiry into the teaching of virtue and excellence, theories on the faculties of mind, the correspondences between the Bible and the medico-philosophical tradition are some of the classical matters in the *Examen de Ingenios*. Nevertheless, nothing has been said about the presence of geographical knowledge in this work. The absence of any critical literature on the subject is clearly justified given that the author nowhere explicitly states an interest in the problems of geographical or cosmographical projects. To advance a geographical interpretation of Huarte's work risks being judged as unreasonable or even, in the author's words, as feigning to see "golden mountains and flying oxen."[12]

However, a quick review of the relationship between medicine and geography, and the role of geographical education in the sixteenth century, may dispel any doubts. Previous studies on history of geography have demonstrated that the teachings of cosmology, cosmography and geography permeated the scholar practice and education in the Renaissance.[13] In Huarte's case, the sources of his *geographical thought* could be traced back to the context of the medico-philosophical[14] tradition to which he is heir. The philosophical and cosmological bases of his geographical thought encompass the Hippocratic Corpus–within which it is worth mentioning *Air, Waters and Places*–, Galen's writings, several Platonic dialogues, especially the *Timaeus*, and Aristotle's *Problemata*.[15] The encyclopaedic model of a universal knowledge proposed by Renaissance universities entails, as a matter of fact, a necessary connection and correspondence among the sciences. In effect, *geographical thought* works within the humanistic sciences; it is a *given* in Renaissance academic culture, a fact that must be highlighted explicitly in critical and interpretative research.

---

[12] In Spanish: "fingiendo montañas de oro y bueyes volando". *Examen de Ingenios,* Op. Cit. p. 688.

[13] Without denying the scope of Physics, Astronomy and the mathematical and geometrical method, it is clear that cultural and scientific activity of that time cannot be reduced to a mathematical-physical model. The history of the meditation on the relationship between World and Human being allows us to see other versions of the history of sciences. This claim was made by Clarence J. Glacken (1967) and more recently by J.-M. Besse (2003).

[14] Cf. Pigeaud (2008: 425-469).

[15] See Iriarte (1939), particularly the list of classical authors and the editions that were used by Juan Huarte in chap. 3.1: "Lecturas y biblioteca del Dr. Huarte". In the same direction, see also L. García Vega (2001).

Within the field of Medicine the presence of such geographical education can be better appreciated. Inspired on the classic Hippocratic method expounded in *Airs, Waters and Places*, Huarte is aware that the *nosological* approach for a correct diagnosis and treatment of diseases of the inhabitants has to be founded on a whole analysis of geographic phenomena such as the orientation, situation and environmental characteristics and properties of a region. The *spatial sense* springs precisely within this relation between medicine and geography. To a considerable extent we can say that the *Examen de Ingenios* displays a medical, moral and spiritual cartography of the Earth, i.e., Huarte's work charts a geographical organization of human beings, their customs and their cultural productions. Indeed, the fundamental idea proposed by Huarte of a *difference of ingenios* among men, i.e. the distinction between the different skills, abilities and diseases forms a perfect analogy with the geographical principle of difference and variety that can be found in all the regions of the World.[16] This is well confirmed by the author when he argues that

> You shall understand that the varieties of men, as well in the composition of the body, as of the wit and conditions of the soule, spring from their inhabiting countries of different temperature, from drinking divers waters, and for not using all of them one kind of food.[17]

Furthermore, during the sixteenth century the geographical matters concerning the concepts, representations and measurement of the Earth constituted much of the scientific and cultural agenda of the time. The encyclopaedic efforts of rearrangement, compilation and review of cosmographical techniques and information were carried out by Theology, Natural Philosophy (i.e. Physics and Medicine), Astronomy, and travel relations. This European *geographical education* set in circulation a tremendous amount of materials for study and pleasure: annotated and translated Classical geographical texts of authors such as Strabo, Pomponius Mela, and Ptolemy, as well as the great Atlases of Mercator,

---

[16] Here there is a defiant relationship to study between the concept of human ability (*Ingenium*) and the concept of properties of a region (*Terrae Ingenia*).

[17] The translation was made by Richard Carew in the English edition of 1594: *The examinations of Mens Wits*. This quotation comes from the chapter 12 where Huarte develops his idea of *difference*. The Spanish text says: "es de saber que la variedad de los hombres, así en la compostura del cuerpo como en el ingenio y condiciones del ánima, nace de habitar regiones de diferente temperatura, y de beber aguas contrarias, y de no usar todos unos mismos alimentos" *Examen de Ingenios,* Op. Cit. p. 506. Huarte is following Galen's *Quod animi mores,* IX.

Ortelius and Braun-Hogenberg, landscape paintings and chorographic prints, were amongst the most popular resources. In this sense, Jean-Marc Besse (2003) insists that the entire intellectual and artistic culture of Europe was mobilized to carry out the formation and dissemination of a new understanding of the Earth as a universal space of human life.[18] That is how geographical and cartographic activities came to be present, circulating and in full bloom in every corner and centre of scientific and cultural production of the time. For instance, during the reigns of Charles V and Philip II–to whom Huarte dedicated his text–Spain developed his renown "mapping sense" through both its own scientific production and the trade of knowledge, maps, engravings, books and astronomical instruments coming mainly from the Low Countries.[19]

Thus, how to portray Huarte as a geographer? Despite the fact that apparently he never made any trip out of Spain, it is important to point out that Huarte travelled through books and representations. Indeed, his working concepts and principles of geography, as already suggested, come from his encyclopaedic medical formation, from the readings of traditional sources and from the widespread geographical culture of his time. Thus, the appearance of the *Forges of Vulcan in Pozzuoli* is to be placed closer to the contents of studies, of the iconic figurations, of the spiritual journey and the rhetorical tradition rather than to the verification and the weary journeys. In this sense, it could be argued that the places referred to in geography are not merely circumscribed to an objective or physical reference; geographical knowledge, especially during the Renaissance, is both an intellectual and visual arrangement of the classical tradition, cultural values, stories, descriptions and images of the Earth. Huarte travelled and visited places around the World from his *Studiolo*. That is why the description of a place, the explicit situation of a region, the orientation of an area, the mention of a climatic division or any other

[18] Besse (2003: 20).

[19] I refer to the study made by R.L. Kagan on the Spanish University (1981); Cf. also Sandman (2007). Here, Sandman affirms that "as an intellectual endeavor, cosmography was part of a university tradition, based fundamentally not only on Ptolemy (both his astronomical and geographical works) but also on Aristotle, Johannes de Sacrobosco, and Pomponius Mela. It included aspects of geography, hydrography, geometry, astronomy, cosmology, and natural history, and throughout the sixteenth century cosmographers in Spain were closely involved not only in astronomy and mathematics, but also in medicine and as official historians and chroniclers." (2007: 1107). On the cultural relationship between Spain and the Low Countries see the overview by D. Buisseret (2007).

geographical reference within a discourse should be understood also as an argument entailing a philosophical background and a rhetorical effectiveness.

Forum Vulcani: the place of Fire

The *Forum Vulcani* appears in the *Examen de Ingenios* within a *Digresión sobre el fuego* (*Digression on Fire*).[20] There, Juan Huarte states–contrary to Classical cosmology assertions–that fire is the heaviest element and that it has its natural place in the centre of the Earth. The Digression, then, is guided by the problem of the four elements' location and topographical arrangement. This discussion is divided as follows: refutation of Aristotle's and Galen's classical arguments,[21] presentation of proofs evidenced by senses and through experience and Huarte's final anthropological solution. Two main theses are debated here, a cosmographical one and a medical one: the first, that fire is not present in the concave part of the moon but is rather in the centre of the world and that it is the heaviest of all the elements; second, that the fire present in the human being does not come from the breath or the pulse but is preserved within him from the moment he is formed in the womb until his death. In the *Digression on fire* we are presented then with two interwoven scales of the argument: on the one hand the world, and on the other hand human nature; in conventional terms, it could be said that macrocosm and microcosm are the two extremes involved in the reasoning.[22]

The *Digression on Fire* emerges at the centre of a medical discussion on "What diligence ought to be used, that children may proone wittie and wise."[23] Here, Huarte asserts that the *Ingenio* or human nature is formed in the mother's womb and not under the influence of the stars,

[20] In the Sub Princeps edition, Baeza 1594, there are three digressions: on fire, on salt, and on the tree of knowledge. The name of d*igression* was given by Charles Vion d'Alibray in his French translation *L'Examen des esprits pour les sciences* (1645). Differences between the 1575 and the 1594 editions were commented by Rodrigo Sanz in his edition, *Examen de Ingenios* (1930).

[21] Huarte is particularly dealing with Aristotle *De Caelo*, II, 293a.

[22] On this classical parallel we refer to the authoritative study by A.O. Lovejoy (1973). See also F. Rico (1970).

[23] *Examination of Mens Wits*. Op. Cit. p. 300. Despite of the fact that this English edition, based on the 1575 version, does not include the Digressions, this medical advice on sexuality, hygiene and reproduction remained unchanged in both editions (i.e. 1575 and 1594). In Spanish: "Donde se ponen las diligencias que se ha de hacer, para que los hijos salgan ingeniosos, y sabios", *Examen de Ingenios,* Op. Cit. p. 642.

> for then the starres do superficially alter the child, giving him heat, coldnesse, moisture, and drouth; but not his substance, wherein the whole life relieth, as do the foure elements, fire, aire, earth, and water, who not only yield to the party composed, heat, cold, moisture, and drainesse, but also the substance which may maintain and preserve the same qualities, during all the course of life.[24]

In this way, deviating from astrology, Huarte situates himself and his theory into an embryological and mundane framework. Huarte leaves aside the considerations about the purity of the Elements to focus instead on the way in which the Elements are perceived on the Earth. The background of this proceeding is a particular interpretation of Aristotle's *Meteora* widespread by the physician Francisco Vallés who was, as already mentioned, Huarte's professor in the University of Alcala. Vallés' main idea was to understand how the mixtures and qualities of the four Elements, as explained by Aristotle, were present in the body, its organs and humours. Based primarily on the processes of digestion (*pepsis*) and putrefaction (*sepsis*), the medical application consisted of establishing a dietary approach to diagnose and treat diseases as well as explaining the changing processes at work in the body.[25] This theory should also explain how the Elements form the body while remaining inside. That is why Huarte recalls in the very beginning of the Digression that–following Galen–the Elements penetrate into the body by the virtue of mixed entities: "The earth commeth lurking in the accustomed meates which we ate, as are flesh, bread, fish, and fruits; the water in the liquors which we drinke, the aire and the fire (Galen saith) are mingled by order of nature, and enter into the body by way of the pulse and of respiration."[26] But this statement motivates Huarte to introduce his Digression while asking: If fire is naturally on high and above the air, how could it be that this element goes down into the world's depths and into the body's cavities to

---

[24] *Examination of Mens Wits*. Op. Cit. p. 300. In Spanish: "porque entonces alteran las estrellas superficialmente al niño, dándole calor, frialdad, humidad y sequedad, pero no sustancia en que restribe toda la vida, como lo hacen los cuatro elementos, fuego, tierra, aire y agua; los cuales no solamente dan al compuesto calor, frialdad, humidad y sequedad, pero también sustancia que le guarde y conserve estas mesmas calidades todo el discurso de la vida" Examen de ingenios", Op. Cit. p. 643.

[25] See Craig (2002) who refers to Galen's treatisses *De Sanitate tuenda* and *De elementis ex Hippocratis sententia.* The work where Francisco Vallés conceived his medical application is *In quartum librum Meteorologicum Aristotelis commentaria* (1588).

[26] *Examination of Mens Wits*. Op. Cit. p. 301

participate in the processes of generation, conservation and corruption of mixed things? Huarte answers his question claiming that:

> ...in truth Galen's opinion is false, as even more so the one expounded by Aristotle in placing the sphere of fire in the concavity of the Moon. It is true that God and Nature never do anything idly or without purpose. If fire is in the concavity of the Moon, it is good for nothing. Then God did not create it, and if He did create it he did not locate it in such a place. And, that in being there it is good for nothing, is a transparent fact running through all the uses that fire can have.[27]

In this line of argument it is worth noting how Huarte uses the Aristotelian principle of "Deus et natura nihil faciunt frustra" (*De caelo,* I, IV) to contradict the traditional Aristotelian theory of the external fire. The solution proposed by Huarte lies, then, in a reconfiguration of the order of elements guided by a radical anthropological teleology.

The first radical solution consists of rearranging the order of the element of fire on the Earth according to experience. This logic of appearances is based on phenomena such as combustion, corrosion and lightning. The first reason that leads Huarte to argue that fire is the heaviest element is the observation–and the particular description of what he observes, we must say–of what happens to a candle's flame: fire goes down and nourishes itself from below, whereas what goes up is a lighter element, namely, the smoke, the excrement. The second reason lies in a classification of materials characteristic of technical knowledge such as pharmaceutics, carpentry and blacksmithing: the compound bodies or mixtures that have a greater amount of fire–Huarte argues–are heavier than the mixtures composed of earth, as happens with various minerals–such as

---

[27] The only two English translations we have of the *Examen de Ingenios* were made by Richard Carew in 1594, that we have used, and by Edward Bellamy in 1698 (*Examen de ingenios; or, The Tryal of Wits*) The reader will notice that the *Digression on Fire* did not appear in either of them. Bellamy's edition included several modifications and the new chapters of the *Sub Princeps*, but he decided to exclude the Digression. In general, critical studies of these digressions and of their editorial fate are a field of research that needs to be undertaken. The translation of the *Digression on Fire* appearing in this paper was made by Mónica Boza. The text in Spanish says: "Pero realmente la opinión de Galeno es falsa, y mucho más la de Aristóteles en poner la esfera del fuego en el cóncavo de la Luna. Porque es cierto que Dios y Naturaleza nunca hacen cosa baldía y sin fin. Estando el fuego en el cóncavo de la Luna, no sirve de nada. Luego Dios no lo crió, y si lo crió no lo puso en tal lugar. Y que no sirva de nada estando allí, es cosa muy clara discurriendo por todos los aprovechamientos que del fuego se pueden tener."*Examen de Ingenios,* Op. Cit. p. 685 and quoting here Aristotle's *Physica* III.

corrosive stones for medicinal use (*lapis corrosivus*)–and the ebony wood that sinks in water. Finally, he points out that the precipitous fall of lightning is nothing but falling fire. After this call for observation–seeming more like a call on imagination of different events in an act of meditation on fire–Huarte concludes as follows:

> …fire is the heaviest thing on earth. Principally on seeing and considering how well things are made coherent when placing fire in this place, and the many inconveniences born of placing it in the concavity of the Moon. The nourishment of fire, the expelling of smoke and the generation of impetus, is carried out without contradiction. Because fire is endowed with the virtue of attracting towards it all things and the cavities of the earth are full of air and water. Holding these three elements (earth, water and air) together, it combines them easily, cooks and alters them, and makes nutriment in order to preserve itself (such as sulphur and saltpetre), and has large paths and breathing holes through which it can emit smoke and obtain ventilation. For which, **evident argument are The Forges of Vulcan in Pozzuoli**, close to Naples, where there are lakes and mountains of fire ever since God created the world. And there must be other similar places due to the roundness of the Earth, where fire is nurtured with thousands of species of minerals suited to its nutrition. And in the same manner that this fire feeds and maintains itself on the outside, we can easily understand what happens in the centre of the Earth, because I do not doubt that these mountains and lakes of fire are of the same gender, and likewise breathing spots for it. [28]

[28] Translated by Mónica Boza. In Spanish: "[…] el fuego es la cosa más pesada del mundo. Mayormente viendo y considerando cuán bien consuenan las cosas poniendo al fuego en este lugar, y cuántos inconvenientes han nacido de ponerlo en el cóncavo de la Luna. La nutrición del fuego, la expulsión del humo, y la generación de los ímpetus, se hace sin ninguna contradicción. Porque el fuego tiene virtud de atraer a sí todas las cosas y las cavidades de la tierra están llenas de aire y de agua. Teniendo junto consigo estos tres elementos (tierra, agua y aire), fácilmente los mezcla, los cuece y altera, y de ellos hace alimento para mantenerse (como es el alcrebite y salitre), y tiene grandes caminos y respiraderos por donde despedir el humo y ventilarse. De lo cual es **evidente argumento las Herrerías de Vulcano en Puzol**, junto a Nápoles, donde aparecen lagos y montañas de fuego dende que Dios crió el mundo. Y de la manera que se ven éstas habrá otras muchas más por la redondez de la Tierra, donde el fuego se mantiene con mil géneros de minerales acomodados a su nutrición. Y de la manera que este fuego se nutre y mantiene acá en lo exterior, entenderemos fácilmente lo que pasa en el centro de la Tierra, porque yo no dudo sino que estas montañas y lagos de fuego son del mismo género, y por ventura respiraderos suyos." *Examen de Ingenios,* Op. Cit. p. 693-694.

This subtlety in reasoning is, as I see it, definitely his strong point. As Huarte states, the *Forges of Vulcan in Pozzuoli* constitutes his *evident argument*, what stands out to convince us, that is, a convincing perception. The argument emerges as a visual device within the reasoning, that is to say, the reader is dealing at the same time with both a representation of a place on the Earth and a cosmological argument. In rhetorical terms that may be called an *argumentum a loco*. The description of the *Solfatara* contains all the Digression's arguments and proves them at one stroke. The occurrence of the *Forum Vulcani* gathers, like a magnetic point, the flame, smoke, minerals and, ultimately, the human everyday experience of fire. Also, it would enable Huarte to present the unity and perpetual mixing of the Elements on the Earth and to introduce the biblical idea of hellfire. From the moment Huarte locates the reasoning in a specific place he proceeds as a geographer. The descriptive method in geography proceeds gathering and locating in a specific place a wide range of information, analysis and features that are linked to it. To recur to the representation of the *Solfatara* to prove his arguments amounts to granting them spatiality, grounding knowledge and thought, and consequently, unfolding a horizon of sense within the represented space.

Huarte uses the volcanic typology of the emblematic landscape of Pozzuoli in order to locate and arrange the meaning of his Digression. The description of the *Forum Vulcani* brought from Classical tradition shapes the argumentation. In addition to the stories that this place recalls, Huarte adds his cosmological, geographical and medical arguments. Huarte's geographical imagination opens up the possibility of *seeing* not only the history but also the reveries indulged by the *Forum Vulcani*. This reminiscence and visual force inherent in a geographical representation would have been emphasized by Abraham Ortelius' suggestive idea *Historiae Oculus Geographia*.[29] Such visual and cognitive geographical operability, which Ortelius refers to, is equally effective in the case of Huarte's meditation. The phlegrean landscape, in which the humanist focuses his attention on, will assist him to chart his reasoning. In this manner *The Digression on Fire* acquires a kind of argumentative high relief quality, a depth and continuity that can be traversed with thought and reading.

Being the *marabilia* that the *Forum Vulcani* is, it allows Huarte to develop, using the literary style of the paradox, a range of cosmological arguments with decisive implications on the conception of Human nature.

---

[29] The phrase appeared in the frontispiece of his *Parergon sive Veteris Geographiae Aliquot Tabulae* (Antwerp, 1592)

In other words, once Huarte finds ground for his arguments, the *Digression on fire* engages in an itinerary of thought in which the teleological implications of the centrality of fire will begin to take shape via analogical correspondences. This is how a geographical place becomes a place for discourse.[30]

The *Forges of Vulcan* take central stage once they have endowed the Digression with a characteristic ground. In fact, the representation of the *Forum Vulcani* establishes a framework of connections such as that shown in the emblem of J. Hoefnagel (Fig. 3).[31] Here, it is possible to see how travellers, workers, other social characters, and means of transportation are gathered in an industrious environment praising the marvellous presence and utility of heat, fire and techniques on the Earth. Likewise, Huarte humanizes the volcanic landscape by reading it from a medico-philosophical perspective, while turning human nature into a volcanic body.

Fig. 3

[30] On this relationship between geographical *situs* and rhetorical *locus* see L. Pernot and his studies on Menander (1993); also the useful and brillant work by C. Jacob and F. Lestringant (1981).

[31] In Georg Braun and Franz Hoegenberg, *Civitatis Orbis Terrarum*, 1593. Vol. 3, p. 58.

According to a system of analogical correspondences, where Plato's *Timaeus* is the fundamental base, decisions taken at the cosmological level are also decisions at the anthropological level. Attributing a centrality to fire and, especially, giving a volcanic chorography to the Digression allow Huarte to traverse the landscape as a physician and as an anatomist in order to meditate on Human nature. In sixteenth century knowledge and culture the analogies between the human being and the world were highly present and widespread. Peter Apian's *Cosmographia*, another European editorial success in the early modern age science, is a well known example showing how the correspondences operate. This text–translated and edited several times in Spanish since 1548–systematized through the implicit connections between cosmography, geography, and chorography, a scale of proportions going from the senses to the entire Universe, that is, from the eye to the cosmic sphere. The most important effectiveness of the Digression occurs in this natural and scientific transit through these orders of magnitude–the Earth and the human body. From a teleological point of view the fire on Earth, and mostly the volcanic fire, has a necessary correspondence with a human inner fire. This is why, after addressing the question of the place of fire on the Earth, Huarte proceeds to deal with the second issue he had posed: how fire is present in the body independently of breathing and pulsation. For Huarte, fire resides within the human being from the beginning of his conception in the mother's womb and does not need to come down from the moon's concavity. The analogy stretches even further: just like volcanoes and lakes of sulphur are sure signs of an inner central fire, the human body also signals the presence an inner fire, a natural and non-accidental heat. Embodied as it is supposed to be, the inner fire is always present in the processes of respiration, reproduction, nutrition, digestion, excretion, and in general, in the consumption of the body and life; and on the other hand, this embodied fire manifests itself in the cognitive and practical skills of the wit, the sparkling *Ingenio*. Human beings, proposes Huarte following Galen, have within themselves not a simple accidental heat but real fire: "a wicker lamp within our body, with its match cord and oil, burning as we see here on the outside."[32]

To finalize on the *Digression on Fire*, Huarte concludes with an erotic theory of *Ingenio*. In the same way that inner fire incites the sexual attraction and the reproduction of life it generates conceits and concepts, it springs in knowledge and lifestyles. Hence, fire and heat manifest themselves in the activities of the understanding and the imagination. The

[32] Here Huarte quotes Galen in his *De placitis,* VIII: "cor ut funiculus est, sanguis ut oleum, pulmo ut organum ir quo est oleum," *Examen de Ingenios* Op. Cit. p. 699.

volcanism brought by the *argumentum a loco* of the Forges of Vulcan in Pozzuoli opens up a philosophical reflection whereby the human being is conceived from his generative capacities, industrious nature, and creative fecundity. In this application we might see the double dimension of the procedure of a geographical description: the position and situation of a place in an ordered spatial system must be completed by the characterization of the properties of that place (*ingenia*). In this double dimension of quantity and quality the geographical knowledge displays its force as intellectual ordering principle. In that sense, the *Digression on fire* is a geographical description since it comprehends from situating fire and its representation in the *Forum Vulcani* to the description of the ingenious stream of mind, the act of the *Ingenio*. The passage from the *Forum Vulcani* to the reflection on human nature is nothing but the systematization of a *moral geography* where the possibilities of creation and human action are placed.[33]

The representations of the *Forum Vulcani* and the idea of an inner and central fire continued drawing interest in natural and moral researches after the *Examen de Ingenios para las ciencias*. Giovani Nardi in his *De igne physica prolusio* (Florence, 1641) and Athanasius Kircher in his *Mundus Subterraneus* (Amsterdam, 1664) are two examples of how these kind of geographical curiosities will continue to be operative describing problems in the natural realm such as combustion, volcanoes' eruptions or the functions and formation of mountains. Representations of the places of fire, as the descriptions of the Etna, the Vesuvius and the Sun were in themselves, will continue to be a *locus* at once geographical and rhetorical allowing consequently medical and moral meditations about the nature and variety of human beings. This anthropological presence in geographical thought is something that stands out. Nevertheless, I claim that it is required to keep questioning about the extent of moral dimension in the geographical knowledge and, also, about the spatial horizon displayed in the discourses on human culture.

This overview of Huarte's *Forum Vulcani* allows us to see the scope and versatility that geographical knowledge presented in the Renaissance both in the encyclopaedic knowledge production and in the process of transmitting and updating Classical learning. Renaissance culture's general

---

[33] In order to engage fully in the debate on moral geography and moralized geography see R. Salomon (1936) who was the first to introduce the concept. See also J. Schulz (1978). More recently G. Mangani (2006) has been referring to a moral cartography, a more radical proposal with the advantage of giving full attention to the context and importance of spiritual practice during the fifteenth and sixteenth centuries.

interest in the world's diversity produced a monumental work of interpretation and critique of historical representations of the Earth that were crystallized in translations, commentaries and compilations of authors that were rediscovered under new light as Strabo, Pomponius Mela, Pliny, Ptolemy, and Macrobius. Thus, the modern geographical education in the Renaissance arose from the study and appropriation of Classical sources. To sum up, the geographical knowledge not only transmits the classical tradition in its contents but also in its argumentative and descriptive mechanisms and devices. As it has been shown, geographical thought is a rhetorical operator appearing dispersedly, albeit effectively in discourses that are not eminently geographical as such laying typologies of places, such as the volcanic *Forum Vulcani*, emblematic places, such as Pozzuoli, and displaying spatial orders to orientate knowledge and action.

References

Amalfitano, P., ed. (1986). *Il destino della Sibilla. Mito, Scienza e storia dei campi flegrei. Atti del convegno internazionale di studi sui Campi Flegrei promosso dalla Fondazione Napoli Novantanove*, Napoli: Bibliopolis.

Apiano, P. (1548). *Libro de la Cosmographia*, Antwerp: Gregorio Bontio.

Besse, J.M. (2003). *Les grandeurs de la terre. Aspects du savoir géographique à la Renaissance*, Lyon: ENS Éditions.

Braun, G. and F. Hogenberg. (1593). *Civitates Orbis Terrarum*, Koln.

Buisseret, D. (2007). "Spanish Peninsular Cartography, 1500-1700". In D. Woodward and J.B. Harley (eds.), *The History of cartography*, (pp. 1069-1094), Chicago: the University of Chicago Press.

Castiglione Minischetti, V. *et al.* (2006). *Le voyage français en Italie des origines au XVIIIe siècle.* (Bibliographie analytique), Fasano: Schena Editore; Paris: Editions Lanore.

Catin, J. (1997 [1568]). *Voyage aux Champs Phlégréens*, ed. by Luigi Monga, Moncalieri: Centro interuniversitario di ricerche sul "Viaggio in Italia".

Craig, M. (2002). "Francisco Vallés and the renaissance reinterpretation of Aristotle's *Meteorologica IV* as a medical text", *Early Science and Medicine* 7 (1): 1-30.

D'Amato, J. (1993). "Cicero's villas in the Phlegrean fields: the development of a historical and cultural myth", *Viator* 24: 385-419.

—. (1996). "The Turris Pharo at capo Miseno: a monument of medieval scholarship", *Viator* 27: 215-260.

García Vega, L. (2001). *Juan Huarte de San Juan*, Madrid: Ed. Del Orto.

Glacken, C. J. (1967). *Traces on the Rhodian shore: nature and culture in Western Thought from ancient times to the end of the 18th century*, Berkeley; Los Angeles: University of California Press.

Huarte, J. (1594). *The examination of Mens Wits. In Which, by discouering the varietie of natures, is shewed for what profession each one is apt, and how far he shall profit therein.* Richard Carew (tr.), London: by Adam Islip, for Richard Watkins.

—. (1645). *L'Examen des esprits pour les sciences.* Charles Vion d'Alibray (tr.), Paris: J. Le Bouc.

—. (1698). *Examen de ingenios; or, The Tryal of Wits. Discovering the great difference of wits among men, and what sort of learning suits best with each genius.* Edward Bellamy (tr.), London: for Richard Sare, at Grays-inn-gate in Holborn.

—. (1930). *Examen de Ingenios*. Edición comparada de la príncipe (Baeza, 1575) Sub-príncipe (Baeza, 1594). Rodrigo Sanz (ed.), Madrid: Biblioteca de filósofos españoles.

—. (1989). *Examen de Ingenios para las ciencias.* Guilermo Serés (ed.), Madrid: Cátedra.

Iriarte, Mauricio. (1939). *El doctor Juan Huarte de San Juan y su Examen de Ingenios. Contribución a la historia de la psicología diferencial*, Madrid: Jeraquía.

Jacob, C. and Lestringant, F. (1981). *Arts et légendes d'espaces : figures du voyage et rhétoriques du monde*, Paris: Presses de l'École normale supérieure.

Kagan, R.L. (1981). *Universidad y sociedad en la España moderna*, Madrid: Tecnos.

Letts, M. (1918). "Some Sixteenth-Century Travellers in Naples", *The English Historical Review*, 33 (130): 176-196.

Lovejoy, A. O. (1973). *The great chain of being. A study of the history of an idea*, Cambridge: Harvard University Press.

Mangani, G. (2006). *Cartografia morale. Geografia, persuasione, identitá*, Modena: F. C. Panini

Ollanda , F. de. (1940). *Os desenhos das antiguallas que vio Francisco d'Ollanda, pintor portugues*, Madrid: Ministerio de Asuntos Exteriores. N. 26 de la Biblioteca del Real Monasterio del Escorial.

Ortelius, A. (1592). *Parergon sive Veteris Geographiae Aliquot Tabulae*, Antwerp: Christophe Plantin.

Pernot, Laurent. (1993). *La rhétorique de l'éloge dans le monde gréco-romain*, Paris: Institut d'études augustiniennes.

Pigeaud, J. (1989). *La maladie de l'âme: étude sur la relation de l'âme et du corps dans la tradition philosophique antique médical*, Paris: Les Belles Lettres.

—. (2008). *Poétiques du corps: aux origines de la médecine*, Paris : Les Belles Lettres.

Rico, F. (1970). *El pequeño mundo del hombre,* Madrid: Castalia.

Salomon, R. (1936). "Opicinus de Canistris", *Studies of the Warburg Institute*, 1 (vol. 1-2): I-348.

Sandman, A. (2007). "Spanish Nautical Cartography in the Renaissance". In D. Woodward and J.B. Harley, *The History of cartography* (pp. 1095-1142, Ch. 40), Chicago: the University of Chicago Press.

Schulz, J. (1978). "Jacopo de' Barbari's View of Venice: Map Making, City Views, and Moralized Geography Before the Year 1500", *Art Bulletin*, 60 (3): 425-474.

Sirpettino, M. (1999). *I Campi Flegrei. Guida Storica*, Napoli: Edizioni Scientifiche Italiane.

Stevens, J. (1726). *New Spanish and English Dictionary*, London.

Strabo. (1988). *Geography*. H.L. Jones (tr.), Cambridge: Harvard University Press; London: W. Heinemann.

Tooley, R.V. (1939). "Maps in Italian atlases of the sixteenth century", *Imago Mundi* 3: 12-47.

Vallés, F. (1588). *In quartum librum Meteorologicum Aristotelis commentaria*, Augusta Taurinorum.

Zevi, F. (1999). "Virgilio e la topografía storica dei Campi Flegrei". In Paolo Amalfitano (ed.), *Il destino della Sibilla. Mito, Scienza e storia dei campi flegrei. Atti del convegno internazionale di studi sui Campi Flegrei promosso dalla Fondazione Napoli Novantanove* (pp. 21-42), Napoli: Bibliopolis.

CONTRIBUTORS

VALERIA ANDÒ

Valeria Andò is professor of Greek Language and Literature at the Università degli Studi di Palermo where she has made all her academic career. As a researcher, she has applied her philological expertise to the study of diverse anthropological and sociological aspects of Greek culture as matrimonial and burial rituals, the social function of wine or the public status of the poets. She has focused as well on cultural studies regarding the Greek feminine world, the social representation of women and the way Greek medicine dealt with women's health and biology–e.g. in Hippocratic texts like *Diseases of Virgins* or *On the Nature of the Woman.* She has published many papers, articles and book chapters on these topics and is the author of the successful book *L'ape che tesse. Saperi femminili nella Grecia antica* (Carocci, 2005, ^{2}2006, ^{3}2008). She has also worked of late on the study of violence in Greek poetry and culture.

JORGE CANO

Jorge Cano is Secretary of the Lucio Anneo Séneca Institute of Classical Studies at Madrid's University Carlos III. His research lines focus on ancient Greek philosophy (pre-Socratics, Plato, Hellenistic philosophy) and medicine. He has recently defended his PhD on Plato's *Timaeus* and its connections with pre-Socratic medical thought. As editor and translator of classical texts for the Spanish market he has published versions of Sophocles (*Oedipus Rex*), Marcus Aurelius (*Meditations*), Plutarch (*Lifes of Lysander and Sulla*), Epicurus (*Letter to Menoeceus*) and Hippocrates (*On the Nature of Man*).

HELEN KING

Helen King is professor of Classical Studies at the Open University. She was previously professor of the History of Classical Medicine and head of the department of Classics at the University of Reading, England. Since her first studies on menstruation in ancient Greece, she has been interested in setting ancient medical thought within its social and cultural context. She has focused on ancient ideas about creation, the role of women, and sacrifice to illuminate Hippocratic gynaecology. She is the author of several book on these issues among which *Hippocrates' Woman:*

Reading the female body in ancient Greece (Routledge, 1998), *Greek and Roman Medicine* (Bristol Classical Press, 2001). She has also worked extensively on the reception of ancient medicine, starting with the sixteenth-century compilation, the *Gynaeciorum libri* and the impact of Hippocratic gynaecology in the period after its publication in Latin by Calvi in 1525. Another monograph, *Midwifery, Obstetrics and the Rise of Gynaecology* (Ashgate, 2007), focuses on uses of classical medicine in the eighteenth century.

SIR GEOFFREY LLOYD

Geoffrey Lloyd has been throughout his University career based chiefly at Cambridge, holding various University and College posts. From 1983 onwards he held a personal Chair in Ancient Philosophy and Science and, since his retirement in 2000, is Senior Scholar in Residence at the Needham Research Institute. He is the author of 19 books and editor of 4, some of which have been translated into French, Italian, Spanish, German, Greek, Romanian, Polish, Slovenian, Japanese, Korean and Chinese. Internationally renown as a historian of ancient science and though, his approach to these research areas has always been informed by a keen interest in anthropology. In the mid 1980's Lloyd turned to the study of Classical Chinese. This has added a broad comparative scope to his more recent work. His most recent book (*Disciplines in the Making: Cross-cultural Perspectives on Elites, Learning and Innovation*, Oxford, 2009) concerns the similarities and differences in the understanding of certain core activities of human experience involved in diverse societies ancient and modern, and the factors that have encouraged or impeded their establishment as learned disciplines.

SARA MACIAS

Sara Macías is currently a post-doc researcher at the Institute of Languages and Cultures of the Mediterranean and the Near East, which is part of the Spanish Superior Council of Scientific Research (CSIC), after a term at the Intituto Lucio Anneo Seneca (University Carlos III). Her research focuses on the Greek, Latin and Semitic origins of European culture. She has published papers on Orphic myths and Orphism, their presence in Greek tragedy and on Euripides' works (*Hecuba*, Fr. 912, Derveni Papyrus).

ANTONI MALET

Antoni Malet is professor of History of Science at the University Pompeu Fabra in Barcelona and director of its Seminar on History of

Science. His research has focused on various aspects of sixteenth and seventeenth-century science, especially mathematics, and also on the influence of Franco's policies and regime in the scientific communities of twentieth-century Spain. He has published numerous papers and chapters of books and given talks about these topics in different universities and countries (France, Italy, Germany, Colombia, Mexico, etc.). Among his most relevant publications we may mention the monograph *From Indivisibles to Infinitesimals. Studies on Seventeenth-Century Mathematizations of Infinitely Small Quantities* (UAB, 1996), the edition of Francesc Santcliment's *Summa de l'art d'Aritmètica (1482)* (EUMO, 1998) and articles like "Renaissance notions of number and magnitude" (*Historia Mathematica* 33(2006): 63-81).

REVIEL NETZ

Reviel Netz is professor of Classics at Stanford University. Netz's main field is the history of pre-modern mathematics. His research involves the wider issues of the history of cognitive practices, e.g. visual culture, the history of the book, and literacy and numeracy. His books from Cambridge University Press include *The Shaping of Deduction in Greek Mathematics: a Study in Cognitive History* (1999, Runciman Award), *The Transformation of Early Mediterranean Mathematics: From Problems to Equations* (2004), and *Ludic Proof: Greek Mathematics and the Alexandrian Aesthetic* (2009).

PAULA OLMOS

Paula Olmos has been based at several Spanish academic institutions (University Carlos III, UNED) as a post-doc researcher in Classics and Philosophy. Her research lines include different aspects of the history of logic, argumentation and rhetoric–particularly in periods with a strong classical inspiration, as the early-modern Spanish humanism or the Latin late antiquity–as well as the contemporary theory of argumentation, in which she takes a rhetorical stance. She has published papers on these issues in journals like *Informal Logic*, *Theoria*, *Res Publica Litterarum* (Salerno), *Renaissance Studies*, *Argumentation* or *Studies in History and Philosophy of Science* (Cambridge). She is the author of a monograph on the Spanish sixteenth-century philosopher Pedro Simón Abril (*Los negocios y las ciencias. Lógica, argumentación y metodología en la obra filosófica de Pedro Simón Abril (ca. 1540-1595)*, CSIC, 2010) and co-editor, together with Luis Vega, of the collective encyclopaedic volume *Compendio de Lógica, Argumentación y Retórica* (Trotta, 2011, ^{2}2012).

SALVATORE DI PIAZZA

Salvatore Di Piazza received his PhD in Philosophy of Language and Mind at the University of Palermo (2008), where he has currently a post-doc position (Department FIERI-AGLAIA). He is also visiting researcher at the Université Libre de Bruxelles (GRAL) and researcher at the Centro di Studi e iniziative culturali "Pio La Torre" in Palermo. He is the author of several articles on issues concerning philosophy of language and semiotics in ancient thought. He has published several books as *Mafia, linguaggio, identità* (Centro di Studi e iniziative culturali "Pio La Torre", Palermo, 2010), which has been translated into French by L'Harmattan, and *Congetture e approssimazioni. Forme del sapere in Aristotele* (Mimesis, Milano, 2012).

AMEDEO ALESSANDRO RASCHIERI

Amedeo Alessandro Raschieri is a lecturer on Latin language and literature in the Dipartimento di Studi Umanistici at the Università di Torino. His research interests include classical, particularly Latin, philology and also ancient geography, a topic he first addressed in his PhD work on Avienus' *Orbis Terrae*, after which he has become Avienus' editor for the electronic library project *Musisque Deoque. Un archivio digitale di poesia latina, dalle origini al Rinascimento italiano*, directed by Paolo Mastandrea. He has also published works on other ancient authors particularly focused on their interest in geography as Cicero (as a frustrated geographer), Rutilius Namazianus, Artemidorus or Dionysius *Periegetes* of Alexandria. He has a book forthcoming on ancient geography and astronomy and its reception (*Geografia e astronomia antiche e moderne: alcune riscoperte umanistiche*, Edizioni dell'Orso).

CRISTIAN TOLSA

Cristian Tolsa is a graduate student in the Department of Greek Philology at the University of Barcelona. His research interests focus on ancient Greek history, art, literature, math, and science and Graeco-Roman Alexandria. He has also been a visitor student at the Humboldt-Universität zu Berlin and Stanford University where he has worked with Reviel Netz. He is working on a PhD on ancient Greek mathematical and musical theory and has taken part in seminars and conferences with papers about mathematical papyri and related topics.

MARIANNA TOMASELLO

Marianna Tomasello has recently defended her PhD at the Università degli Studi di Palermo under the supervision of Valeria Andò with a thesis on the relationship between musical theory and its concept of "concord" and natural, political and ethical thought in Plato and also in the lyric poets, under the title *La tranquillità all'unisono. Il modello della consonanza tra scienza della natura, politica e riflessione sul linguaggio dalla Lirica tardoantica a Platone.*

LUIS VEGA

Luis Vega is professor of logic, theory of argumentation and history of logic in the department of Logic, History and Philosophy of Science at the Spanish Distance Learning University (UNED). He is the author of two books on the history of proof and demonstration, focused on ancient Greece (*La trama de la demostración*, Alianza, 1990) and medieval Europe (*Artes de la razón. Una historia de la demostración en la Edad Media*, UNED, 1999). He has been for the last seven years the head and principal investigator of an international research group on contemporary theory of argumentation funded by the Spanish Ministry of Science and Innovation whose efforts have produced the collective work co-edited by him and Paula Olmos, *Compendio de Lógica, Argumentación y Retórica* (Trotta, 2011, ^{2}2012). He has a book forthcoming on the changing historical traditions in the theoretical treatment of fallacies (*La fauna de las falacias*, Trotta).

MARIO VEGETTI

Mario Vegetti is now emeritus professor of ancient philosophy after a long career chiefly based at the Università degli Studi di Pavia. His research has covered many different aspects of ancient thought but especially focuses on its scientific-epistemological and ethical-political implications. He has edited and translated works by Hippocrates and Galen and Aristotle's biological works. He has been the coordinator of the collective work in three volumes *Introduzione alle culture antiche* (Boringhieri, 1992). Among his chief publications we may mention *Tra Edipo e Euclide* (Il Saggiatore, 1983), *L'etica degli antichi* (Laterza, 1996), *Quindici lezioni su Platone* (Einaudi, 2003, Premio di filosofia di Siracusa), *Dialoghi con gli antichi* (Academia Verlag, 2007), *Un paradigma in cielo. Platone politico da Aristotele al Novecento* (Carocci, 2009, Premio di filosofia di Castiglioncello). Many of his books have been translated into English, French or Spanish.

ANDRÉS VÉLEZ

Andrés Vélez is a graduate student of the programme Histoire et Civilisations at the EHESS in Paris under the supervision of Prof. Yves Hersant. He is working on a PhD on the concept of "ingenio" (wit) in sixteenth-century natural philosophy, a topic which he has already approached from a rhetorical point of view in his book *Figuras del Ingenio. Estudio sobre la fuerza de la Analogía y la Ironía* (Editorial UPB, 2006), the fruit of his postgraduate studies in Colombia as part of the research project on "Rhetoric and Humanistic Tradition" (Universidad Pontificia Bolivariana, July 2002-Novenber 2004) and co-organizer of the "Tenth Muse Seminar on Poetry" (Universidad Pontificia Bolivariana, 2005).